Thermal Comfort in Built Environment: Challenges and Research Trends

Thermal Comfort in Built Environment: Challenges and Research Trends

Editors

Nianping Li
Yingdong He

MDPI

Basel • Beijing • Wuhan • Barcelona • Belgrade • Novi Sad • Cluj • Manchester

Editors

Nianping Li
Hunan University
Changsha
China

Yingdong He
Hunan University
Changsha
China

Editorial Office
MDPI
St. Alban-Anlage 66
4052 Basel, Switzerland

This is a reprint of articles from the Special Issue published online in the open access journal *Buildings* (ISSN 2075-5309) (available at: https://www.mdpi.com/journal/buildings/special_issues/04DYW2F443).

For citation purposes, cite each article independently as indicated on the article page online and as indicated below:

Lastname, A.A.; Lastname, B.B. Article Title. *Journal Name* **Year**, *Volume Number*, Page Range.

ISBN 978-3-7258-1225-7 (Hbk)
ISBN 978-3-7258-1226-4 (PDF)
doi.org/10.3390/books978-3-7258-1226-4

Contents

buildings

MDPI

Article

High-Performance Glazing for Enhancing Sustainable Environment in Arid Region's Healthcare Projects

Esam M. H. Ismaeil [1,2,*] and Abu Elnasr E. Sobaih [3,4,*]

1 Civil and Environmental Department, College of Engineering, King Faisal University, Al-Ahsaa 31982, Saudi Arabia
2 Architecture and Urban Planning Department, Faculty of Engineering, Port Said University, Port Said 42526, Egypt
3 Management Department, College of Business Administration, King Faisal University, Al-Ahsa 31982, Saudi Arabia
4 Hotel Management Department, Faculty of Tourism and Hotel Management, Helwan University, Cairo 12612, Egypt
* Correspondence: emohamed@kfu.edu.sa (E.M.H.I.); asobaih@kfu.edu.sa (A.E.E.S.)

Abstract: The integration of sustainability rating systems in healthcare projects and healthcare building envelope specifications is a growing concern in the construction industry, especially in the arid region. The external facade of healthcare buildings is one of the most significant contributors to the energy cost and comfort level of healthcare buildings in such a region. This study undertook a comprehensive comparison analysis of an adaptive model of high-performance glazing (HPG) specifications for patient rooms in a case study inside Saudi Arabia based on multi-criteria, including the LEED Healthcare rating system. The study used a technical comparative analysis for three onsite glazing models with HAB software v6.0 based on specifications of specialist manufacturer organizations for glazing window performance, climatic conditions, and the region's culture. Significant results in the case study project were achieved in energy saving and sustainability ranking in the healthcare rating system, providing new specification guidelines for HPG applications in healthcare buildings located in an arid region, and cultural environment considerations.

Keywords: healthcare rating system; high-performance glazing (HPG); green hospitals; sustainable environment; arid region

Citation: Ismaeil, E.M.H.; Sobaih, A.E.E. High-Performance Glazing for Enhancing Sustainable Environment in Arid Region's Healthcare Projects. *Buildings* **2023**, *13*, 1243. https://doi.org/10.3390/buildings13051243

Academic Editors: Yingdong He and Nianping Li

Received: 19 April 2023
Revised: 30 April 2023
Accepted: 5 May 2023
Published: 9 May 2023

1. Introduction

A green hospital recognizes the connection between physical and mental human health and the environment to demonstrate its governance, strategy, and operations [1–3]. A green hospital building enhances patient well-being, promotes public health, reduces environmental impact, contributes to the elimination of disease burden, and aids the curative process, while utilizing natural resources in an efficient, environment-friendly matter [4,5]. Green hospital rating systems are considered in both construction and operation and provide a therapeutic environment in which the overall design of the building contributes to the healing process and reduces the risk of healthcare-associated infections rather than simply being a place where treatment takes place [5,6]. The healthcare planning and design process needs to be correspondingly broad enough to include not only the issues surrounding the treatment of disease but also the promotion of health and prevention of disease—essentially the creation of a safe and therapeutic care environment [7–9]. Therefore, the design of a green hospital considers lighting, indoor air quality, passive and active measures, clean and green interior building materials, and the landscape. Table 1 illustrates sustainability rating tools and organization for healthcare buildings [9,10].

Identifying precisely the external glazing elements specifications in healthcare building facades supports architects and contributes to achieving a green hospital. They also help in achieving sustainability conditions and elements, which include indoor air quality and using natural materials and resources without any harmful content to reduce CO_2 monitoring. They also help in providing desirable and better daylight transmission; contributing with new energy-saving technology; reducing solar heat gain in hot climates and preventing heat loss in cold climates; supporting safety benefits needs; enhancing occupant comfort, improving the productivity of occupants; and supporting patient recovery, healing, and connectivity to the external environment view. Figure 1 illustrates the benefits of using glazing elements in hospitals facades [8,9]. The processes and practices of sustainability focus on the green healthcare structures' design, justify architectural approaches to interior design in a healthcare setting, and improve the quality of services to the hospital occupants [10,11].

Table 1. Sustainability rating tools and organization for healthcare buildings.

Issues and Elements for Rating		International Organization Rating Green Healthcare Building
Construction	Operation	BREEAM
Integrated Design	Integrated Operation	LEED
Sustainable Sites	Sustainable Education	GREEN STAR
Water Efficiency	Sites Management	DGNB
Energy + Atmosphere	Transportation	FGI
Materials + Resources	Facilities Management	WHO "The World Health Organization"
Environmental Quality	Chemical Management	
	Waste Management	LEED
Innovation + Design Process	Environmental Services	WELL
	Food Service	WELL
	Environmentally Preferable Purchasing	EPA
	Innovation in Operation	ISO

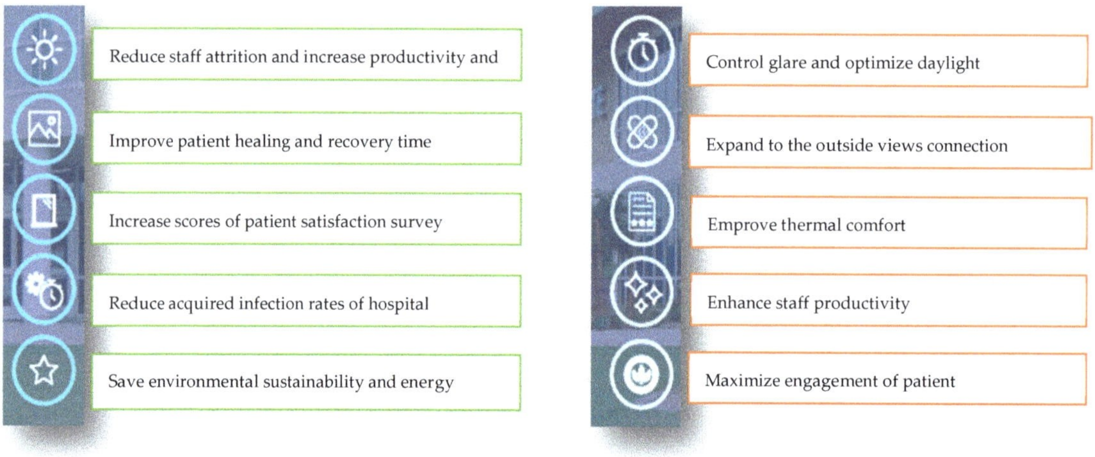

Figure 1. Benefits of using glazing elements in hospitals facades.

Designing green and healthy environments for patients, staff, and visitors' physical and mental well-being in hospital buildings is a significant concern for all international and local organizations [12,13]. Enhancing the formulation of the requirements for healthcare architectural design elements and the building system positively affects the relationship between the indoor environmental quality (IEQ) performance and the overall inpatient satisfaction [14,15]. Green building rating systems (GBRSs) are adopted worldwide to investigate the most effective glazing types in hospital patient room buildings, which affect the performance of IEQ, occupants' satisfaction in a hospital ward, heating, and cooling energy needs [6,16–18]. The healthcare façades, which include HPG elements with special specifications, can achieve a better balance between occupants' satisfaction and the building energy demand, well-being, privacy in clinic diagnosing, daylight transmission, occupant comfort inside the hospital spaces, energy saving, and a reduction in CO_2 [19–24].

1.1. High-Performance Glazing (HPG) Elements Specifications and Sustainability

The term 'high-performance building' is commonly applied to both buildings and their facades, and often to the materials that they comprise, such as high-performance glass [25,26]. The exterior enclosures of high-performance sustainable facades use the minimum possible amount of energy to maintain a comfortable interior environment and to create a healthy and productive environment [27,28]. A high-performance building optimizes the integration of all high-performance attributes through a life cycle basis, including energy conservation, durability, safety, accessibility, and the operation processes' cost [29,30]. Therefore, the attributes for determining the performance of the building facades include: first, environmental impacts of the building facade, including energy consumption and resulting emissions over the operations phase of the building lifecycle, as well as lasting impacts; second, the building occupant security and safety using the façade's system; third, durability, which is the fundamental aspect of performance and sustainability for all building systems; fourth, the economic or cost benefit, and the efficiency of all facade systems and materials; fifth, human comfort, which affects the productivity, e.g., thermal, acoustic, daylight, visuality, and connection to the natural environment. The benefits of high-performance skins provide the occupants with fresh-air exchanges, collecting solar energy through various technologies, harvesting rainwater for cistern storage and water heating, and providing daylight and views to occupants while minimizing glare [31–33].

The technical description of high-performance glass provides energy efficient, resistant glass, safety, and glass with electrically charged interlayers [34]. The innovations in glass and glass coating technology provide control over solar radiation, maintain the virtually neutral appearance, and provide a high light transmission [35]. The benefits of using HPG include an energy saving of 35–40% more than conventional glass, a payback period from 3 to 4 years, and enhancing the occupant comfort and productivity by accessing daylight [36]. The common parameters and values to measure the internal comfort status for the whole glass in HPG include [13,16]: first, visible light transmittance, which varies between 0 and 1; second, the solar radiation admitted through a window, which is addressed with the solar heat gain coefficient (SHGC), expressed between 0 and 1; and third, the U-factor, which measures the resistance heat flow from the building to outside, where its ratings are between 0.15 and 1.20 [17,18].

1.2. Types of High-Performance Glass

The different types of high-performance glass include: first, insulated double and triple-glazed glass, which consist of two or more glass panes with an air space filled with inert gas, e.g., argon and krypton, which resists the heat flow and can reduce the U-value by 0.2–0.3 W/m^2K [21]; second, tinted glass, which reduces the solar heat gain coefficient and visible light transmittance using common colors, e.g., blue, green, gray, and bronze [21,22]; third, reflective glass coatings, which usually consist of thin metallic layers with a variety of colors, e.g., silver, gold, and bronze, applied on the outer surface of the glazing, where the reflective coatings enhance the transmittance of clear glass from 89% to more than 96% and the reflectivity from 8% to less than 2% [22]; fourth, triple-silver-glass coatings, where a third silver layer is added to enable a high light transmission and low solar factor [21,23]; fifth, low-iron glass or ultra-clear glass, which reduces the iron content to produce beautifully transparent results, with a high safety, transmittance, aesthetics, and environmental friendliness [22–24]; sixth, vacuum-insulated glazing, which is used to minimize conduction and convection heat losses [25]; seventh, solar control low-E glass, which is a special oxide-coated glass that transfers a lower amount of heat in the building, reducing the glare of light entering and the cost of artificial lights, with excellent thermal insulation properties and an emissivity of 0.84, which means that it absorbs 84% of long-wave radiation and reflects 16% [26,27]; eighth, smart glass and laminated glass or switchable transparent glass (STG), which are used as smart PDLC film to change between clear and opaque using voltage control [27,28].

1.3. Compatibility Factors of Glazing Elements with Sustainability Rating System

There are many advantages in the composition of glazing elements in all types of healthcare building facades because of their aesthetic specifications and positive technical environmental effects commensurate with the modern standard in healthcare construction projects, which aims toward being green healthcare projects with a high ranking in achieving sustainability requirements and patient-healing needs [30,31]. The attainment of these requirements and needs can be illustrated and classified in a compatibility matrix by reviewing the most important standards required by the USGBC (LEED Healthcare Guideline 2009), Guideline for Design and Construction of Health Care Facilities FGI "Facility Guidelines Institute", and the use of technical specifications of glazing elements from specialized manufacturer companies, e.g., Saint-Gobain, Guardian, and Pilkington, and aluminum companies such as Choco and Technal [32]. The domains of the environmental performance of sustainable healthcare construction include: site selection, alternative transportation, water conservation, energy efficiency, recycled materials, renewable systems, low-emitting materials, natural daylight, waste generation reduction, and local organic food [33,34].

The Leadership in Energy and Environmental Design (LEED) system is widely recognized as a green building standard in the construction industry, and provides a framework for healthy-environment, high-efficiency, low-carbon, and cost-saving green buildings. The LEED rating system is an evaluation system used for the building's environmental performance and sustainability measurements, which are based on an points-earning system for meeting certain criteria in categories such as the buildings' indoor environmental quality, water conservation, and energy efficiency [32]. The LEED rating system has an emphasis on energy efficiency and renewable energy, using energy-efficient equipment and systems such as high-efficiency HVAC systems, LED lighting, solar panels, and analytics data platforms, and using renewable energy sources such as geothermal, wind, or hydroelectric power. The LEED certification has four levels: Certified (40–49 points), Silver (50–59 points), Gold (60–79 points), and Platinum (80+ points), through a continuous review process that includes a pre-certification review, a construction review, and a post-construction review [33].

The LEED for Healthcare rating system is provided for inpatient and outpatient healthcare facilities and its content is specific to designing strategies relevant to healthcare environments. LEED-certified sustainable hospitals contain specific criteria values as optimal green healthcare environments that influence the patients' health, well-being, and recovery periods. The base of the standards is classified into categories of location and transportation, sustainable site, water efficiency, energy and atmosphere, materials and resources, indoor environmental quality, innovation and design process, and regional priority [33,34]. Improving the high-performance glazing (HPG) specifications in healthcare projects can achieve more points in the LEED rating system and improve sustainability in the three main sustainable categories of the whole project.

First: in the energy and atmosphere (EA) category, HPG can help to achieve more points in EA Credit 1 (1–24 points) and optimize the energy performance to achieve increasing levels by demonstrating a percentage improvement in the proposed building performance [37,38]. Moreover, regarding EA Credit 2 (1–8 points), which concerns onsite renewable energy, the intent for this category is to encourage increasing on-site renewable energy self-supply levels by using on-site renewable energy systems to offset building energy costs. Figure A1 (Appendix A) shows a technical performance comparison between clear and tinted glass [38]. The HPG can contribute to these two credits by obtaining the best energy-efficient glazing, which combines thermal insulation and solar with increasing the SHGC value from 0.20 to 0.25 as required in EN 410, NFRC 200, and NFRC 300 (American Standard) [35]. Moreover, it helps in achieving a combination of solar control and low emissivity by installing, as an example, high-tech HPG such as transparent conductive oxide (TCO) as cover plates, electricity generation, and hot water, including certain types of solar materials, e.g., crystalline silicon photovoltaics, thin-film photovoltaics, concentrated solar power applications, and solar thermal collectors. Figure A2 (Appendix A) shows crystalline, amorphous, and coating type used to generate power to reduce air conditioning (AC) loads and CO_2 emissions [36].

Second: in the indoor environmental quality (IEQ) category, HPG can help to achieve more points in IEQ Credit 2 (1–2 points), which concerns the acoustic environment, in order to provide building occupants with an indoor healing environment free of intrusive or disruptive levels of sound to meet the 2010 FGI Guidelines on control site exterior noise [37,38]. HPG can contribute to noise reduction according to noise reduction coefficients (NRCs), traffic loads, and the thickness of glass layers as follows: traffic, 30–50 db reduction; non-traffic, 35–40 db reduction [36,39]. HPG also helps to achieve more points in IEQ Credit 4 (1–4 points), which concerns low-emitting materials, by reducing the quantity of indoor air contaminants that are odorous, irritating, and harmful to the well-being and comfort of patients. Therefore, HPG can protect a healthy indoor environment from pollution, characterized by a level of formaldehyde and volatile organic compound (VOC) emissions close to 0 in materials and systems, e.g., coated glass, tempered glass, the sealant, and the PVB layer in laminated glass [40]. HPG also contributes to IEQ Credit 5 (1 point)—the control of potentially hazardous indoor chemical and pollutant sources to minimize their exposure to building occupants—by using silver ions in the upper layers of the glass, which destroy bacteria and germs by disabling their metabolism and disrupting their division mechanism, particularly in warm and moist conditions, where the glass coating supports the antibacterial process as shown in Figure A3 (Appendix A) [24,41,42]. Furthermore, HPG helps to achieve points in IEQ Credit 6.2 (1 point), which indicates the controllability of systems' thermal comfort, by providing a high-level-of-thermal-comfort system and individual thermal comfort controls for the patient room [37–39]. HPG can achieve a high level of insulation by using double or triple-insulating glazing and coated glass in the range of approximately 8″ c to 15″ c (IRUV coating as an example) [25,43,44], resulting in a reduction in the solar factor, increase in the thermal insulation value, a 25% energy-saving value, 70%:80% visible transmittance, a U-value as low as 0.9, and a 2:5 year payback [23,37]. Figure A4 (Appendix A) illustrates one of the job references used: transparent coating tested in a lab with a 15% reduction in insulation [42,45]. HPG contributes to both IEQ Credit 8.1 (1 point)—daylight—and IEQ Credit 8.2 (1 point)—daylight and views—by providing building occupants with a connection between the outdoor environment and indoor spaces through daylight and views that reach regularly occupied areas of the building [46,47], achieved by calculating the window-to-floor area ratio (WFR), window-to-wall area ratio (WWR), and window head height. HPG can achieve a wide variety of visible light transmittance, ranging from 10% to 83% [48]. Therefore, the glass connects building occupants with the outdoor views and reduces the use of electrical lighting [45]. The design of the window glazing could serve to be compatible with dim electrical light systems as well as allow for the achievement of a suitable angle and WFR ratio. On average, hospitals require an ambient light of 500 to 1000 Lux for visual comfort. Figure A5 illustrates the different types of integrated window and curtain wall structures with thin and photovoltaic film in visible light transmittance (LT). Figure A5 (Appendix A) shows different types of integrated window-and-curtain-wall structures with thin and photovoltaic film in visible light transmittance [46].

Third: in the materials and resources (MRs) category, HPG can help to achieve more points in MR credit 3 (1–4 points), which concerns sustainably sourced materials and products, by using the recycled content value and the cost of the assembly, as well as materials and products whose source/manufacture must not exceed 500 miles (800 km) from the project location [46,49]. The manufacturing measurements and local raw materials of HPG in Saudi Arabia comply with the current edition of the ASTM E966 Standard [50,51]. All glass factories are within 800 km from inside SA, and they have a minimum of 30% recycled glass [49,52]. Figure 2 shows the locations of the glass manufacturer for the case study far 239 km from the case study location.

Case study Building

Processing plant

The location of glass

Figure 2. Locations of manufacturing processes that produce glass for the case study.

The study presents an evaluation process for HPG specifications guidelines that could be applied in public and private healthcare projects in arid regions to have a positive impact regarding the external environment, meet the patients' healing needs, indoor patient comfort, caregiver performance, and patient and staff satisfaction, and provide excellent solutions for growing convergence that fosters public health and environmental sustainability.

This study undertook a full analysis of compatibility factors of HPG elements with sustainable rating system credits and international standards in order to update tender specifications and achieve distinguished results in energy saving, indoor air quality, and other sustainability goals, using King Faisal University (KFU) teaching hospital as a case study. This study examined energy conservation and the sustainable rating level in a public healthcare project in a desert campus area based on numerical analysis with HAP software v6.0, considering LEED for Healthcare as parameters for a sustainable rating system, in order to evaluate three HPG alternatives in the case study hospital in terms of their effectiveness in improving the hospital efficiency. The study adapted new specifications for HPG, energy saving, and more than 13 points in LEED for Healthcare that could be earned in the case study hospital. These significant results were accomplished according to partnerships between several agencies (e.g., the campus technical team, the national energy service, private glass manufacturers, and municipal governments). The study presents new applicable HPG specification guidelines to be addressed in the HPG local construction regulations and codes for public and private healthcare projects in arid regions. This will support gaining sustainability points in the rating system and linking positive external environment impacts with patients' healing needs, indoor patient comfort, caregiver performance, and satisfaction, which are excellent solutions for growing convergence that foster public health and environmental sustainability.

2. Methodology

The technical analysis and numerical evaluation of the external glass composite, e.g., curtain walls, windows, and doors in healthcare construction projects, are considered as essential processes for improving energy performance and adapting the external and internal environment toward supporting sustainability, comfortability, and well-being of all patients, staff, and visitors. This study addressed the benefits of improving energy performance, sustainability value, and culture needs in the inpatient ward in a healthcare case study construction project according to the technologies used to improve the performance of the facades and international sustainability rating system.

The study performed technical analysis, with HAP software [53] for energy simulation, LEED Healthcare rating system, and physical monitoring with technical team of experts and decision makers for three actual high-performance external window alternative samples in ongoing teaching hospital facades as a case study in Saudi Arabia between 2020 and 2021 before installing the facade windows in the actual site. The authors of the study, with an expert team that included a project manager, an aluminum manufacturer specialist in specifications, and a mechanical engineer specialist with HAP software, as well as interviews with specialist medical staff, established, as an expert judgment of the analysis processes within, the following stages:

1. Gathering all available data and explaining the compatibility matrix between external glazing elements and LEED Healthcare sustainability credits as a sustainability international rating system related to healthcare projects. These include (A) energy and atmosphere regarding optimizing energy performance and onsite renewable energy; (B) indoor environmental quality (IEQ) in the acoustic environment, indoor chemical and pollutant source control, low-emitting materials, daylight and views, and thermal comfort system controllability; and (C) materials and resources (MRs) regarding sustainability sourced materials and products.
2. Interviews with experienced medical staff in analyzing the impact of the connection between indoors and outdoors through the external building envelope on patient healing and well-being in public hospitals.
3. The project aluminum manufacturer installing three actual samples of glazed windows as per bidding dimensions: one as per tender specification, and the other two samples each having different glass specifications, including transmittance (LT), reflectance (in), shading coefficient SC, U value, reflectance (out), privacy, and color, in order to make a physical evaluation by expert judgment committee.
4. Conducting technical and numerical comparisons using HAP software for the three samples to identify the compatible sample with sustainability standards, the required privacy degree in the patient room according to environment culture, and natural light for improving the well-being of the patient.

The study adopted technical analysis for the energy performance and sustainability rating in the window glazing case study for improving the tender specifications. The study undertook numerical and technical comparison for the three actual alternative samples installed by the aluminum supplier before installation in the site as a procedure in order to achieve sustainability in one of the important external façade elements, which have a direct effect on inpatient healing, privacy, comfortability, and satisfaction. The study submitted a pioneer example of updating HPG window specifications for healthcare construction projects in arid areas to be addressed in regulations and codes of desert healthcare construction projects. Figure 3 shows the study method flowchart.

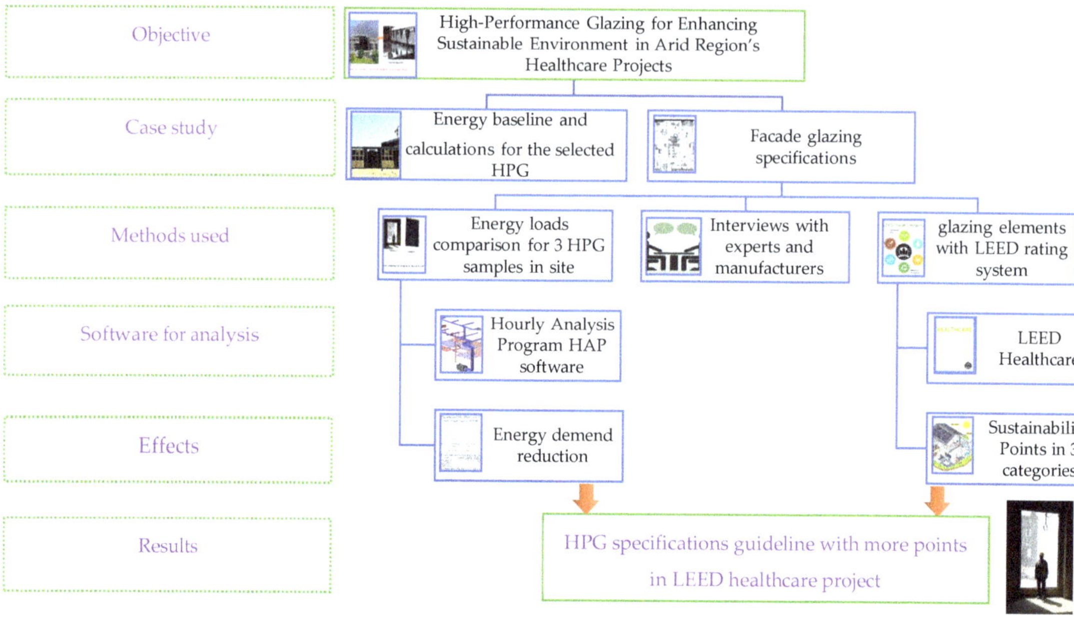

Figure 3. The study method flowchart.

Applying High-Performance Glazing HPG in King Faisal Teaching Hospital Facades

The public teaching Hospital located at King Faisal University (KFU), Saudi Arabia, consists of five buildings (main building, outpatient, physiotherapy, oncology, and multi-story parking). Building 02, which is the outpatient hospital, was particularly designed for complete examination in clinics and laboratories to test and determine exact illness results for diagnosis. The built area of outpatient building B02 is 45,000 m^2, and its skeleton structure is composed of four levels with 29 clinics on three floors and staff parking on the basement floor. Building B02 contains 448 m^2 of structural glazing elements for the whole building, and the total load for cooling reaches approximately 2841 kw, where the tender specifications of glazing elements include non-intensive clinic dimension of 1.8 m $\times$ 3.0 m (5.40 m^2), intensive clinic dimension of 1.8 m $\times$ 2.4 m (4.32 m^2), 2 units of sensitive medical clinics with dimension of 0.6 m $\times$ 2.4 m (2.88 m^2), with insulating clear double-glazed bronze color, tempered, 24 mm thick 6 mm external pane, 6 mm internal pane, and 12 mm sealed air space in between panes, and cooling load for each window of 5.5 kw. Figure 4 illustrates the outpatient ward façade and Figure 5 illustrates the outpatient building B02 inpatient ward plan [54].

Project tender specifications of glass and glazing structure were designed to have its final composition be 24 mm thick insulating double-glazed, 6 mm thick bronze color float and tempered glass for the external pane, and 6 mm thick clear float and tempered glass for the internal pane, with 12 mm distance of sealed air space in between panes. In addition, the shading coefficient was designed to be SHGC 0.3%, with protective UV value of 2.5 W/m^{2}K, without mentioning any other conditions for maximum relative heat gain (RHG) or daylight visibility transmittance (LT). These specifications was limited regarding the achievement of sustainability requirements for the patient healing, staff, employees, and visitors. Therefore, the authors and KFU experts committee built the methodology based on choosing the suitable specifications from a comparison of alternatives for three real samples manufactured by the approved aluminum supplier according to different specific specifications, e.g., daylight transmittance (LT), glare reflectance (in), shading coefficient (SHGC), protective U-value, glare reflectance (out), privacy, and color.

Figure 4. Outpatient ward façade.

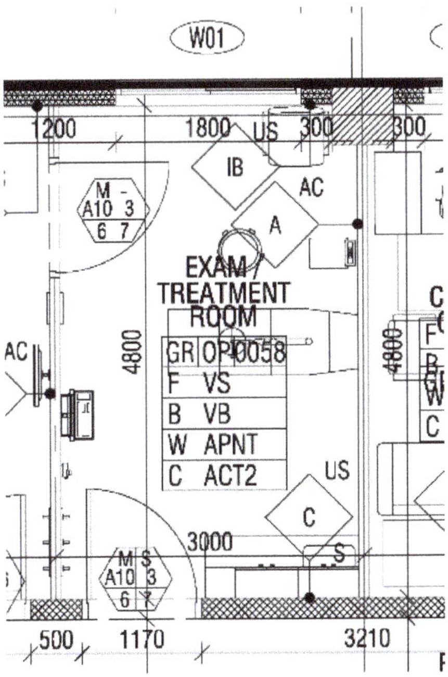

Figure 5. Outpatient ward plan.

Table 2 illustrates the specification comparison sheet for the following: Sample one: daylight visibility transmittance 18%, glare reflectance agreed (in) 13%, shading coefficient 0.2, U-value 1.74, glare reflectance agreed (out) 13%; Sample two: daylight visibility transmittance 27%, reflectance (in) 11%, shading coefficient 0.3, U-value 1.8, reflectance (out) 8%; Sample three: daylight visibility transmittance 4%, glare reflectance (in) 49%, shading coefficient 0.18, U-value 2.4 W/m^{2}K; glare reflectance (out) 15%. The committee made a practical comparison on site in order to select the suitable alternative based on achieving the degree of sustainability requirements within compatibility matrix between glazing elements and LEED sustainability credits.

Table 2. Specifications comparison sheet for the three real samples.

Specification	Tender Project Specification	Sample 1	Sample 2	Sample 3
Transmittance (LT)	-	27%	18%	4%
Reflectance (in)	-	11%	13%	49%
Shading Coefficient (SC)	0.3	0.3	0.2	0.18
U-value	2.5	1.8	1.7	2.4
Reflectance (out)	-	8%	13%	15%
Privacy	-	Less privacy	Medium	More privacy
Color	-	Grey	Between grey and brown	Brown

Figure 6 illustrates real alternative samples erected on site in the outpatient building façade to facilitate physical and realistic comparison with the international standard recommended by the international manufacturer, such as Guardian, Obeikan, and Pilkington, to support the authors and KFU-authorized committee in choosing applicable high-ranking sample and achieve the following advantages: compatibility between capabilities of glazing elements with new technology manufactured from internationally experienced glass manufacturers such as Onyx, AGC Obeikan, and Guardian, and LEED for Healthcare 2009 guide as sustainability requirements standard mentioned in Table 2, which focuses on: (A) energy and atmosphere (EA) regarding optimizing performance of energy and renewable energy onsite; (B) indoor environmental quality (IEQ) in the acoustic environment, concerning indoor chemical and pollutant source control, low-emitting materials, daylight and views, and thermal comfort systems; (C) materials and resources (MRs) regarding sustainably sourced materials and products, all considering the international standard, codes, and requirements in glazing elements and green hospitals such as European Norm EN 410 and EN 673, as well as American standards such as NFRC 100/200/300 [55–57]. The results from using HAB software program to calculate energy saving in cooling load provide decision maker with quantitative applicable alternatives approach to enhance the glazing windows' tender specifications and save energy for cooling system in outpatient building regarding the project life cycle.

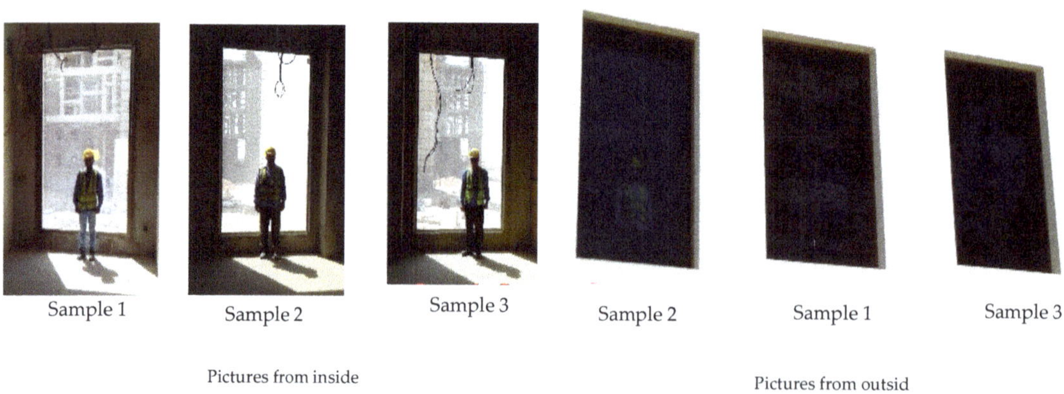

Figure 6. Proposed alternative samples in case study site for outpatient ward.

3. Results and Discussions

After using HAP software to perform the analysis for the alternative specification and comparing the alternative specification with both sustainability requirements in LEED for Healthcare 2009 and requirements in glazing elements and green hospitals, in the study, the KFU-authorized committee and author chose sample two as HPG with a shading coefficient of SHGC 0.21%; a protective UV-45 value of a maximum of 1.74 W/m^{2}K, considering the season of summer and expecting to control the relative heat gain with a maximum RHG of 149 W/m^2; and a transmittance (LT) of 18% with a glare reflectance agreed to be 13% on both the inside and outside. These choices were made according to the following results.

Sample two follows international organizations that focus on sustainability and green healthcare construction and have codes and credits, e.g., European Norm (EN), The National Fenestration Rating Council (NFRC), the glass manufacturer ONYX, and the glass manufacturer AGC Obeikan. The technology used for the installation of glass coating involved a transparent microscopically thin coating on the glass to reduce the heat gain into the building [57–59]. Table 3 illustrates a comparison between tender specifications and sample two specifications, which assure that sample two specifications are compatible with the glass international standard requirements and achieve significant results.

Table 3. Technical comparison factors for choosing alternative sample with tender specifications.

Glass Thickness 6t + 12air + 6t	U-Value	Shading Coefficient SC	Transparency Lt	Reflectance (in) %	Reflectance (out) %	Room Cooling Load KW	Energy Saving (1–10)
Tender	2.5	0.3	N/A	N/A	N/A	3.4	5
Sample two	1.74	0.21	18	13	13	3.1	7

The HPG specifications of sample two support the needed cooling load reduction, enhance the air flow inside the inpatient room, and compensate for the tender specification shortage. Fully detailed descriptions for zone sizing and energy modeling are illustrated in Figures A6 and A7 (in Appendix B). Figure A6 shows a zone sizing results summary for the fan coil of maximum tender specifications, and Figure A7 shows a zone sizing results summary for HAP software of the fan coil of sample two. The result of the comparison between the two instances of zone sizing indicates a 13% reduction in the cooling load and 15% reduction in the air flow needed for cooling energy after applying sample two. The other advantages are illustrated in Figure 7, which shows significant results of a comparison of elements, e.g., U-value, shading coefficient (SC), cooling load, and airflow, for the two instances of sizing.

Sample two selected from the expert staff and authors as HPG through the site visit achieved an optimum connection between the privacy quality for the patient in the diagnosing process and a good feeling resulting from the outdoor daylight and outside landscape. The colors of sample two as HPG matched the façade's finish, and included two colors: medium beige and dark brown as seen in Figure 5. Sample two as HPG is consider as the guideline in glazing elements in healthcare construction, which adapt with healing, engineering requirements, and satisfaction for the patient. Since the project is under construction, the study proved that the cost of the supply and application of sample two, for the whole building, does not exceed the tender price and does not make the contract unbalanced. Sample two as HPG can contribute to achieving a sustainable 19 points in LEED for Healthcare 2009 [60] according to the following.

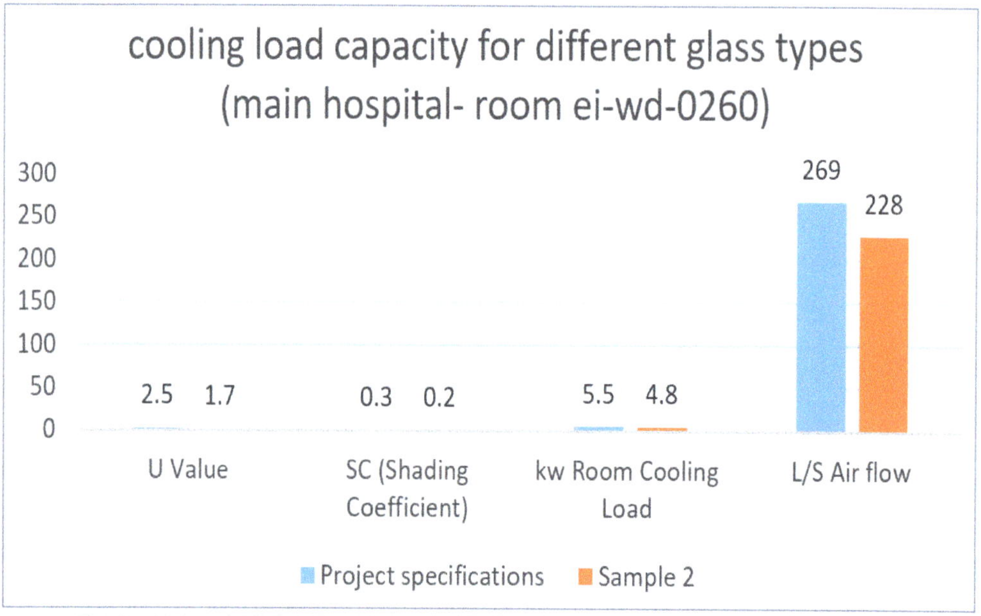

COOLING LOAD CAPACITY FOR DIFFERENT GLASS TYPES (MAIN HOSPITAL- ROOM EI-WD-0260)					
SN	GLASS	U Value	SC (Shading Coefficient)	B01 Room Cooling Load	Air flow
1	Project specifications	2.5	0.3	5.5 KW	269 L/S
2	Sample 2	1.7	0.2	4.8 KW	228 L/S
	U VALUE IN (W/M2.K)				
	KW (kilowatt)				
	L/S (LITER PER SECOND)				

Figure 7. Comparison analysis regarding energy saving, U-value, SC, cooling load, and air flow results.

3.1. Energy and Atmosphere (EA) Category

The study analysis achieved 5 points in optimizing the energy performance credits: regarding the cooling loads required for the air conditioning of the two study rooms mentioned in the table for both the hospital building and the outpatient building, it is clear that changing the quality of the glass prescribed by the specifications of the project to lower U-values and SHGC values leads to reduce the cooling load by 13%, in energy consumption.

3.2. Indoor and Environment Quality (IEQ) Category

The study analysis achieved 2 points in the acoustic environment: sample two as HPG can reduce the external noise by 45 db; 1 point in low-emitting materials: the final material of sample two as HPG is a low-E origin brand that contains a non-VOC; 5 points in daylight and views: the final material of sample two as HPG provides a light transmittance of 18% from a 4.32 m^2 window area, which is compatible with the international sustainability standard, providing a wide vision and connection to the external landscape, as well as preserving the needed privacy for the patient within the diagnosing process; 1 point in indoor chemical and pollutant source control: the final material of sample two was manufactured as antibacterial glass; 1 point in thermal comfort system controllability: sample two as HPG can result in energy saving—as an example, the cooling load reduction reaches approximately 13%, which affects the total energy reduction for the whole building.

3.3. *Materials and Resources (MRs) Category*

The study analysis achieved 4 points in sustainability regarding sourced materials and products: all components of sample two as HPG that were raw material components came from mines less than 800 km away; furthermore, they are highly capable of being recycled and are fire-resistant. The study with numerical analysis using HAP software achieved significant advanced results regarding retrofit processes for specific façade elements in the ongoing project, focusing on achieving a high ranking in the sustainability rating system, which supports creating a healthy environment for patient wellbeing in healthcare projects located in difficult desert environments. The study supports the decision maker with robust results in energy saving based on the specifications comparison between actual samples carried out by an expert committee. The selected specifications can be addressed in the local Saudi construction codes and regulations for healthcare construction projects, and present pioneering actual specifications for high-performance glazing adopted with an international sustainable rating system.

4. Conclusions

High-performance glazing (HPG) elements are inevitable for all healthcare construction projects and international glass manufacturer investments, which contribute through their research centers to glass quality improvement, green hospitals, and all sustainability standards. This study was based on the installation of three actual samples in a healthcare case study project that were manufactured and erected on site for a comparison process carried out by an expert committee according to achieving results in sustainability impacts and energy saving. The study drew up these recent improvements and classified them according to LEED for Healthcare 2009 in the form of a practical and applicable compatible elements matrix that contributes to supporting healthcare sustainability credits, green hospitals requirements, and the patient's well-being. The study used this compatible elements matrix for a technical analysis comparison process to explore the sustainable impacts and ranking of the three actual HPG sample specifications in the healthcare case study project. The study used HAP software as an energy design, zone sizing, and comparison tool for the technical specifications of the three actual samples to register the energy impacts and savings. The selected sample specification achieved a 13% reduction in the cooling load and a 15% reduction in the air flow needed for cooling energy, contributing with 19 potential points in the LEED Healthcare rating system in three categories (energy and atmosphere (EA), indoor and environment quality (IEQ), materials and resources (MRs)), compatible with EN and the NFRC standard. The selected sample supports the high privacy level needed for patient healing, achieves a contract cost balance, and matches its color with that of the facade elements. The selected sample specifications could be addressed in local healthcare construction project codes and regulations as HPG window specifications in such an environment.

5. Limitations and Future Research Directions

The current study focused on a technical analysis for high-performance glazing and the LEED for Healthcare rating system as a retrofit approach toward energy conservation and asset management inside public organizations buildings, i.e., hospitals, using an analysis approach toward energy consumption inside a university hospital as a case study. The results may be limited to other public spaces of the same context. The study also opens the door for future research studies on public organization building in relation to sustainability, such as the indoor air quality. Future research could also address the conflicting goals of the window manufacturers, such as a greater living comfort at the expense of poorer recyclability, in order to ensure the production of HPG. Additionally, the economic impacts of new technology in sustainable building features can be another interesting area of research.

Author Contributions: Conceptualization, E.M.H.I. and A.E.E.S.; methodology, E.M.H.I.; software, E.M.H.I.; validation, E.M.H.I. and A.E.E.S.; formal analysis, E.M.H.I.; investigation, E.M.H.I. and A.E.E.S.; resources, E.M.H.I. and A.E.E.S.; data curation, E.M.H.I.; writing—original draft preparation, E.M.H.I. and A.E.E.S.; writing—review and editing, E.M.H.I. and A.E.E.S.; visualization, E.M.H.I.; supervision E.M.H.I.; project administration, E.M.H.I.; funding acquisition, E.M.H.I. and A.E.E.S. All authors have read and agreed to the published version of the manuscript.

Funding: This work was supported by the Deanship of Scientific Research, Vice Presidency for Graduate Studies and Scientific Research, King Faisal University, Saudi Arabia [Project No. GRANT3368].

Institutional Review Board Statement: Not applicable.

Informed Consent Statement: Not applicable.

Data Availability Statement: Data available upon request from the first author.

Conflicts of Interest: The authors declare no conflict of interest.

Appendix A

U Factor summer, SHGC & RHG

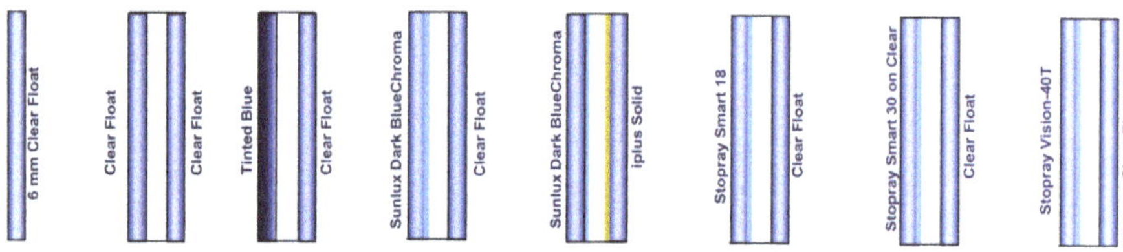

	6 mm Clear Float	Clear Float / Clear Float	Tinted Blue / Clear Float	Sunlux Dark BlueChroma / Clear Float	Sunlux Dark BlueChroma / iplus Solid	Stopray Smart 18 / Clear Float	Stopray Smart 30 on Clear / Clear Float	Stopray Vision-40T / Clear Float
LT	=39	=81	=52	=14	=12	=18	=30	=40
LR int	=3	=14	=13	=15	=12	=16	=15	=15
U-Factor	=5.27	=2.75	=2.75	=2.30	=1.56	=1.32	=1.4	=1.29
SHGC	=0.87	=078	=0.44	=0.22	=0.18	=0.15	=2	=0.20
RHG	=671	=584	=337	=180	=141	=121	=0.22	=154

Figure A1. Technical performance comparison between clear and tinted glass.

Crystalline blind to generate 100W/H/M²

Transparent coating generates 4–28 w/h/m²

Using crystalline glass to generate 100w/h/m²

Thin film as generates 25–50w/h/m²

Figure A2. Using crystalline and amorphous material and coating to generate power.

Observed reduction bacteria after 24 hours results

99% fewer bacteria in:

S. Aureus:

E. Coli:

P. Aeruginosa:

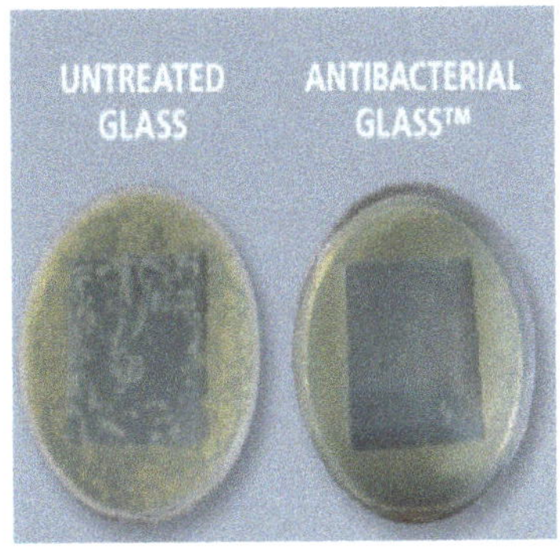

Figure A3. Coating glass supports the antibacterial glass process.

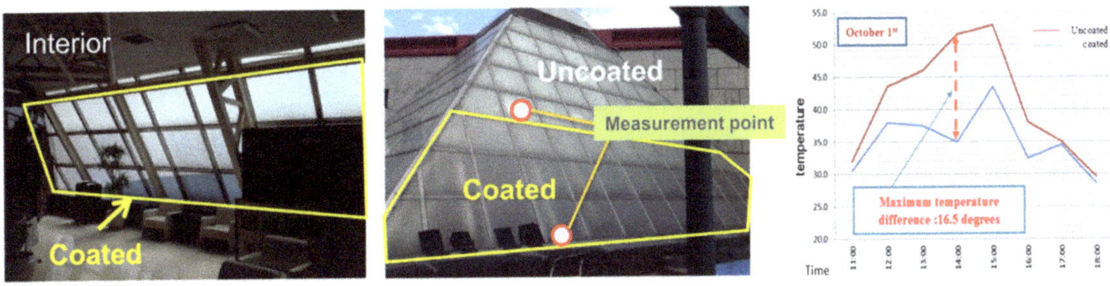

Figure A4. Job reference with transparent coating tested in lab.

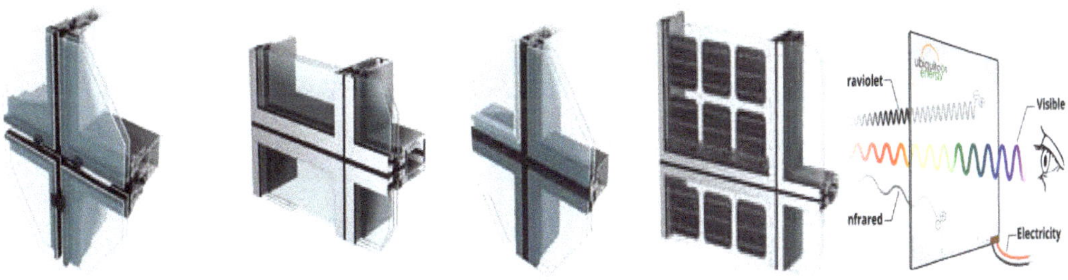

Structural glazing | Unitized curtain wall | Structural fixed | Photovoltaic and thin film integrated with curtain wall

Figure A5. Different types of integrated window-and-curtain-wall structures with thin and photovoltaic film in visible light transmittance.

Appendix B

Zone Sizing Summary for FAN COIL (PROJECT SPECIFICATIONS
Project Name: Glass load comparison (B01)
Prepared by: ZFP

05/06/2021
09:40AM

Air System Information

Air System Name . **FAN COIL (PROJECT SPECIFICATIONS**
Equipment Class .... **TERM**
Air System Type **4P-FC**

Number of zones **1**
Floor Area .... **29.0** m²
Location .... **AL AHSA, Saudi Arabia**

Sizing Calculation Information

Zone and Space Sizing Method:

Zone L/s .... **Sum of space airflow rates**
Space L/s .... **Individual peak space loads**

Calculation Months .... **Jan to Dec**
Sizing Data .... **Calculated**

Zone Sizing Data

Zone Name	Maximum Cooling Sensible (kW)	Design Air Flow (L/s)	Minimum Air Flow (L/s)	Time of Peak Load	Maximum Heating Load (kW)	Zone Floor Area (m²)	Zone L/(s-m²)
Zone 1	3.5	269	269	Jul 1500	0.8	29.0	9.29

Terminal Unit Sizing Data - Cooling

Zone Name	Total Coil Load (kW)	Sens Coil Load (kW)	Coil Entering DB / WB (°C)	Coil Leaving DB / WB (°C)	Water Flow @ 8.0 °K (L/s)	Time of Peak Load
Zone 1	5.5	3.7	25.1 / 19.1	13.6 / 13.1	0.16	Jun 1600

Terminal Unit Sizing Data - Heating, Fan, Ventilation

Zone Name	Heating Coil Load (kW)	Heating Coil Ent/Lvg DB (°C)	Htg Coil Water Flow @15.0 °K (L/s)	Fan Design Airflow (L/s)	Fan Motor (BHP)	Fan Motor (kW)	OA Vent Design Airflow (L/s)
Zone 1	0.8	20.8 / 23.4	0.01	269	0.000	0.000	52

Space Loads and Airflows

Zone Name / Space Name	Mult.	Cooling Sensible (kW)	Time of Load	Air Flow (L/s)	Heating Load (kW)	Floor Area (m²)	Space L/(s-m²)
Zone 1							
EI-WD-0260 (PROJECT)	1	3.5	Jul 1500	269	0.8	29.0	9.29

Hourly Analysis Program v4.50 — Page 1 of 1

Figure A6. Zone sizing summary for tender specifications of fan coil.

Zone Sizing Summary for FAN COIL (SAMPLE 2)

Project Name: Glass load comparison (B01)
Prepared by: ZFP

05/06/2021
09:48AM

Air System Information

Air System Name .... FAN COIL (SAMPLE 2)
Equipment Class .... TERM
Air System Type 4P-FC

Number of zones 1
Floor Area .... 29.0 m^2
Location .... AL AHSA, Saudi Arabia

Sizing Calculation Information

Zone and Space Sizing Method:

Zone L/s .... Sum of space airflow rates
Space L/s .... Individual peak space loads

Calculation Months .... Jan to Dec
Sizing Data .... Calculated

Zone Sizing Data

Zone Name	Maximum Cooling Sensible (kW)	Design Air Flow (L/s)	Minimum Air Flow (L/s)	Time of Peak Load	Maximum Heating Load (kW)	Zone Floor Area (m^2)	Zone L/(s-m^2)
Zone 1	3.0	228	228	Jul 1500	0.7	29.0	7.86

Terminal Unit Sizing Data - Cooling

Zone Name	Total Coil Load (kW)	Sens Coil Load (kW)	Coil Entering DB / WB (°C)	Coil Leaving DB / WB (°C)	Water Flow @ 8.0 °K (L/s)	Time of Peak Load
Zone 1	4.8	3.1	25.4 / 19.6	13.9 / 13.4	0.14	Jul 1500

Terminal Unit Sizing Data - Heating, Fan, Ventilation

Zone Name	Heating Coil Load (kW)	Heating Coil Ent/Lvg DB (°C)	Htg Coil Water Flow @15.0 °K (L/s)	Fan Design Airflow (L/s)	Fan Motor (BHP)	Fan Motor (kW)	OA Vent Design Airflow (L/s)
Zone 1	0.7	20.7 / 23.4	0.01	228	0.000	0.000	52

Space Loads and Airflows

Zone Name / Space Name	Mult.	Cooling Sensible (kW)	Time of Load	Air Flow (L/s)	Heating Load (kW)	Floor Area (m^2)	Space L/(s-m^2)
Zone 1							
EI-WD-0260 (SAMPLE 2)	1	3.0	Jul 1500	228	0.7	29.0	7.86

Hourly Analysis Program v4.50   Page 1 of 1

Figure A7. Zone sizing summary for fan coil when choosing alternative sample two.

References

1. Koytcheva, M.K.; Sauerwein, L.K.; Webb, T.L.; Baumgarn, S.A.; Skeels, S.A.; Duncan, C.G. A systematic review of environmental sustainability in veterinary practice. *Top. Companion Anim. Med.* **2021**, *44*, 100550. [CrossRef]
2. Danilov, A.; Benuzh, A.; Yeye, O.; Compaore, S.; Rud, N. Design of healthcare structures by green standards. *E3S Web Conf.* **2020**, *164*, 05002. [CrossRef]
3. Karliner, J.; Guenther, R. *Global Green and Healthy Hospitals, A Comprehensive Environmental Health Agenda for Hospitals and Health Systems around the World*; Health Care Without Harm: Reston, VA, USA, 2022.
4. Billanes, J.D.; Ma, Z.; Jørgensen, B.N. The bright green hospitals case studies of hospitals' energy efficiency and flexibility in philippines. In Proceedings of the 2018 8th International Conference on Power and Energy Systems (ICPES), Colombo, Sri Lanka, 21–22 December 2018; pp. 190–195.
5. Hensher, M.; Mcgain, F. Health Care Sustainability Metrics: Building A Safer, Low-Carbon Health System: Commentary examines how to build a safer, low-carbon health system. *Health Aff.* **2020**, *39*, 2080–2087. [CrossRef] [PubMed]
6. Balo, F.; Sua, L.S.; Polat, H. Green Hospitals and Sustainability: Case of Companion House of a Research Hospital. In *Green Energy and Infrastructure*; CRC Press: Boca Raton, FL, USA, 2020; pp. 63–92.
7. de Fátima Castro, M.; Mateus, R.; Bragança, L. Healthcare building sustainability assessment tool-sustainable effective design criteria in the Portuguese context. *Environ. Impact Assess. Rev.* **2017**, *67*, 49–60. [CrossRef]
8. Kim, S.; Osmond, P. Analyzing green building rating tools for healthcare buildings from the building user's perspective. *Indoor Built Environ.* **2014**, *23*, 757–766. [CrossRef]
9. Caixeta, M.C.B.F.; Fabricio, M.M. Physical-digital model for co-design in healthcare buildings. *J. Build. Eng.* **2021**, *34*, 101900. [CrossRef]
10. Minaei, M.; Aksamija, A. Performance-based facade framework automated and multi-objective simulation and optimization. In Proceedings of the 11th Annual Symposium on Simulation for Architecture and Urban Design, Virtual Event, 25–27 May 2020; pp. 1–8.
11. Mengjie, S.; Fuxin, N.; Ning, M.; Yanxin, H.; Shiming, D. Review on building energy performance improvement using phase change materials. *Energy Build.* **2018**, *158*, 776–793.
12. Miedema, E.; Lindahl, G.; Elf, M. Conceptualizing health promotion in relation to outpatient healthcare building design: A scoping review. *Health Environ. Res. Des. J.* **2019**, *12*, 69–86. [CrossRef] [PubMed]
13. William, M.; El-Haridi, A.; Hanafy, A.; El-Sayed, A. Assessing the energy efficiency and environmental impact of an egyptian hospital building. *IOP Conf. Ser. Earth Environ. Sci.* **2019**, *397*, 012006. [CrossRef]
14. Chías, P.; Abad, T. Green hospitals, green healthcare. *Int. J. Energy Prod. Manag.* **2017**, *2*, 196–205. [CrossRef]
15. Nimlyat, P.S. Indoor environmental quality performance and occupants' satisfaction [IEQPOS] as assessment criteria for green healthcare building rating. *Build. Environ.* **2018**, *144*, 598–610. [CrossRef]
16. Sadek, A.H.; Mahrous, R. Adaptive glazing technologies: Balancing the benefits of outdoor views in healthcare environments. *Sol. Energy* **2018**, *174*, 719–727. [CrossRef]
17. Lee, Y.; Choi, J.; Yoon, H.; Kim, S.K. Perception and Attitude on Augmented Reality Smart Glass for Healthcare Convergence Simulation. *J. Korea Converg. Soc.* **2021**, *12*, 369–377. [CrossRef]
18. Alameh, K.; Vasiliev, M.; Alghamedi, R.; Nur-E-Alam, M.; Rosenberg, V. Solar energy harvesting clear glass for building-integrated photovoltaics. In Proceedings of the 2014 11th Annual High Capacity Optical Networks and Emerging/Enabling Technologies (Photonics for Energy), Charlotte, NC, USA, 15–17 December 2014; pp. 210–213.
19. Trajanoska, B.; Doncheva, E. Influence of structural glass on working environment quality and healthcare benefits. *J. Environ. Prot. Ecol.* **2019**, *20*, 468–473.
20. Mikhailov, L.; Mikhailova, S.; Ismailova, G.; Yersayin, R.; Kenes, N.; Lavrishev, O.; Nikulin, V. Using solar energy by a smart window for the needs of urban residents. *Mater. Today Proc.* **2020**, *25*, 64–66. [CrossRef]
21. Naqash, M.T.; Formisano, A.; De Matteis, G. Aluminium framing members in facades. In *Key Engineering Materials*; Trans Tech Publications Ltd.: Baech, Switzerland, 2016; pp. 327–332.
22. Bontekoe, E.; Van Sark, W.; Van Leeuwen, J. Building-Integrated Photovoltaics. In *Designing with Photovoltaics*; CRC Press: Boca Raton, FL, USA, 2020; pp. 127–163.
23. Lee, J. Noise reduction and air behaviors in ventilated single-glazed façade with glass fiber-based shading louvers and compact silencers. In Proceedings of the INTER-NOISE and NOISE-CON Congress and Conference, Chicago, IL, USA, 26–29 August 2018; Institute of Noise Control Engineering: Reston, VA, USA, 2018; pp. 162–170.
24. Hodapp, R.T. LEED-EB: Leadership in Energy and Environmental Design for Existing Buildings. In *Managing Human and Social Systems*; CRC Press: Boca Raton, FL, USA, 2020; pp. 401–411.
25. Shahin, H.S.M. Adaptive building envelopes of multistory buildings as an example of high performance building skins. *Alex. Eng. J.* **2019**, *58*, 345–352. [CrossRef]
26. Mirmozaffari, M.; Alinezhad, A. Window analysis using two-stage DEA in heart hospitals. In Proceedings of the 10th International Conference on Innovations in Science, Engineering, Computers and Technology (ISECT-2017), Dubai, United Arab Emirates, 17–19 October 2017; pp. 44–51.
27. Sok, E.; Sanders, H. New developments in dynamic glass: Towards a new era in sustainable construction. In Proceedings of the Challenging Glass Conference Proceedings, Gent, Belgium, 16–17 June 2016; pp. 133–142.

28. Pradinuk, R. Incentivizing the daylit hospital: The green guide for health care approach. *Health Environ. Res. Des. J.* **2009**, *2*, 92–112. [CrossRef] [PubMed]
29. Benzidia, S.; Makaoui, N.; Bentahar, O. The impact of big data analytics and artificial intelligence on green supply chain process integration and hospital environmental performance. *Technol. Forecast. Soc. Chang.* **2021**, *165*, 120557. [CrossRef]
30. Cesari, S.; Valdisseri, P.; Coccagna, M.; Mazzacane, S. Energy savings in hospital patient rooms: The role of windows size and glazing properties. *Energy Procedia* **2018**, *148*, 1151–1158. [CrossRef]
31. Huisman, E.R.C.M.; Morales, E.; van Hoof, J.; Kort, H.S.M. Healing environment: A review of the impact of physical environmental factors on users. *Build. Environ.* **2012**, *58*, 70–80. [CrossRef]
32. Golbazi, M.; Aktas, C.B. Analysis of credits earned by LEED healthcare certified facilities. *Procedia Eng.* **2016**, *145*, 203–210. [CrossRef]
33. Balabel, A.; Alwetaishi, M. Toward Sustainable Healthcare Facilities: An Initiative for Development of "Mostadam-HCF" Rating System in Saudi Arabia. *Sustainability* **2021**, *13*, 6742. [CrossRef]
34. Kocman, D.; Števanec, T.; Novak, R.; Kranjec, N. Citizen science as part of the primary school curriculum: A case study of a technical day on the topic of noise and health. *Sustainability* **2020**, *12*, 10213. [CrossRef]
35. Lee, A.D.; Shepherd, P.; Evernden, M.C.; Metcalfe, D. Optimizing the architectural layouts and technical specifications of curtain walls to minimize use of aluminium. *Structures* **2018**, *13*, 8–25. [CrossRef]
36. Kang, H.; Hong, T.; Jung, S.; Lee, M. Techno-economic performance analysis of the smart solar photovoltaic blinds considering the photovoltaic panel type and the solar tracking method. *Energy Build.* **2019**, *193*, 1–14. [CrossRef]
37. Wagdy, A.; Sherif, A.; Sabry, H.; Arafa, R.; Mashaly, I. Daylighting simulation for the configuration of external sun-breakers on south oriented windows of hospital patient rooms under a clear desert sky. *Sol. Energy* **2017**, *149*, 164–175. [CrossRef]
38. Eisazadeh, N.; de Troyer, F.; Allacker, K. Integrated energy, daylighting, comfort and environmental performance analysis of window systems in patient rooms. *Archit. Sci. Rev.* **2022**, *65*, 319–337. [CrossRef]
39. Waheed, Z. The British Council Lahore's Green and LEED-certified Library Building. In *Green Behavior and Corporate Social Responsibility in Asia*; Emerald Publishing Limited: Bingley, UK, 2019; pp. 17–25.
40. Sathishkumar, R.; Tamilselvan, S.; Magesh, C.J.; Venkatapathy, K.; Vimalan, M.; Levanya, G. Synthesis, crystal growth, optical, thermal, mechanical and dielectric properties of mono (bis (2-(4-butylphenyl) imino) methyl) phenoxy) zinc (II) dichloride (MBPMP) single crystal as a nonlinear optical (NLO) material. *J. Mater. Sci. Mater. Electron.* **2019**, *30*, 17504–17513. [CrossRef]
41. Cuce, E.; Cuce, P.M. Vacuum glazing for highly insulating windows: Recent developments and future prospects. *Renew. Sustain. Energy Rev.* **2016**, *54*, 1345–1357. [CrossRef]
42. Gatland, S., II; Ihab, E.; Aldo, G.; Yacine, D. Measuring the Impact of a High-Performance All-Glass Building on the Indoor Acoustic Environment and the Occupants Perception of Health, Satisfaction and Productivity. In Proceedings of the INTER-NOISE and Noise-Con Congress and Conference, Chicago, IL, USA, 26–29 August 2018; Institute of Noise Control Engineering: Reston, VA, USA, 2018; pp. 1801–1812.
43. Antoniadou, P.; Symeonidou, M.; Kyriaki, E.; Giama, E.; Boemi, S.-N.; Chadiarakou, S.; Papadopoulos, A.M. High performance building façades for Zero Energy Buildings in Greece: State of the art and perspectives. *IOP Conf. Ser. Earth Environ. Sci.* **2020**, *410*, 012036. [CrossRef]
44. Mayhoub, M.M.G.; El-Sayad, Z.M.T.; Ali, A.A.M.; Ibrahim, M.G. Assessment of green building materials' attributes to achieve sustainable building façades using AHP. *Buildings* **2021**, *11*, 474. [CrossRef]
45. Silenzi, F.; Priarone, A.; Fossa, M. Hourly simulations of an hospital building for assessing the thermal demand and the best retrofit strategies for consumption reduction. *Therm. Sci. Eng. Prog.* **2018**, *6*, 388–397. [CrossRef]
46. Ramana, M.V.; Saboor, S. A novel glazing system filled with hydrogel granules: Energy saving, diurnal illumination, color rendering, and CO_2 emission mitigation prospective. *Energy Sources Part A Recovery Util. Environ. Eff.* **2021**, 1–16. [CrossRef]
47. Shan, M.; Hwang, B.G. Green building rating systems: Global reviews of practices and research efforts. *Sustain. Cities Soc.* **2018**, *39*, 172–180. [CrossRef]
48. Moghtadernejad, S.; Chouinard, L.E.; Mirza, M.S. Design strategies using multi-criteria decision-making tools to enhance the performance of building façades. *J. Build. Eng.* **2020**, *30*, 101274. [CrossRef]
49. Saeli, M.; Piccirillo, C.; Parkin, I.P.; Ridley, I.; Binions, R. Nano-composite thermochromic thin films and their application in energy-efficient glazing. *Sol. Energy Mater. Sol. Cells* **2010**, *94*, 141–151. [CrossRef]
50. Atzeri, A.M.; Gasparella, A.; Cappelletti, F.; Tzempelikos, A. Comfort and energy performance analysis of different glazing systems coupled with three shading control strategies. *Sci. Technol. Built Environ.* **2018**, *24*, 545–558. [CrossRef]
51. Jin, Q.; Overend, M. A comparative study on high-performance glazing for office buildings. *Intell. Build. Int.* **2017**, *9*, 181–203. [CrossRef]
52. Rose, A.A. Glazing Performance in the Patient Care Setting. Master's Thesis, Graduate School of Clemson University, Clemson, SC, USA, 2017.
53. Zaphar, S.; Sheworke, T. Computer program for cooling load estimation and comparative analysis with hourly analysis program (HAP) software. *Int. J. Latest Technol. Eng. Manag. Appl. Sci.* **2018**, *7*, 53–61.
54. Zuhair Fayez Partnership Consultants. *Design, Kfu Campus Supervisor Teaching Hospital 400 Beds Bidding Document and Monthly Report*; Zuhair Fayez Partnership Consultants: Riyadh, Saudi Arabia, 2015.

55. Raj, S.S.R.; Dhas, J.; Edwin, R.; Jesuthanam, C.P. Challenges on machining characteristics of natural fiber-reinforced composites–A review. *J. Reinf. Plast. Compos.* **2021**, *40*, 41–69. [CrossRef]
56. Ciraulo, R.; Nubbe, V.; Wedekind, S.; Jean-Michel, C.; Stanley, J. *Commercial Building Fenestration Market Study Prepared for: The National Fenestration Rating Council August 2021*; Lawrence Berkeley National Laboratory (LBNL): Berkeley, CA, USA, 2021.
57. Sherif, A.; Hanan, S.; Rasha, A.; Ayman, W. Energy efficient hospital patient room design: Effect of room shape on the window-to-wall ratio in a desert climate. In Proceedings of the 30th International PLEA Conference: Sustainable Habitat for Developing Societies: Choosing the Way Forward-Proceedings. PLEA-Passive and Low Energy Architecture, Hong Kong, China, 16–18 December 2014; pp. 352–360.
58. Wilson, H.R.; Elstner, M. Spot landing: Determining the light and solar properties of fritted and coated glass. In Proceedings of the Challenging Glass Conference Proceedings, Delft, The Netherlands, 17–18 May 2018; pp. 203–212.
59. Hart, R.; Selkowitz, S.; Curcija, C. Thermal performance and potential annual energy impact of retrofit thin-glass triple-pane glazing in US residential buildings. In *Building Simulation*; Tsinghua University Press: Beijing, China, 2019; pp. 79–86.
60. Xuan, X. Effectiveness of indoor environment quality in LEED-certified healthcare settings. *Indoor Built Environ.* **2016**, *25*, 786–798. [CrossRef]

buildings

MDPI

Article

Exploring Energy Retrofitting Strategies and Their Effect on Comfort in a Vernacular Building in a Dry Mediterranean Climate

Andrea Lozoya-Peral [1], Carlos Pérez-Carramiñana [1,*], Antonio Galiano-Garrigós [1], Ángel Benigno González-Avilés [1] and Stephen Emmitt [2]

1 Departamento de Construcciones Arquitectónicas, Escuela Politécnica Superior, Universidad de Alicante, 03690 Alicante, Spain; andrealozoyaperal@gmail.com (A.L.-P.); antonio.galiano@ua.es (A.G.-G.); angelb@ua.es (Á.B.G.-A.)

2 Department of Architecture and Civil Engineering, University of Bath, Bath BA2 7AY, UK; s.emmitt@bath.ac.uk

* Correspondence: c.perez@ua.es; Tel.: +34-615388644

Abstract: This research explores the energy behaviour of a traditional house on the Mediterranean coast of south-eastern Spain. The objective of the work is to determine the optimal passive strategies for rehabilitating a traditional house, improving its energy savings and comfort, considering the characteristics of the warm semi-arid Mediterranean climate. The main novelty of this article is that it demonstrates that the limits imposed by current regulations, based on globalised climate strategy approaches, undermine the energy efficiency capacity that passive solutions in vernacular architecture already employed. The methodology used consists of a systematised multi-objective study of various energy rehabilitation strategies. Four strategies were studied: raising the thermal insulation of enclosures, improving thermal insulation and solar control glazing with movable shading devices, increasing the size of windows and introducing the use of natural ventilation enhanced by ceiling fans. The results show that simultaneous improvement of these parameters reduces cooling and heating requirements by up to 87%, reducing the energy consumption of air conditioning systems. Indoor temperatures are also maintained within the comfort limits set by regulations for 91% of hours per year without the need for air conditioning systems. This results in a passive energy-efficient and comfortable house almost all year round. This work offers an alternative solution to the comfort standards of current Spanish regulations and demonstrates the need to adapt Fanger's analytical method for comfort estimation. The research concludes that the comfort criteria of current energy regulations should be modified to better adapt the design criteria to the dry Mediterranean climate.

Keywords: energy efficiency; thermal comfort; nearly zero energy buildings; sustainable rehabilitation; warm semi-arid dry Mediterranean climate; simulation

Citation: Lozoya-Peral, A.; Pérez-Carramiñana, C.; Galiano-Garrigós, A.; González-Avilés, Á.B.; Emmitt, S. Exploring Energy Retrofitting Strategies and Their Effect on Comfort in a Vernacular Building in a Dry Mediterranean Climate. *Buildings* **2023**, *13*, 1381. https://doi.org/10.3390/buildings13061381

Academic Editors: Nianping Li and Yingdong He

Received: 30 March 2023
Revised: 13 May 2023
Accepted: 23 May 2023
Published: 26 May 2023

1. Introduction

It is established that buildings are responsible for 29% of the world's energy use and contribute to 21% of CO_2 emissions [1]. In the EU, the share of energy consumption of residential buildings reached 40% [2] at the beginning of the century and 25.7% in 2016 [3]. The Sustainable Development Goals stress the importance of reducing the environmental impact of urban environments [4] and improving the energy efficiency of buildings [5,6]. These objectives entail a progressive increase in European energy efficiency regulatory requirements, such as European Directive 2010/317EU [7] and subsequently Directive (EU) 2018/844 [8], and consequently the Spanish regulations have also been revised [9,10]. These regulations have greatly increased the thermal insulation requirements for new and renovated buildings [11,12], with a particular focus on reducing the thermal transmittance and airtightness of building envelopes. However, the creation of airtight and highly insulated thermal envelopes is not the best construction response in hot and semi-arid climates with high annual insolation such as the dry Mediterranean climate (BShs). The dry Mediterranean climate is characterised by a lower annual temperature oscillation than other European climates, with mild winters and not excessively hot summers, and a high level of sunshine throughout the year. In recent years, there has been an increase in tropical nights throughout the year [13,14]. In Europe, this type of climate is not very representative because it is found exclusively in south-eastern Spain and southern Greece [15]. This specificity implies designing a particular architecture that is different from the rest of the European climate zones, which is missing in the Spanish regulations. The current literature and building energy efficiency regulations do not specifically analyse the advantages of high annual insolation and natural ventilation of houses in a dry Mediterranean climate (BShs). The originality of this work is in studying, with priority to comfort and passive solutions, the optimal strategies for rehabilitating a traditional Mediterranean house with the most balanced intervention, improving its energy efficiency and avoiding air conditioning systems.

Adapting design criteria to climatic characteristics is one of the key actions to meet the regulatory requirements for nearly zero energy buildings (NZEB) [16] and to reduce greenhouse gas emissions from buildings [17]. Additionally, these design criteria should incorporate optimal construction solutions to achieve the lowest energy and material consumption over the entire life cycle of the building. To this end, the architectural design and construction systems should be adapted to the characteristics of the local climate. This means using design strategies that do not need energy to work in accordance with the Kyoto Pyramid [18] to reduce energy consumption and increase user comfort with less use of air conditioning systems.

In the case of the dry Mediterranean climate of south-eastern Spain, the optimal passive design solutions are those that in winter take advantage of the "free" heating from high solar radiation and in summer protect the interior from the sun and dissipate excess heat from inside the building via ventilation. Therefore, in addition to improving the insulation of the building envelope [19–21], adjustable solar shading systems [22] and natural ventilation should be used to avoid overheating due to the unwanted greenhouse effect in the warmer months [23]. However, current Spanish and European energy efficiency regulations focus on ensuring highly thermally insulated and airtight building envelopes. These regulations do not facilitate the application of some passive systems typical of the dry Mediterranean climate [24], where vernacular architecture is defined by a less-tight envelope that promotes natural ventilation.

They also do not consider the advantages of natural ventilation for the thermal comfort of users in this climate, as proposed in the bioclimatic diagrams of Givoni [25,26] or Oligay [27] and analysed in many studies [28–36]. Many researchers have studied the influence of natural ventilation on the indoor thermal environment in residential buildings as well [37–40].

This standard sets much more restrictive operating temperature limits and does not consider the positive effect of air velocity in achieving comfort even at higher temperatures than allowed [41]. Current Spanish energy efficiency regulations and national energy certification software [42,43] do not allow for modelling the effect of natural ventilation, despite its advantages of dissipating heat in summer and lowering the temperature and relative humidity of the air [44–47]. It should be noted that there are studies on this type of climate in other parts of the world, but their architectures are not regulated by European standards and have other variations in their design and construction systems [48,49], making comparisons misleading.

The novelty of this research is to demonstrate that, in a BShs climate, only the combined action of improving thermal insulation, increasing the size of windows, increasing the solar control of glazing and using natural ventilation supported by ceiling fans can achieve energy-efficient and comfortable houses without the use of air conditioning. This work proposes an alternative solution to the current energy efficiency regulations in Spain and Europe.

This research analyses a traditional house located in the south-east of Spain. The purpose is to demonstrate the possibility of achieving energy-efficient and comfortable homes all year round in a BShs climate without using conditioning systems. The aim was to optimise energy performance and improve user comfort with minimal intervention, which is compatible with, and respectful of, the original building. The methodology consists of the study and analysis of the hygrothermal behaviour of a traditional house and a systematised study of the most influential design parameters for optimising energy performance and improving indoor comfort. The research modelled and analysed the constructive design that optimises the energy performance of the building to meet the requirements of current regulations [50,51] and improve comfort by adapting to the characteristics of the local climate [52–54]. The research analyses passive design retrofit strategies such as increased thermal insulation [55] and natural cross ventilation [56] to improve the energy efficiency, indoor hygrothermal quality [57] and thermal comfort [58] of traditional buildings. The target is to identify simple passive design strategies that can strongly influence the energy performance of a building in this climate. Other research has demonstrated the benefits of improving the insulation of traditional dwellings in Mediterranean climates in terms of the comfort of their users [59]. However, this research is not specific to the dry Mediterranean BShs climate. It also does not analyse the beneficial effect of the greenhouse effect in winter caused by increasing the size of windows, nor the use of natural ventilation in summer. Other research examines the influence of thermal inertia of traditional envelopes with the incorporation of novel high-thermal-insulation envelope systems with dynamic envelope ventilation strategies [60]. These solutions were discarded because of their aesthetic impact on traditional houses. The present research aims to demonstrate that adequate natural ventilation of the interior of the house with more windows sufficiently improves thermal performance in summer. This research does not consider other scenarios based on active systems, e.g., the use of renewable energies through thermal or photovoltaic panels that can be integrated into the existing fabric [61]. Our purpose is to provide an alternative to traditional mechanical methods of air conditioning [62] through bioclimatic design strategies. Sustainability criteria were considered, reflected using traditional and autochthonous materials, minimum intervention and alteration of the original building, and lower economic cost [63].

A traditional Mediterranean house was selected as a case study and we used the opinions of its occupants through analytical surveys of the Predicted Mean Vote (PMV) and Predicted Percentage Dissatisfied (PPD) indices for estimating comfort using the Fanger method. Many investigations apply the PMV and PPD indices according to the Fanger model and demonstrate the need to correlate the temperature with the relative humidity of the indoor environment to ensure that the human comfort zone is reached [64,65]. This work analyses the applicability of the Fanger model in this BShs climate zone, compares the results with survey data, and proposes some corrections. This paper also proposes modification of the Givoni diagrams and the Spanish regulations on comfort to adapt them to the BShs climate.

This study contributes to a better understanding of the thermal performance of vernacular buildings. It is hoped that it will slow the current degradation and disuse of these dwellings and prevent unsuitable thermal renovations and retrofits, thus helping to safeguard the architectural and cultural heritage of this region of Spain. The usefulness of this research is to quantitatively evaluate the improvement in energy efficiency and thermal comfort produced by the proposed passive systems for the vernacular architecture of south-eastern Spain with a dry Mediterranean climate. This research demonstrates the inefficiency of over-insulating buildings in a BShs climate. It also contributes to questioning the limitations of Spanish energy efficiency regulations regarding passive systems such as natural ventilation. In this way, Pfafferot [66] drew beneficial conclusions from the use of night ventilation in subtropical, Mediterranean and desert climates. The demonstration of the validity of economical passive systems adapted to vernacular traditions can contribute to a more sustainable energy rehabilitation compatible with the built heritage of the area and protecting the culture of the "orchard of Alicante".

2. Materials and Methods

The case study building is a traditional country house located near the city of Elche in the province of Alicante (Spain). It is located on the outskirts of the city in an area designated for agricultural plantations. Its geometry and construction method are common in the vernacular architecture of this area. The footprint is rectangular, 9.70 m long and 6.00 m wide, and the building is oriented south–north like most of the vernacular houses in the area. As is common in this typology, it has a main hall that runs longitudinally from east to the west, where the main activity of the house takes place, with a living–dining room and a kitchen. On the side, there are two bedrooms and a bathroom (Figure 1).

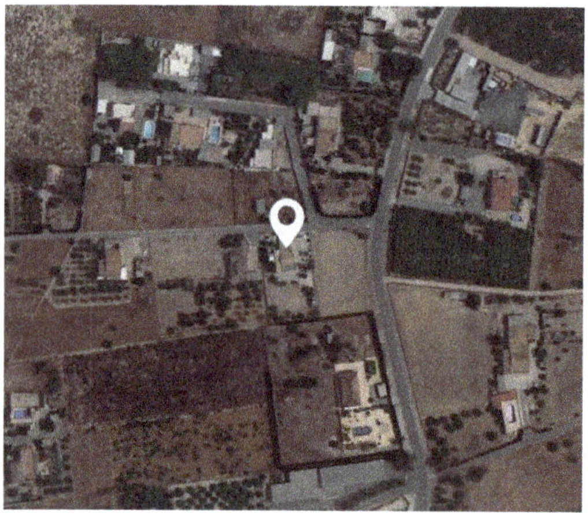

(a)

(b)

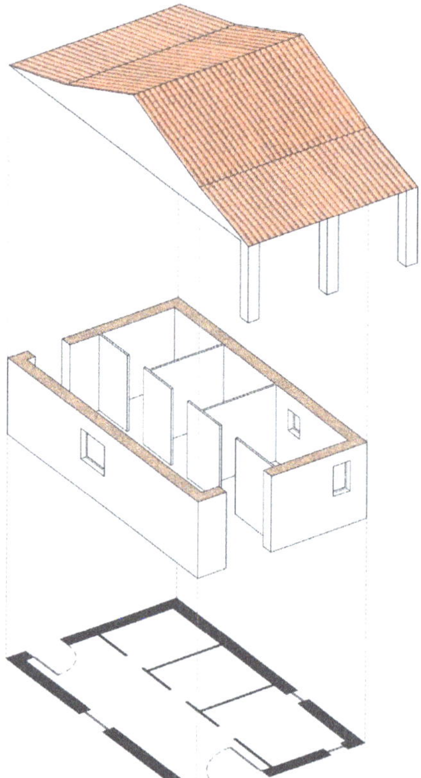

(c)

Figure 1. Case study house: (**a**) location, (**b**) external image, (**c**) floor plan.

This house was selected because it is an example of the traditional Mediterranean autochthonous architecture of south-eastern Spain. This building typology is characterised by local materials, a layout that responds to the old needs of the agricultural sector, and its typical orientation. The construction and spatial characteristics also serve as an example of other similar traditional architectural typologies in this part of Spain. This building typology is currently experiencing a gradual process of abandonment due to its age and insufficient (thermal) comfort. This leads to fabric deterioration and the disappearance of the cultural and architectural heritage of the area [67].

The methodology used in this work consisted of two phases. The first was an investigation of the current hygrothermal conditions of the house, supported by an occupant survey of thermal comfort. This took place over a 12-month period. Second, a systematized and parametric study of different alternative passive design retrofit solutions was conducted. A comparative analysis of five proposed retrofit interventions and their effect on building energy performance and comfort considering the specific conditions of the local climate was performed.

Phase 1. Three types of measurements were made in situ. First, a thermographic survey was carried out to establish the performance of the building's thermal envelope. Second, using a thermal transmittanceflowmeter, the thermal transmittance of the envelope of the enclosures that are part of the house was measured. Third, by using a hot-wire probe, the behaviour and stratification of temperatures inside the spaces was measured simultaneously with air velocity. An important part of this study was the evaluation of user perception of comfort in spaces and their thermal adaptation. This data was captured through a questionnaire survey that collected the occupants' perceptions for twelve months. The questionnaire was completed by the eight members of the family who were residents or regular visitors to the property. A procedure was defined in all visits to ensure that the survey was completed according to pre-established conditions. These conditions were that the measurements were to be taken at the same time and the data recording instruments should be connected five minutes before taking the measurement. An important part of the protocol was that air conditioning could not be turned on for the whole day so as not to interfere with the interior conditions.

At the same time, using a weather station, a climate file was created that collected outdoor temperature and humidity for a period of 12 months, with two daily readings every 3 days.

The instrumentation used for data collection consisted of different teams. The thermographic evaluation was carried out by using a Testo 868 camera. The evaluation of thermal transmittance, humidity and temperature, and air speed was carried out by using a Testo 435-2 multifunctional instrument calibrated with standard settings.

Phase 2. This research has a systematized approach and is based on the four rehabilitation interventions that have the greatest influence on the energy efficiency of this kind of building in this climate. The proposed modifications were:

- Improving the thermal insulation of opaque enclosures, evaluating the optimal thickness of the thermal insulation (Modification 1);
- Improving window characteristics by reducing the thermal transmittance of the glazing (Ug) (Modification 2a), which reduces the thermal transmittance of the windows (Uw). The solar control factor of glass (g) (Modification 2b) and total solar transmittance of the glazing with an activated movable shading device (g_gl;sh;wi) (Modification 2c) were also improved;
- Increasing the proportion of glazed area on the walls (Window Wall Ratio WWR) (Modification 3), analysing the impact of window size and proportion over the total of the façade;
- Use of natural ventilation and the advantages of using ceiling fans to improve cross ventilation and redistribute the air and its temperature (Modification 4). The effect of natural ventilation on the operation of the building was analysed, and we performed computational fluid dynamics (CFD) analysis to determine the circulation speed of indoor air and its influence on the interior conditioning of the spaces.

And finally, a solution was proposed with all improvements implemented simultaneously.

This study analyses the influence of each of these interventions on interior temperatures, interior comfort and energy efficiency (Figure 2).

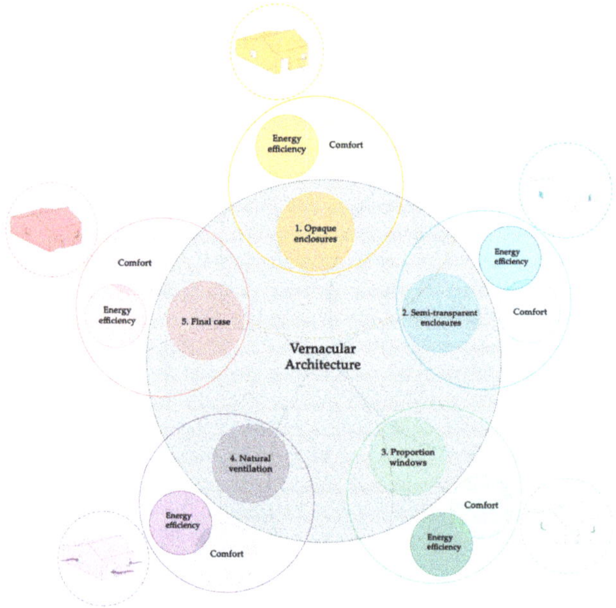

Figure 2. Methodology diagram of the study.

Design Builder software (version v.7.0.1.006, DesignBuilder Software Limited, Stroud, UK) was used for modelling and simulation. This is a recognized software in the process of evaluating the energy performance of buildings that offers a wide variety of results and high customization. It uses the EnergyPlus calculation engine [68], which allows one to accurately define the geometry and construction of a building, in addition to defining use profiles that can include passive ventilation systems.

This software was also used in the development of the CFD analysis, in which the location of the building, the climatic conditions in which it is located, including the prevailing winds, and its construction characteristics were considered.

In a first adjustment, a heat transfer coefficient of 20 W/m2·K was considered, given the absence of insulation and a reference temperature equal to room temperature, since we did not intend to introduce the factor of conditioning systems in the analysis. The parameters used in air configuration in the analysis were:

- Maximum air speed equal to the wind speed at the input factor;
- Static gauge pressure equal to 0 at the air outlet.

The CFD simulation carried out was of the linear static type that omits the influence of external wind in the calculations. A convergence limit was defined in the calculation when the fluctuation or deviation was below the established level. At the same time, a large-enough analysis zone was established such that artificial variations in air acceleration did not occur with a variable fluid material environment and several iterations greater than 500 [69–71].

The calculation method used to evaluate the performance of the building was carried out in accordance with current Spanish regulations. Gains and losses through the building envelope by conduction were considered, to which the transmission of solar energy through the openings with their solar factor corrections and shading elements was added or subtracted. In addition, the loss of air through the gaps, the influence of ventilation and the internal loads generated by the occupants and the equipment were considered.

For all simulations, the climatic data were considered on the same days on which the measurement campaign was carried out, including temperature and relative humidity.

For the analysis of comfort, indoor temperatures measured in situ and those calculated virtually were compared with Givoni's bioclimatic diagrams. The estimation of the thermal sensation of users was also calculated with the Predicted Mean Vote (PMV) and the Predicted Percentage Dissatisfied (PPD) using the analytical Fanger method.

For the thermal simulations, the climate database of the El Altet weather station (Alicante) was used. This geographical area corresponds to climate B4 according to Spanish legislation [72]. It can be fitted into climate BShs (dry Mediterranean climate within a warm semi-arid climate) in the Köppen climate classification [73] (see Figure 2). The climate in this area is defined by mild winters and warm summers, and temperatures allow for comfortable conditions during a large part of the year, as the average annual temperature is 18.3 °C. The most complicated part of the year to allow comfort occurs in the summer when the average temperature is 27 °C and only falls to 22 °C during the nights. However, this situation is compensated by the sea breeze, as the location is near the coast. This climate is also defined by high annual sunshine, high relative humidity and few episodes of rain that mostly happen in the autumn, with storms and torrential downpours (Figure 3).

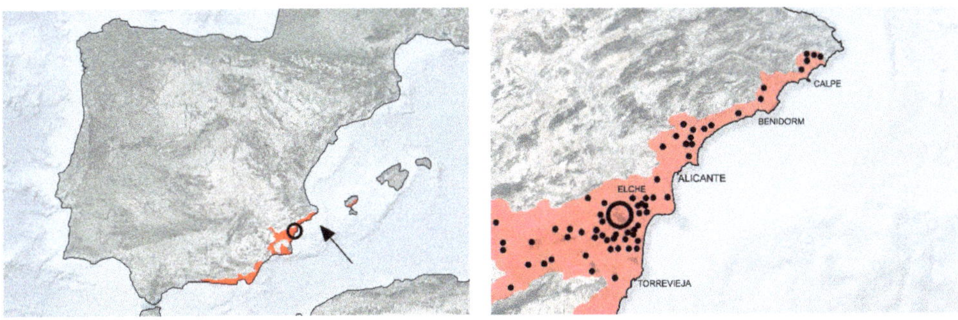

(**a**) (**b**)

Figure 3. (**a**) BShs climate zone (red) according to the Köppen–Geiger Climate Classification for the Iberian Peninsula, the Balearic Islands and the Canary Islands of the Agencia Estatal de Meteorología del Ministerio para la Transición Ecológica del Gobierno de España; (**b**) location map of the house (large circle) and main locations of this housing typology (small dots) (own elaboration).

The computer model of the house considered its situation, dimensions, orientation, geometry, location of glazing and construction of the thermal envelope (Figure 4 and Table 1).

Table 1. Original House Construction Characteristics.

Original House Construction	Thickness (cm)	Thermal Conductivity (W/m·K)	Thermal Resistance (m^2·K/W)	U (W/m^2·K)	g
Roof:				0.67	
Ceramic tile	1	1.00			
Air chamber	4		0.16		
Wattle	2	0.063			
Wooden beam	15	0.18			
Walls:				1.75	
Lime mortar	1	0.55			
Adobe and straw wall	40	1.10			
Lime mortar	1	0.55			
Lower slab:				2.56	
Soil	10	0.40			
Window:				Uw = 5.25	
Glass (87% of the window)				Ug = 5.70	g = 0.85
Frame (13% of the window)				Uf = 2.22	

Frame absorptivity = 0.75; frame air permeability = 100.00 $m^3/h \cdot m^2$.

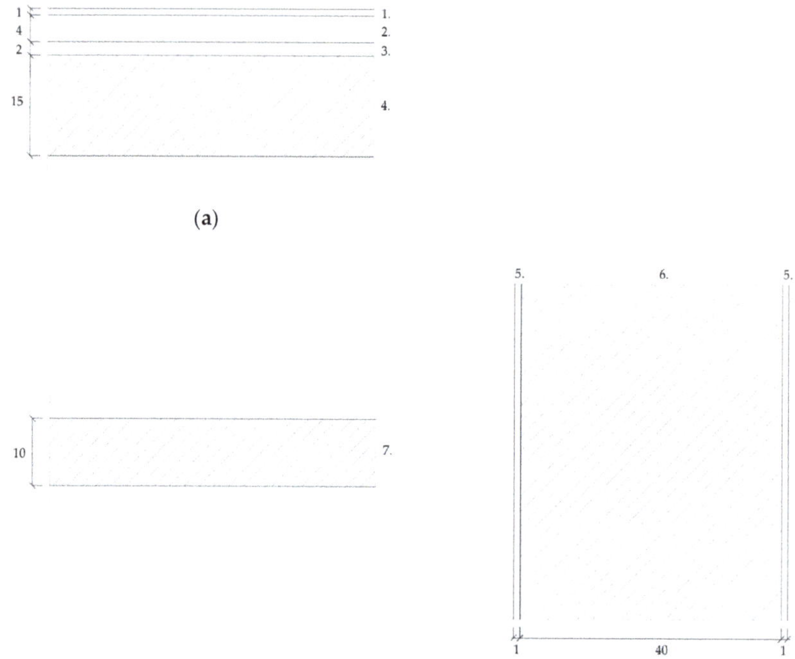

(a)

(b) (c)

Figure 4. Construction of the house: (**a**) roof section; (**b**) floor slab section; (**c**) façade wall construction. Materials: 1.: ceramic tile; 2.: air chamber; 3.: wattle; 4.: wooden beam; 5.: lime mortar; 6.: adobe and straw wall; 7.: soil. (dimensions in cm).

To introduce the modified construction and geometric parameters, the parameter variations were introduced individually and progressively to allow for the evaluation of their effect on the energy performance and comfort of the house (Figure 5 and Table 2).

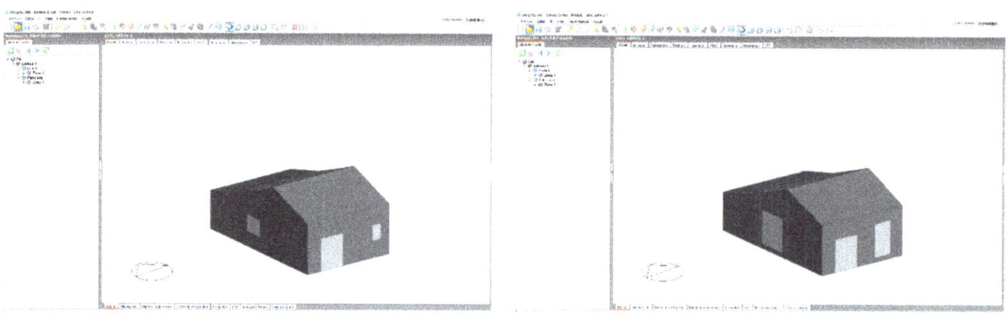

(a) (b)

Figure 5. Design Builder models: (**a**) original house; (**b**) modified house with 10% windows–wall ratio.

Table 2. Original house and proposed house modifications construction and geometrical characteristics.

	Thermal Insulation Thickness (cm)	Ug (W/m^2·K)	g	$g_{gl;sh;wi}$	Percentage of Windows (% with Respect WWR)	Natural Ventilation
Original house	0	1.75	0.85	0.65	2	No
Modification 1	11	1.75	0.85	0.65	2	No
Modification 2a	0	5.60–0.60	0.85	0.65	2	No
Modification 2b	0	1.75	0.75–0.25	0.65	2	No
Modification 2c	0	1.75	0.85	0.53–0.14	2	No
Modification 3	0	1.75	0.85	0.65	10	No
Modification 4	0	1.75	0.85	0.65	10	Yes
Final solution	11	1.60	0.65	0.14	10	Yes

Indicators and parameters of CTE DB-HE: Ug (W/m^2·K): thermal transmittance of glass; g: solar factor of glass; $g_{gl;sh;wi}$: total solar energy transmittance of the glazing with the movable shading device activated.

3. Results

3.1. Energy Efficiency

The calculations performed for the original house yielded winter heating needs above 39.07 KWh/m^2·year and summer cooling needs reaching 10.35 KWh/m^2·year (Figure 6). Therefore, the overall annual energy need (Heating + Cooling) reached 49.42 KWh/m^2·year.

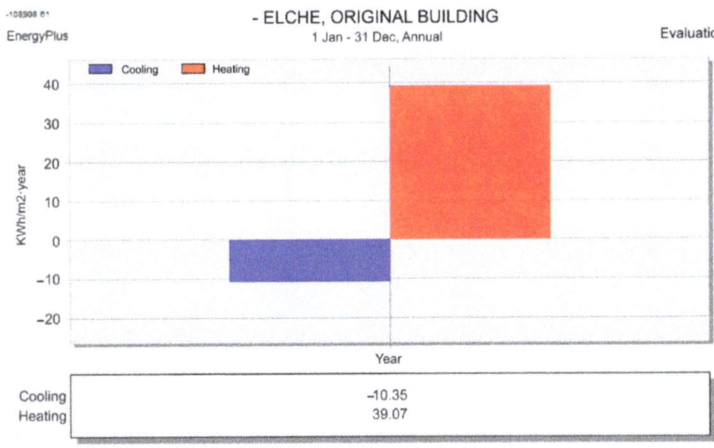

Figure 6. Simulation results of the calculations of heating and cooling demands of the original building with the original construction features.

The results offered by this simulation indicate that an improvement in the thermal insulation of the opaque part of the envelope (Modification 1) by introducing 11 cm of hemp insulation reduces the heating energy need in winter by 29% but increases the cooling energy need in summer by 6%. It was also found that the impact generated by an increase in the thickness of the thermal insulation was reduced if the thickness of the insulation was increased, also generating an increase in the cooling demand (Figure 7a). In this case, the total annual energy demand (Heating + Cooling) is only reduced by 15% (Figure 6b). The use of artificial insulating materials would reduce the thickness of insulation needed to achieve the same overall annual energy requirement, such as expanded polystyrene (EPS) (7.7 cm), polyurethane foam (7.2 cm) or outdoor vacuum insulation panels (VIPs) (5.3 cm).

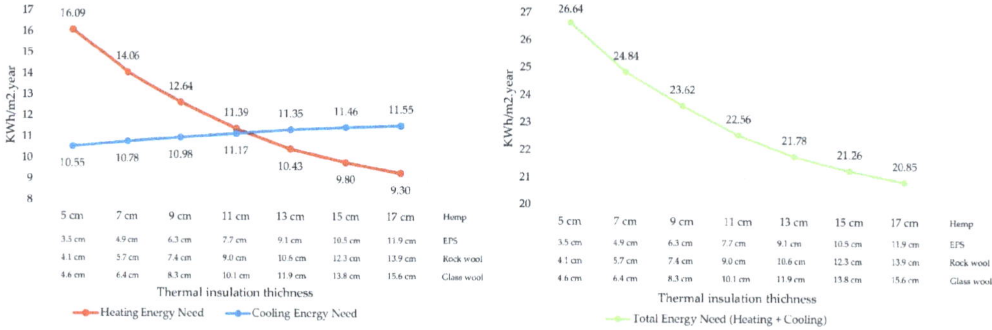

(a) (b)

Figure 7. Improvement in thermal insulation in the opaque parts of the building enclosures: (**a**) cooling and heating energy needs; (**b**) total energy need.

The yielded results show that improving the quality of the glazing by increasing its thermal insulation (Modification 2a) only offers an improvement of 2% and worsens summer cooling needs by 0.5% (Figure 7a), providing an overall improvement of 1.5% (Figure 8b).

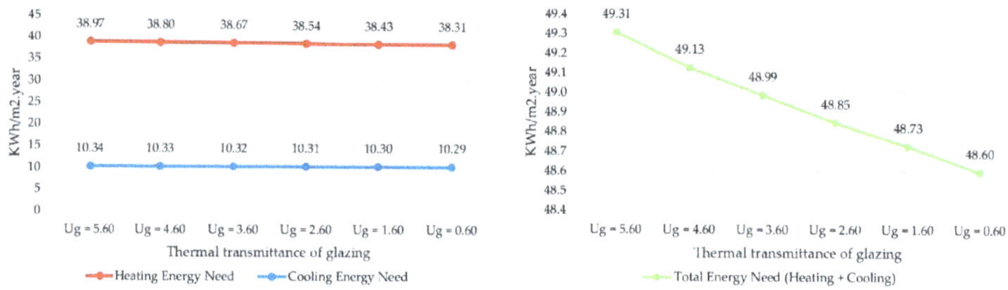

(a) (b)

Figure 8. Improvement in thermal insulation in glazing (Ug): (**a**) cooling and heating energy needs; (**b**) total energy need.

If an improvement in the solar factor (g) is introduced in glazing (Modification 2b), the cooling energy needs in summer are reduced by 2.5% and the heating needs in winter are increased by 2.3% (Figure 9a). As a result, this intervention increases the overall energy need (Heating + Cooling) by 1.3% (Figure 9b).

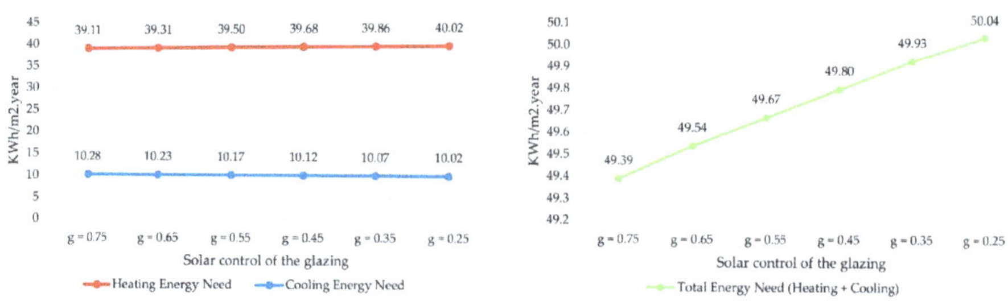

(a) (b)

Figure 9. Improvement in solar control of glazing (g): (**a**) cooling and heating energy needs; (**b**) total energy need.

The incorporation of movable solar shading (Modification 2c), using it only during the summer months, reduces summer cooling needs by 10% and does not worsen winter heating needs (Figure 10a). However, the overall energy need (Heating + Cooling) is only reduced by 2%. (Figure 10b).

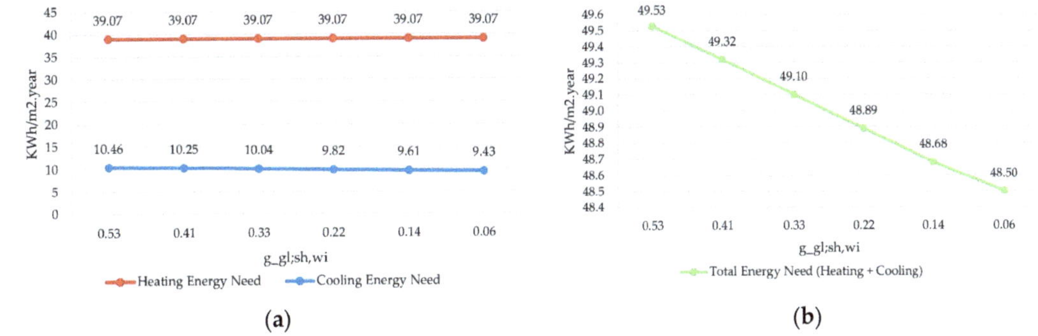

Figure 10. Improvement in glazing total solar energy transmittance by introducing movable shading devices (ggl;sh;wi): (**a**) cooling and heating energy needs; (**b**) total energy need.

Increasing the window–wall ratio (WWR) without adding solar shading (Modification 3) reduces winter heating needs by almost 18% but greatly increases summer cooling needs by more than 40% (Figure 11a). Consequently, the overall energy need (Heating + Cooling) can increase by up to 20% (Figure 11b).

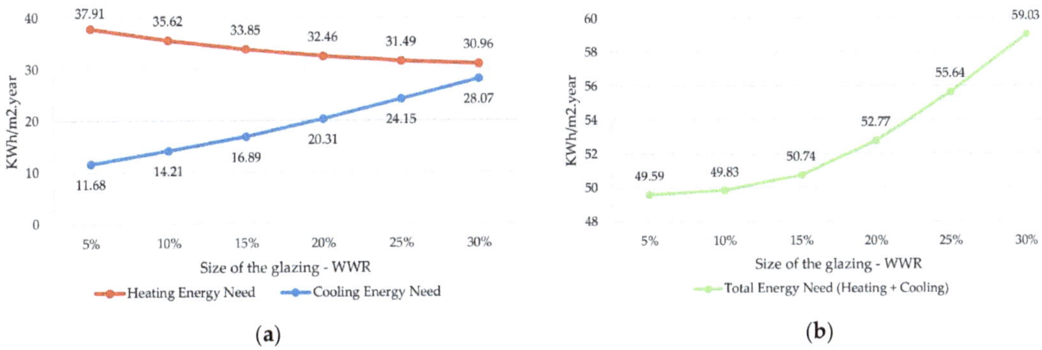

Figure 11. Modification of window–wall ratio (WWR): (**a**) cooling and heating energy needs; (**b**) total energy need.

Increasing the natural ventilation of the original dwelling in the summer (Modification 4) does not affect the winter heating needs and reduces the summer cooling needs by 7% (Figure 12a). This means a reduction in the overall energy need for heating and cooling by 2% (Figure 12b).

The results obtained show that simultaneous modification of the selected parameters allows a reduction in winter heating needs by up 2.02 KWh/m^2·year and summer cooling needs by up to 4.42 KWh/m^2·year (Figure 13). These results imply that it is possible to reduce the energy need for heating by up to 95% and the energy need for cooling by 57%, with a reduction in the global energy need of 87%, improving the annual energy performance of the house more than any of the calculated solutions that modified the parameters individually.

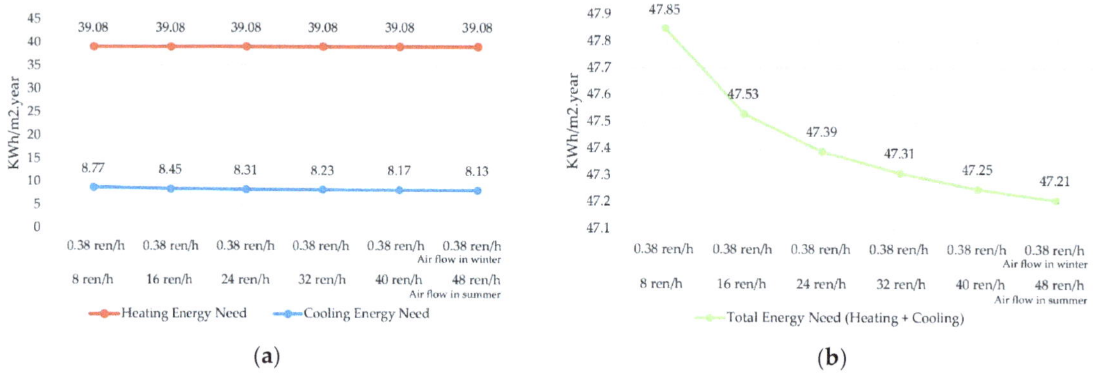

Figure 12. Modification of natural ventilation (ren/h): (**a**) cooling and heating energy needs; (**b**) total energy need.

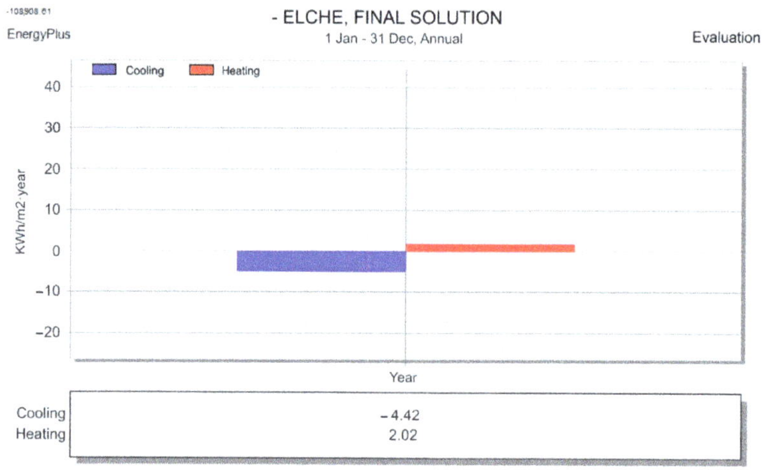

Figure 13. Simulation results of the calculations of heating and cooling demands, implementing all the proposed improvements simultaneously.

3.1.1. Indoor Temperatures

The results obtained from the on-site measurements of the original building show that indoor temperatures in the house, without using air conditioning systems, are below 21 °C approximately 67% of the year (equivalent to 5856 h). Temperatures are below 19 °C for approximately 52% of the year (equivalent to 4598 h) and below 12.4 °C for approximately 3% of the year (equivalent to 238 h). On the other hand, temperatures are above 25 °C for 14% of the year (equivalent to 1219 h) and above 27 °C for approximately 7% of the year (equivalent to 625 h), with more than 5% of hours above 28 °C (equivalent to 430 h). Temperatures did not exceed 32 °C throughout the year. Consequently, for approximately 19% of annual hours (equivalent to 1664 h), indoor temperatures were maintained between 21 °C and 25 °C, and for approximately 41% of annual hours (equivalent to 3592 h), indoor temperatures were maintained between 19 °C and 27 °C. (Figure 14).

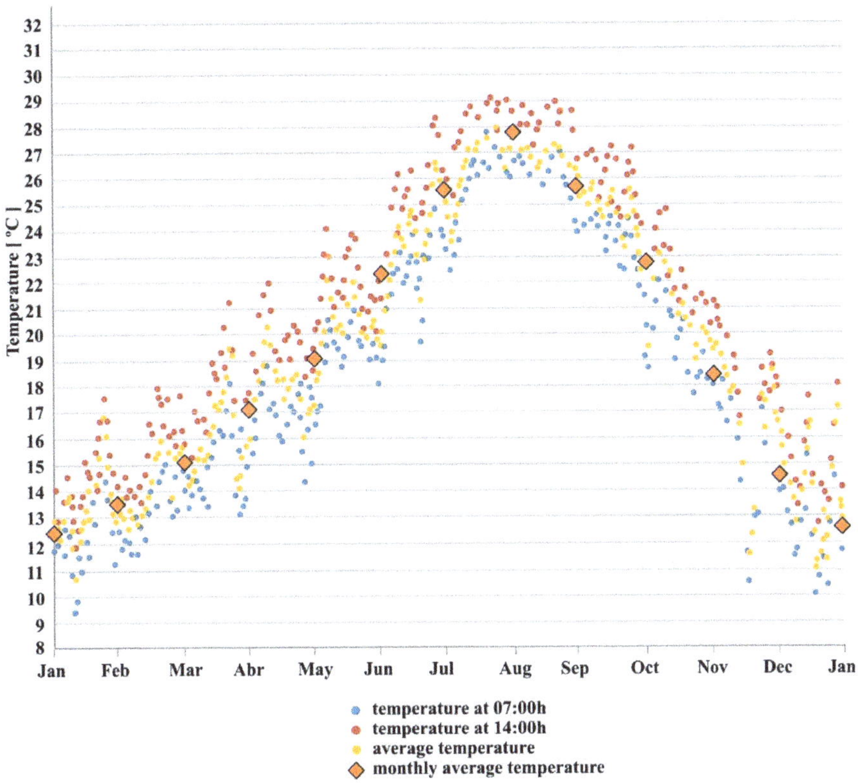

Figure 14. On-site interior temperature measurements in the original building without using air conditioning systems.

Improving the thermal insulation of opaque enclosures reduces annual hours below 21 °C to 39% (3430 h) and annual hours below 19 °C to 23% (2075 h) and also eliminates hours below 12.4 °C. However, this intervention increases the annual hours above 25 °C to 40% (3547 h) and the annual hours above 27 °C to 29% (2569 h) (Figure 15a). Improved glazing insulation and solar shading on windows maintains annual hours below 21 °C at 67% (5784 h), maintains annual hours below 19 °C at 52% (4574 h), and slightly reduces hours below 12.4 °C to 2% (205 h). This intervention slightly reduces the annual hours above 25 °C to 12% (1042 h) and maintains the annual hours above 27 °C at 7% (619 h) (Figure 15b). The increase in WWR keeps annual hours below 21 °C at 68% (6000 h), keeps annual hours below 19 °C at 55% (4854 h), and keeps hours below 12.4 °C at 3.7% (327 h). This intervention increases the annual hours above 25 °C to 19% (1700 h) and maintains the annual hours above 27 °C at 7% (642 h) (Figure 15c). Finally, the use of natural ventilation in summer reduces the annual hours above 27 °C to 5% (438 h). This intervention decreases the hours above 28 °C to only 1% (70 h) (Figure 15d).

With regard to natural ventilation, the simulation results of the indoor air circulation calculations by means of CFD show that cross ventilation achieves comfortable temperatures with adequate air velocities, mostly below 0.10 m/s, in accordance with Spanish regulations. (Figure 16a,b).

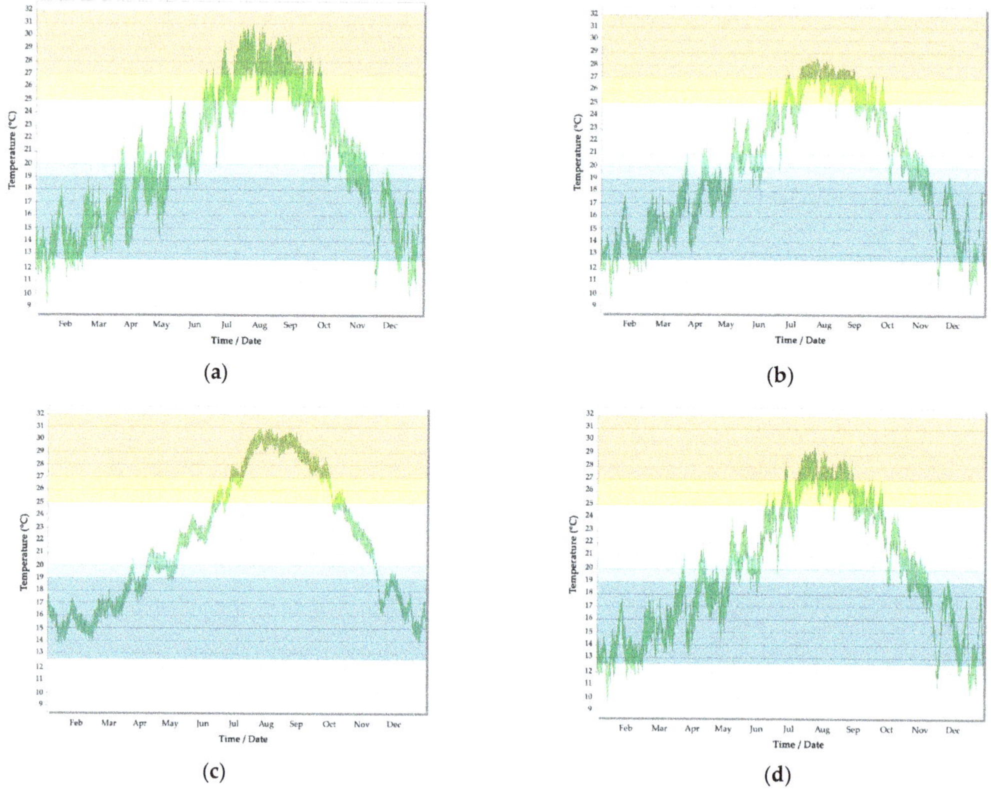

Figure 15. Indoor temperatures considering the proposed energy renovation improvements individually: (**a**) modification of thermal insulation thickness of opaque enclosures (Modification 1); (**b**) modification of the thermal characteristics of the glazing (Modification 2); (**c**) modification of size of the glazing by modifying the size ratio between glass and opaque enclosures (WWR) (Modification 3); (**d**) use of natural ventilation in summer (Modification 4).

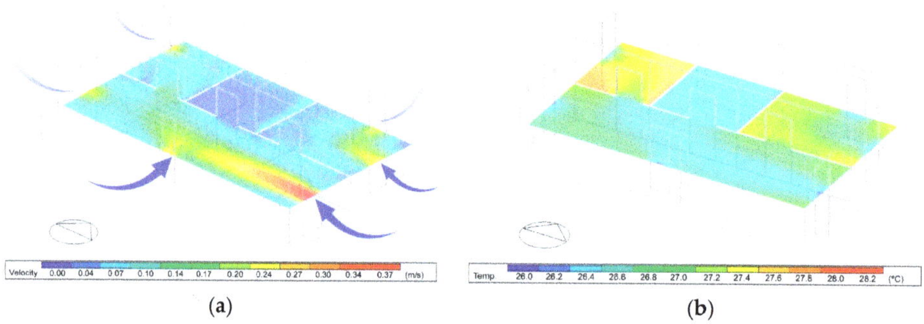

Figure 16. Simulation results of the indoor air circulation calculations by means of CFD using natural cross ventilation: (**a**) air velocity results; (**b**) air temperature results.

At this point in the research, a new scenario was proposed with the incorporation of ceiling fans in each room. This measure has been studied in several studies as a means of improving energy efficiency, air quality [74] and sensation of thermal comfort [75,76], even up to 29° [77]. The simulation shows that the complementary use of ceiling fans allows a more controlled redistribution of air. They achieve more homogeneous temperatures and velocities throughout the dwelling and below 0.10 m/s (Figure 17a,b). It is interesting to recall that Roles et al. [78] concluded that ceiling fans can lead to energy savings of 15–18%.

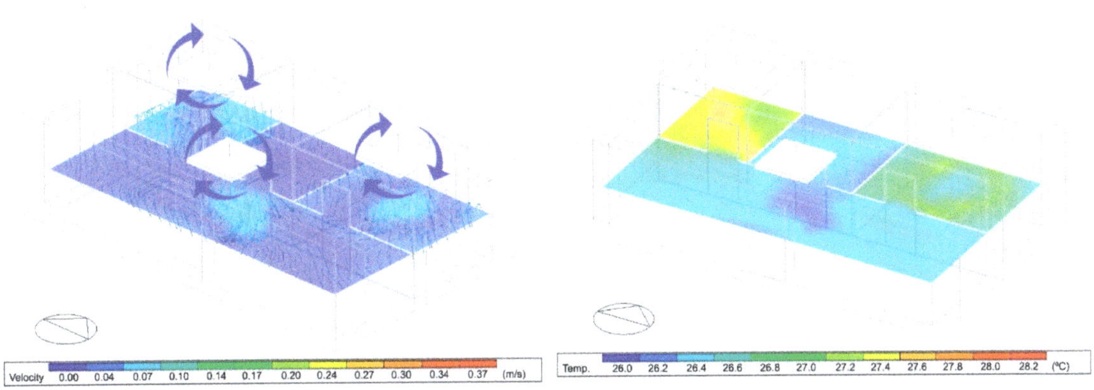

Figure 17. Simulation results of the indoor air circulation calculations by means of CFD using natural ventilation supported by ceiling fans: (**a**) air velocity results; (**b**) air temperature results.

The set of improvements incorporated in the final solution reduces annual hours below 21 °C to 19% (1681 h) and annual hours below 19 °C to just 3% (285 h), and eliminates hours below 12.4 °C. On the other hand, annual hours above 25 °C are increased to 20% (1840 h), but annual hours above 27 °C are reduced to 6% (552 h) and hours above 28 °C are drastically reduced to less than 1% (50 h) (Figure 18).

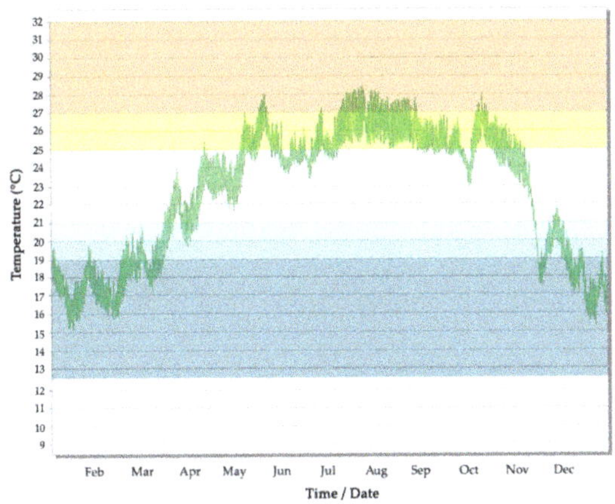

Figure 18. Indoor temperatures considering all proposed energy retrofit improvements in the final solution: modification of thermal insulation thickness of opaque enclosures, modification of the thermal characteristics of the glazing, modification of size of the glazing by modifying the size ratio between glass and opaque enclosures and using natural cross ventilation.

3.1.2. Thermal Comfort

The temperatures obtained from the on-site measurements of the original building in winter are far from the comfort zone, but they are within the comfort zone with internal gains, with an indoor temperature of 9.5 °C on an extreme winter day (Figure 19) and 13.5 °C on the standard day (Figure 20). Improving the insulation of the opaque enclosures shows that temperatures are very close to the comfort zone, increasing from 9.5 °C to 13.8 °C on an extreme winter day and from 13.5 °C to 16.8 °C on the standard day. The rest of the proposed solutions hardly change the temperatures if applied individually. However, the application of all measures together does substantially improve temperatures, bringing them very close to the comfort zone on an extreme winter day and within the comfort zone on the standard day.

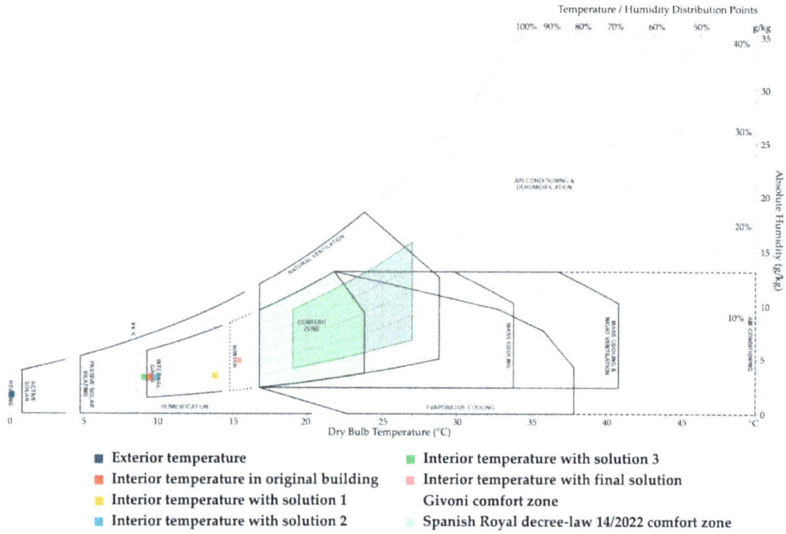

Figure 19. Givoni chart with exterior and interior temperature measurements for winter extreme day.

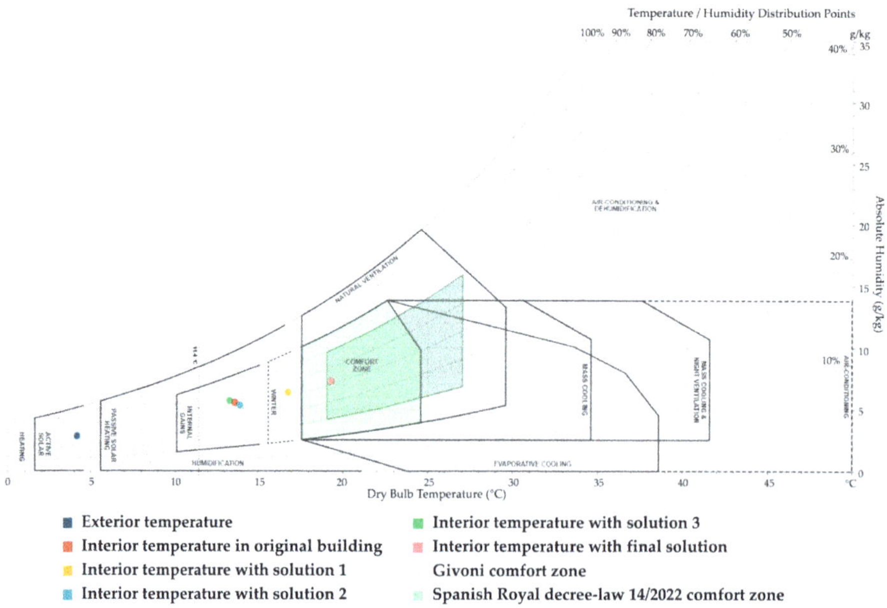

Figure 20. Givoni chart with exterior and interior temperature measurements for winter typical day.

In spring and autumn, indoor temperatures are outside the comfort zone on the coldest days but within the comfort zone with internal gain utilisation (Figure 21). However, temperatures are within the comfort zone on the hottest days (Figure 22). The indoor temperature is 13.8 °C on the coldest day and 25.0 °C on the hottest day. By improving the insulation of opaque enclosures, indoor temperatures remain within the comfort zone throughout spring and autumn. Again, the combined application of all the proposed measures offers the best results.

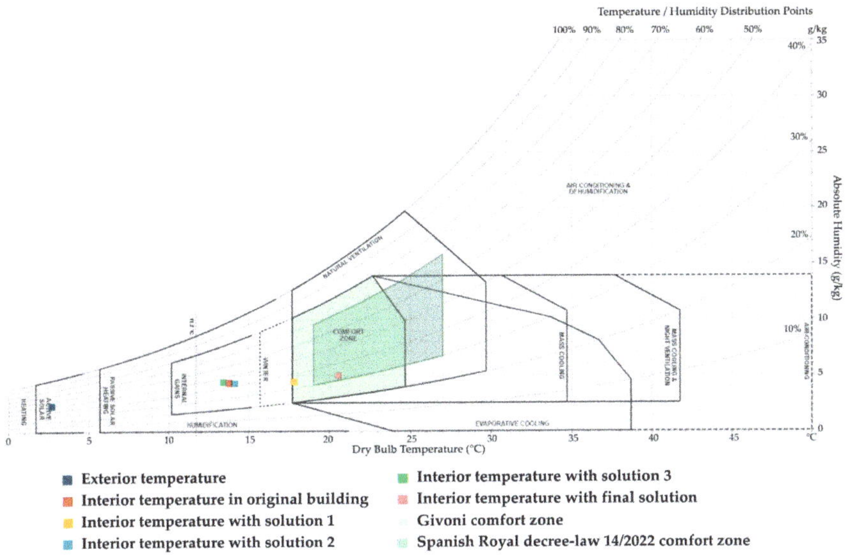

Figure 21. Givoni chart with exterior and interior temperature measurements for spring/autumn cold day.

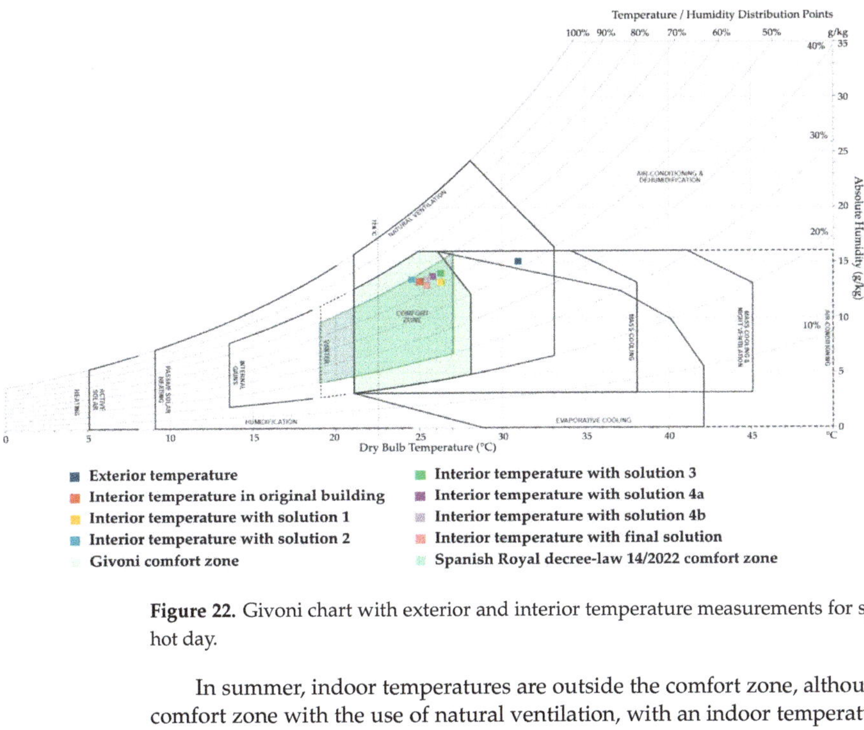

Figure 22. Givoni chart with exterior and interior temperature measurements for spring/autumn hot day.

In summer, indoor temperatures are outside the comfort zone, although within the comfort zone with the use of natural ventilation, with an indoor temperature of 27.2 °C on the average day (Figure 23) and 29.1 °C on an extreme summer day (Figure 24). By improving the insulation of opaque enclosures, temperatures move away from the comfort zone, increasing from 27.2 °C to 28.9 °C on the typical summer day and from 29 °C to 30.7 °C on an extreme day. The rest of the proposed solutions do not improve house temperatures if applied individually. However, the application of all measures together improves temperatures, bringing them within the comfort zone on the standard summer day and bringing them very close to the comfort zone on an extreme day.

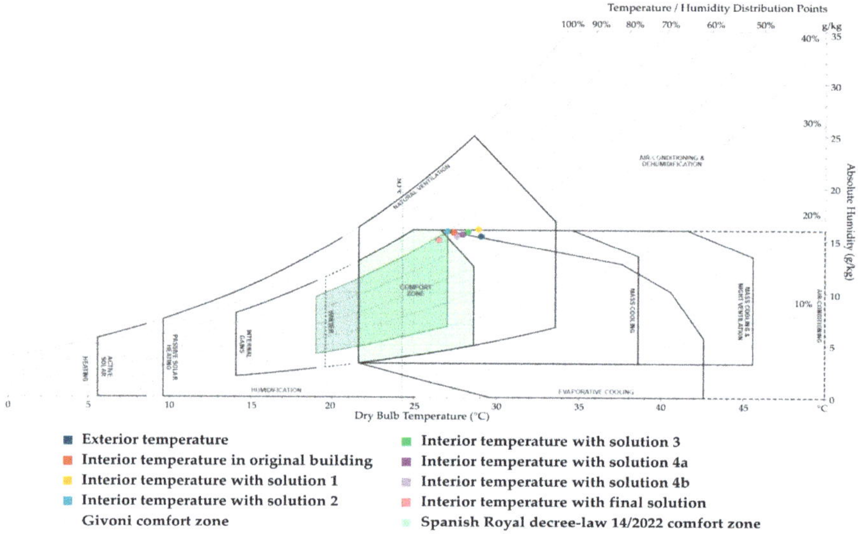

Figure 23. Givoni chart with exterior and interior temperature measurements for summer typical day.

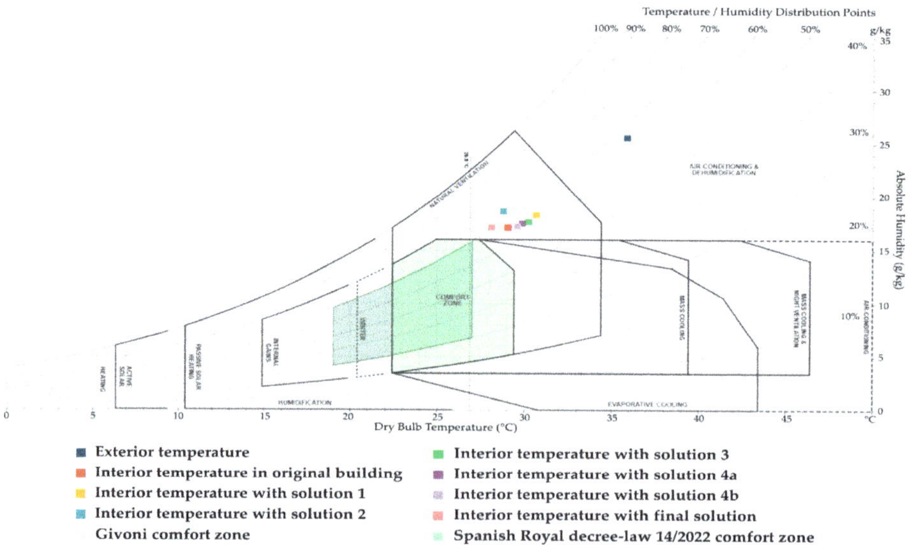

Figure 24. Givoni chart with exterior and interior temperature measurements for summer extreme day.

The results of the comfort surveys conducted with the users of the dwelling show that for all occupants of the dwelling, the average thermal sensation inside the house was perceived as cold or cool in December and January and hot or very hot in July and August. The average thermal sensation remained cool or cold for 87.5% in February and cool for 50.0% in November. Additionally, it remained hot for 25% of occupants in June and September. During the months of April, May and October, the average thermal sensation was considered slightly cool to slightly hot for most occupants (87.5%). The hottest month was August, with 62.5% of occupants considering it very hot, and the coldest month was January, with 87.5% of occupants considering the wind chill to be cold (Figures 25 and 26a).

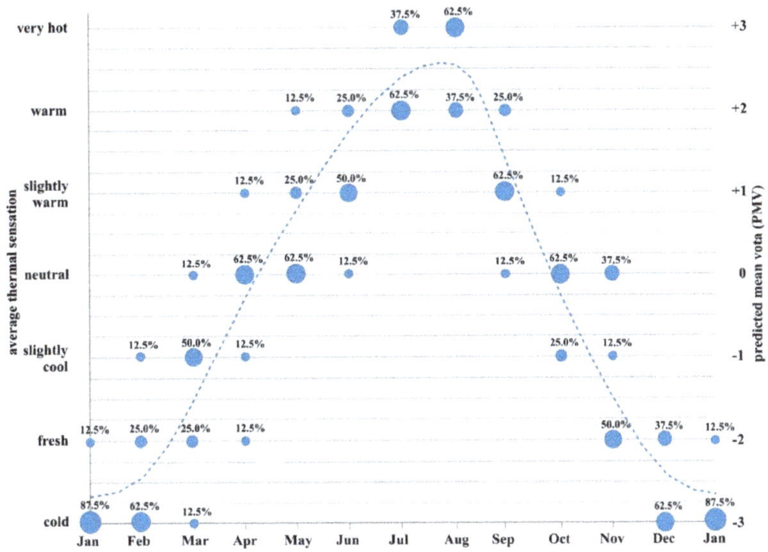

Figure 25. Results of user surveys on indoor thermal sensation.

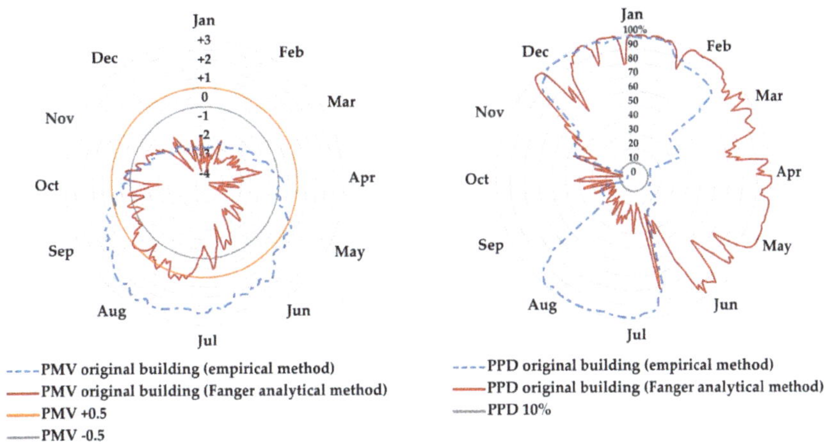

(**a**) (**b**)

Figure 26. Empirical method by means of surveys of housing users, calculated analytically for the original building using the parametric Fanger comfort equation for the calculation of the estimated average vote (PMV) and by means of the equation for calculating the percentage of unsatisfied people (PPD): (**a**) thermal sensation (PMV); (**b**) percentage of dissatisfied (PPD).

The occupants were dissatisfied with thermal comfort in the winter, specifically December and January (100%). They considered the thermal sensation unpleasant because they felt it was too cold. In February, the percentage dropped to 88%, in November to

50%, in March to 37.5%, and in April to 12.5%. In the summer, all occupants considered the thermal sensation unpleasant because they felt it was too hot in July and August. In June and September, the percentage of dissatisfied people dropped to 25%, and in May, the percentage dropped to 12.5%. In October, the percentage of dissatisfied occupants was 0% (Figure 26b). Analytical calculations of wind chill estimation in the house according to the Fanger method show PMV values below −1 from November to June. However, values below −2 were reached across the board from December to May, and even values of −3 on many days in January and February. During the summer, values above +1 were not reached (Figure 26a). In line with these data, the results of the PPD calculations show values close to 100% from December to May and values close to 10% from July to October. The results show two very different times of the year, with much greater discomfort in the cold months than in the warm months (Figure 26b).

By increasing the insulation of opaque enclosures, the results for estimating the thermal sensation improve in winter but worsen in summer. In winter, the PMV results go from average values below −3 in the original house to average values above −2.5 with the improved insulation. In summer, results go from average values around +0.5 to average values above +1. PPD values improve by 5–10% on average in winter, but worsen in summer by 20% on average, from PPD values below 35% in the original house to values above 55% in the modified house. The remaining construction improvements, applied individually, hardly cause any change in the results of the PMV and PPD calculations according to the Fanger method. However, the final solution with the application of all measures together substantially improves wind chill estimation (PMV) and percentage of dissatisfied (PPD) both in winter and summer. In winter, the PMV results go from average values below −3 in the original house to average values above −1.5 with the final solution and remain between −0.5 and +0.5 from May to November. On the other hand, PPD values improve by 20–50% on average all year round (Figure 27a,b).

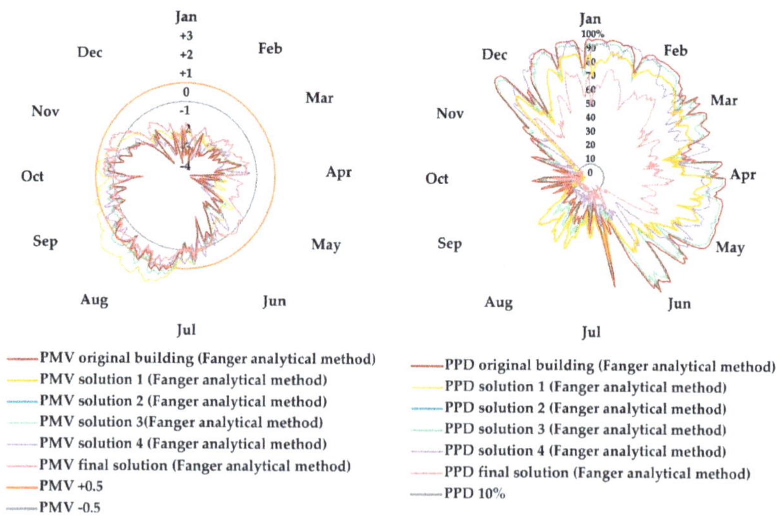

(a) (b)

Figure 27. Fanger method calculated analytically with the proposed energy rehabilitation improvements using the parametric Fanger comfort equation for the calculation of the estimated average vote (PMV) and by means of the equation for calculating the percentage of unsatisfied people (PPD): (**a**) wind chill estimation (PMV); (**b**) percentage of dissatisfied (PPD).

4. Discussion

4.1. Energy Efficiency

In relation to improvements in energy efficiency, the analysis of the yielded results shows that the improvement of the thermal insulation of the opaque parts of the building's enclosures considerably reduces the energy need for heating in winter (−29%) but slightly increases the energy need for cooling in summer (+6%). The reason for this performance can be found in the greenhouse effect inside the house, which also increases (Figure 28). For this reason, in this warm climate with mild winters, over-insulating façades and roofs does not improve the annual energy performance of the house. In the traditional houses of the Spanish Mediterranean, improving the thermal insulation and solar factor of glazing or solar shading has very little influence on the overall energy need (heating and cooling). This is because the size of the windows is very small. Increasing the window–wall ratio (WWR) does have an important influence on the overall energy need because the heat gains from solar radiation are greatly increased. This positively reduces the winter heating needs but progressively increases the cooling needs if the solar protection of the glazing is not improved, because it increases the greenhouse effect in the summer. Therefore, when increasing WWR, increased glass insulation is necessary to reduce heating needs in the winter, but it is also very important to improve the solar shading of windows in the summer. Regarding the increase in natural ventilation in summer, our calculations show that this is a particularly positive measure because it reduces cooling needs in summer but does not increase heating needs in winter. This improvement is most optimal if natural ventilation is used on warm days in spring and autumn. Finally, this research confirms that the best results are obtained by performing a complex multi-factor analysis that allows several parameters to be modified at the same time, reducing the overall annual need (cooling + heating) by up to 87%. The reduction is even greater with the latest regulation on urgent energy saving measures approved by the Spanish government, because it widens the setpoint temperature range [79].

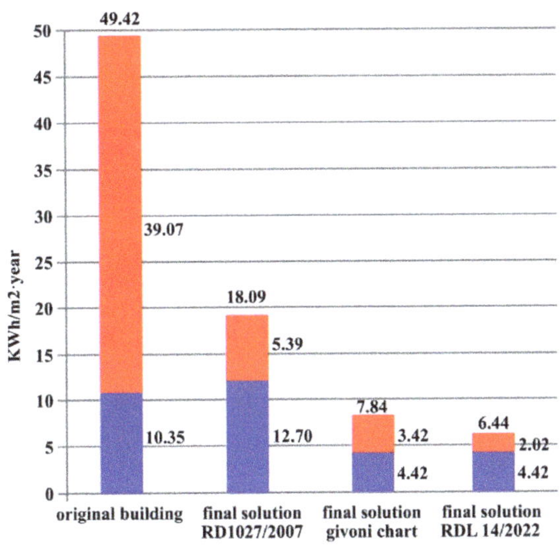

Figure 28. Simulation results of the calculations of heating and cooling demands using the software "Design Builder", comparing the results of the original house and the house with the proposed set of improvements simultaneously, in accordance with RD 1027/2007, with Givoni chart and with RDL 14/2022.

Through the obtained savings in energy consumption, it is possible to obtain a recovery period of 16 years that is below the limits defined by European Regulation (EU) 244/2012 [80], which develops Directive 2010/31/EU [81].

For the calculation of the Net Present Value (NPV), the global cost at the financial level was considered with a calculation period of 30 years [82], according to the values indicated in Annex A of the UNE-EN 15,459 [83] standard. In the calculations, a useful life of the windows of 20 years and the insulation of 50 years were also considered (Figure 29). In addition, the recovery of the investment obtained shows that it would not be necessary to introduce additional costs for maintenance.

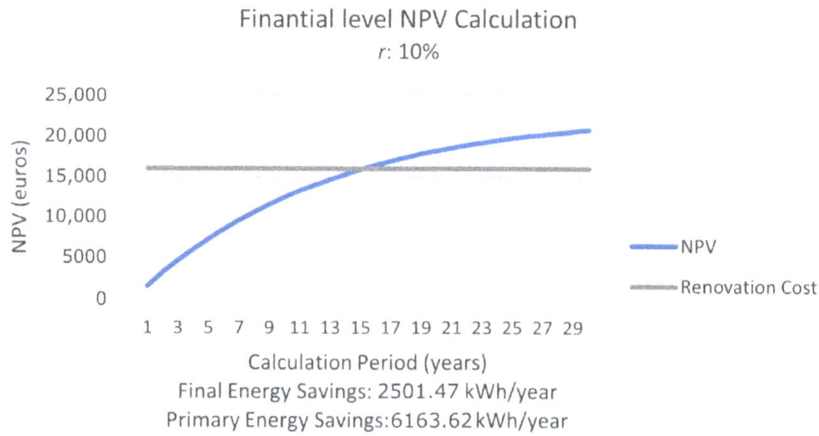

Figure 29. Financial-level Net Present Value (NPV) calculation.

4.2. Indoor Temperatures

The on-site measurements of the original building show that the indoor temperatures of the house are outside the comfort limits set by the current Spanish standard [36] 59% of hours per year. The comparative analysis of the results obtained with the simulations of the different energy improvement proposals shows that none of them individually managed to sufficiently improve the thermal performance of the house. Improving the thermal insulation of opaque envelopes beneficially increases indoor temperatures in winter, but very detrimentally increases temperatures in summer. This is due to the increased greenhouse effect inside the house during the summer because the increased thermal insulation hinders the thermal dissipation of internal heat gains. The improvement of glazing characteristics has very little influence on indoor temperatures due to the small size of the windows of the original house. However, with the increase in WWR, if glazing insulation and solar protection are not improved, there will be an increase in colder hours in winter and a very significant increase in excessively hot hours in summer. Additionally, the incorporation of only natural ventilation in the original house does not improve indoor temperatures much either because there is little greenhouse effect due to the small size of the windows. The results of the simulations carried out show that only the combined application of all the proposed improvements allows an increase in winter temperatures and a reduction in summer temperatures. This joint measure does have a very beneficial effect on indoor temperatures in the house, reducing the annual hours outside the comfort limits set by current Spanish regulations to 9% (Figure 30).

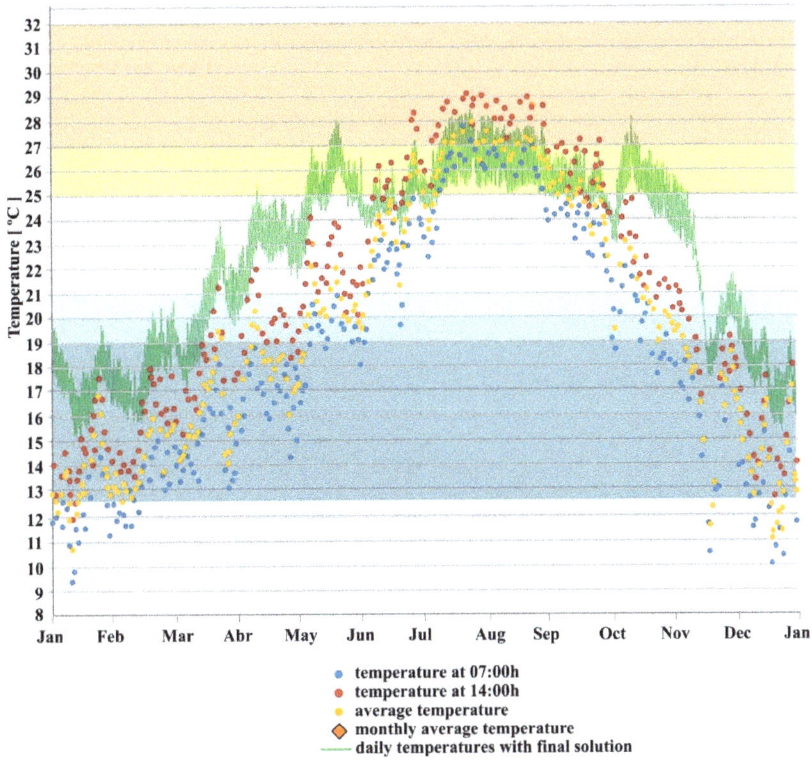

Figure 30. On-site interior temperature measurements in the original building and interior temperatures simulated considering final solution. 3. Thermal Comfort.

In relation to thermal comfort, the comparative analysis of the results obtained applied to the Givoni bioclimatic graph shows that in summer, the original house only requires natural ventilation to create a comfortable indoor climate. In winter, the original dwelling is far away from the comfort zone and requires the joint improvement of the thermal insulation of the walls and the glazing to achieve an indoor environment that is close to the comfort zone. No single solution achieves comfortable conditions. In spring and autumn, the original house stays within the comfort zone on warm days, but on cold days it does not, and needs improved insulation to maintain comfort (Figures 18–23). However, this increase in insulation is clearly detrimental to summer comfort, as it increases the greenhouse effect inside the house. Therefore, only the combined application of all the proposed energy improvements can improve comfort all year round. This is because, in winter, improved insulation helps retain heat inside the house. In addition, the increased size of the glazing helps capture more solar radiation, while in summer, cross ventilation helps dissipate the heat inside the house, and movable sunscreens prevent heat from entering through solar radiation.

The analysis of the results obtained allows us to consider that the current energy saving regulations in Spain do not sufficiently consider the specific characteristics of the BShs climate. The regulations limit the minimum operating temperature in summer and the maximum operating temperature in winter, as well as the range of permissible relative humidity. However, they do not consider it necessary to establish an appropriate relationship between temperature and relative humidity depending on the time of year to ensure adequate comfort. Therefore, it may be considered more appropriate to propose a mixed solution that combines the psychometric diagrams of Givoni with the temperature and relative humidity limits imposed by Spanish regulations, specifically for each season of the year (Figures 31 and 32).

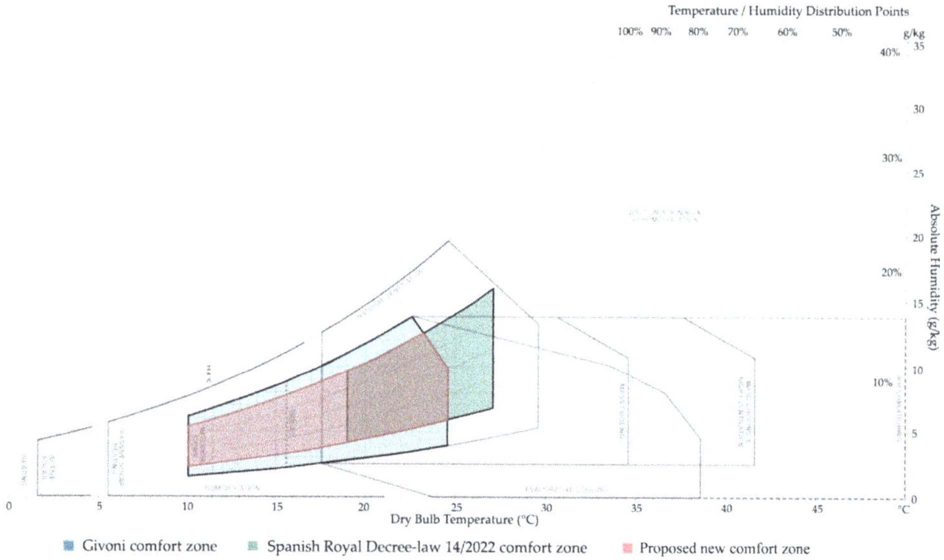

Figure 31. Proposed diagram in winter.

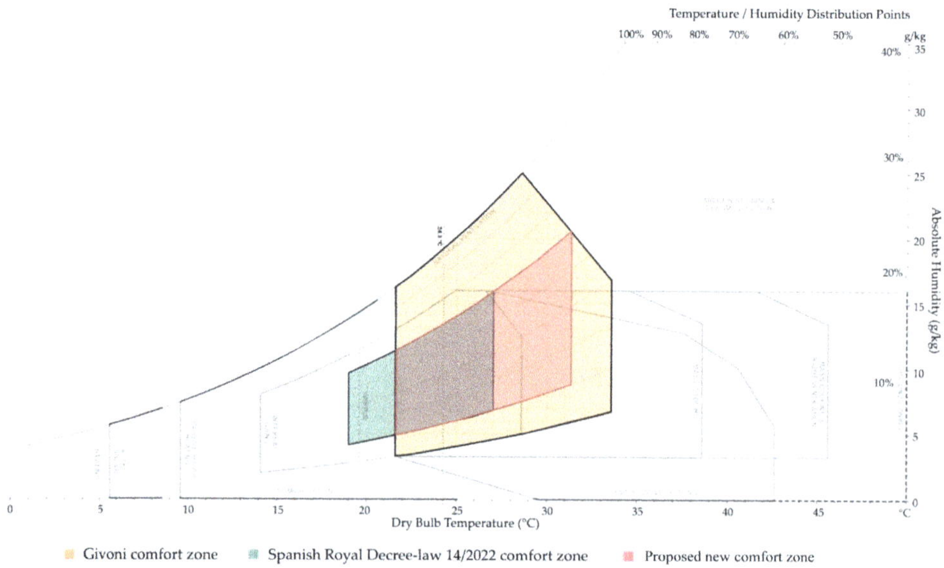

Figure 32. Proposed diagram in summer.

From the comparative analysis of the results of the application of the Fanger method, the values of the analytical calculations for estimating the wind chill in the original house are much lower than those obtained in the surveys carried out. The PMV values obtained using the parametric Fanger comfort equation are less than −0.5 most of the year, with many days in January and February with cold wind chill (−3), and only exceed +0.5 from late July to early September (Figure 33a). However, our surveys show values below

−0.5 from November to March and values above +0.5 from May to September, with the annual number of days with cold or cool and hot or very hot wind chill being balanced. Similarly, PPD values obtained analytically are above 80% from December to the end of June and below 30% from July to November (Figure 33b). However, our surveys show discomfort among users both in the cold months, from December to February, and in the warmer months, July and August. These differences between the results of the analytical calculations and the surveys show that the Fanger method does not sufficiently reflect the discomfort caused by excessive summer heat in the BShs climate. It also does not reflect the smaller temperature oscillation between the annual minimum and maximum temperatures in this climate, which results in mild and increasingly shorter winters. Considering the existing deviation between the results of the analytical calculations and the surveys with respect to the original dwelling, a proportional modification of the results obtained in the calculations of the final solution is proposed (Figure 33a,b).

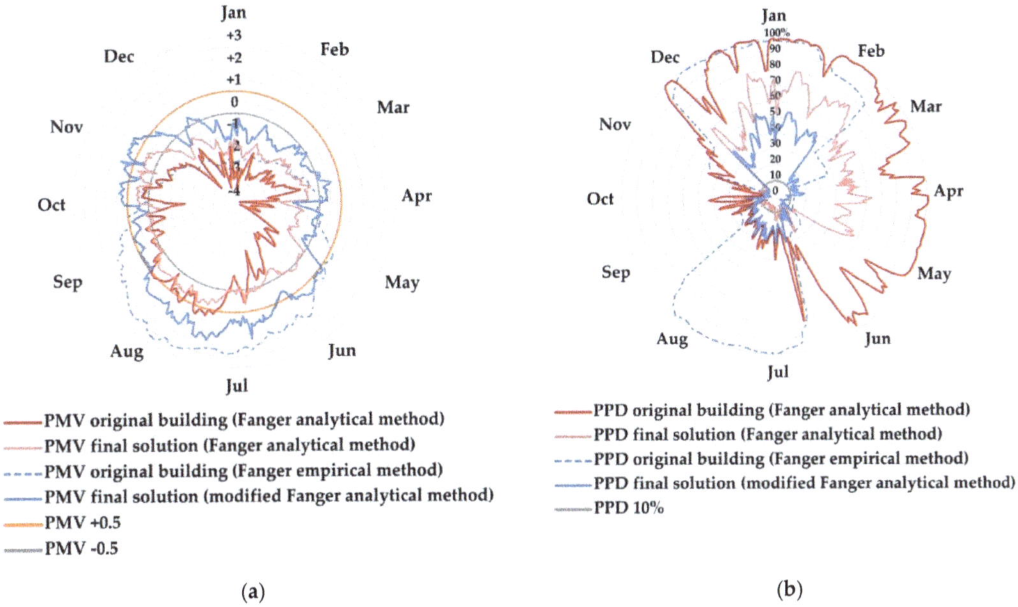

Figure 33. Proportional modification of the results obtained in the calculations of the final solution using Fanger method calculated for the original building and with the final energy rehabilitation: (**a**) wind chill estimation (PMV); (**b**) percentage of dissatisfied (PPD).

5. Conclusions

This research demonstrates the value of using design and energy retrofitting criteria that are adapted to the specific climate of the area rather than following national guidance. The results reinforce the importance of designing for a specific climate and ensuring retrofitting is conducted holistically and simultaneously to suit the local climate. Each intervention was found to influence the thermal behaviour of the building because of its specific climatic context. This implies that retrofitting buildings to improve their energy performance to comply with European and national regulations needs to be questioned. As shown in this paper, there are specific areas of Europe that require a contextualised approach rather than a standardised one.

The best solution, in the case of a dry Mediterranean climate, is to take advantage of passive opportunities to heat the house through solar radiation in winter ("free heat") and protect the building from solar radiation ("unwanted heat") in summer.

In a BShs climate, it is necessary to act simultaneously and improve thermal insulation together with increased glazing size, control of solar incidence on windows and the use of natural ventilation with the support of ceiling fans in summer to achieve energy-efficient and comfortable houses without the use of air conditioning. Partial improvements show limited results that are only effective for certain parts of the year.

Energy efficiency regulations in Spain, which are based on European regulations, are wrong to prioritise in a generalised manner improvements in the insulation and airtightness of dwellings. Furthermore, in contrast to other research and current regulations that defend the benefits of natural ventilation only at night, this work shows that in BShs climates it is also advisable to use daytime ventilation supported by ceiling fans.

An excess of insulation on walls and glazing has a detrimental effect by increasing the demand for energy for cooling in summer. This translates into increased cooling costs. Our analysis of the results shows that combined solutions of improving insulation, increasing the WWR with mobile sun protection systems and improving natural ventilation with the support of ceiling fans in the summer season are more optimal.

The use of flexible solutions has been a permanent feature in vernacular architecture and is confirmed in this study. In this way, solar radiation in winter can be adequately used to reduce the demand for heating energy and at the same time reduce the solar incidence in summer when the sun increases the demand for energy for cooling.

Our research shows that this solution allows cooling and heating requirements to be reduced by up to 87%, reducing the operational needs of air conditioning systems and their energy consumption. If air conditioning systems are used, interior temperatures rise by an average of 5 °C in winter and drop by an average of 2 °C in summer, reducing the annual hours outside the comfort limits established by current Spanish regulations from 59% to 9%. This helps the house stay within the comfort zone of the Givoni diagram all year, with a comfortable thermal sensation during 67% of the year according to the Fanger method.

The main limitation associated with this research was the practical challenge of measuring indoor temperatures in an inhabited house with the air conditioning switched off. In this case, it was necessary to adapt the data collection to the needs of the building users, which meant that measurements were taken by a researcher when it was convenient for the occupants, rather than using automatic data collection systems. The air conditioning was not used so as not to artificially influence the thermal behaviour of the building on the days the measurements were taken. There was no control of occupant behaviour on the days the measurements were not taken, and the researchers acknowledge that the temperatures recorded may have been affected a little by the occupants' behaviour on the "non- measurement" days. Despite this limitation, our data set provides a valuable source to inform the modelling exercises. Another constraint was the geometric and constructive conditions that must be maintained in the traditional house typology studied. These circumstances prevented changes in the size of the windows and the use of overhangs and other solar protection elements due to their aesthetic impact.

This research demonstrates the importance of modifying the comfort criteria of energy regulations to better adapt the design criteria to the particularities of the dry Mediterranean climate. This research offers an alternative solution to the comfort standards of the current Spanish regulations. We consider it appropriate to correlate temperatures and relative humidity based on psychrometric diagrams specifically adapted to the temperature limits established in Spain. Our research also demonstrates the need to adapt Fanger's analytical method of comfort estimation to the particularities of the dry Mediterranean climate to bring the analytical results closer to the actual statistics of the users. This will require further research.

To conclude, this research demonstrates that buildings are more energy-efficient and comfortable the more they are adapted to their local climate. It proves the maxim that there is no cheaper energy than that which is not needed, helping to reinforce the importance of passive design principles and question the applicability of national codes and guidance.

Author Contributions: Conceptualization, C.P.-C., Á.B.G.-A. and A.L.-P.; methodology, C.P.-C., Á.B.G.-A. and A.L.-P.; software, C.P.-C. and A.L.-P.; validation, C.P.-C., A.G.-G., Á.B.G.-A. and S.E.; formal analysis, A.L.-P.; investigation, C.P.-C., A.L.-P., A.G.-G. and Á.B.G.-A.; resources, A.L.-P.; data curation, C.P.-C. and A.L.-P.; writing—original draft preparation, C.P.-C. and A.L.-P.; writing—review and editing, C.P.-C., A.L.-P., Á.B.G.-A. and S.E.; visualization, C.P.-C. and A.L.-P.; supervision, C.P.-C., A.G.-G., Á.B.G.-A. and S.E.; project administration, C.P.-C.; grant acquisition, Colegio Territorial de Arquitectos de Alicante (CTAA). All authors have read and agreed to the published version of the manuscript.

Funding: This research received an external grant from the Colegio Territorial de Arquitectos de Alicante (CTAA).

Data Availability Statement: The data presented in this study are available on request from the corresponding author.

Conflicts of Interest: The authors declare no conflict of interest.

References

1. Di Foggia, G. Energy efficiency measures in buildings for achieving sustainable development goals. *Heliyon* **2018**, *4*, e00953. [CrossRef]
2. Balaras, C.A.; Droutsa, K.; Argiriou, A.A.; Asimakopoulos, D.N. Potential for energy conservation in apartment buildings. *Energy Build.* **2000**, *31*, 143–154. [CrossRef]
3. Tsemekidi Tzeiranaki, S.; Bertoldi, P.; Diluiso, F.; Castellazzi, L.; Economidou, M.; Labanca, N.; Ribeiro Serrenho, T.; Zangheri, P. Analysis of the EU Residential Energy Consumption: Trends and Determinants. *Energies* **2019**, *12*, 1065. [CrossRef]
4. Zakari, A.; Khan, I.; Tan, D.; Alvarado, R.; Dagar, V. Energy efficiency and sustainable development goals (SDGs). *Energy* **2022**, *239*, 122365. [CrossRef]
5. Scrucca, F.; Ingrao, C.; Barberio, G.; Matarazzo, A.; Lagioia, G. On the role of sustainable buildings in achieving the 2030 UN sustainable development goals. *Environ. Impact Assess. Rev.* **2023**, *100*, 107069. [CrossRef]
6. Hafez, F.S.; Sa'di, B.; Safa-Gamal, M.; Taufiq-Yap, Y.H.; Alrifaey, M.; Seyedmahmoudian, M.; Stojcevski, A.; Horan, B.; Mekhilef, S. Energy Efficiency in Sustainable Buildings: A Systematic Review with Taxonomy, Challenges, Motivations, Methodological Aspects, Recommendations, and Pathways for Future Research. *Energy Strategy Rev.* **2023**, *45*, 101013. [CrossRef]
7. European Union. *Directiva 2010/31/UE del Parlamento Europeo y del Consejo de 19 de mayo de 2010 Relativa a la Eficiencia Energética de los Edificios (Refundición)*; European Union: Maastricht, The Netherlands, 2010.
8. European Union. *Directiva (UE) 2018/844 del Parlamento Europeo y del Consejo, de 30 de mayo de 2018, por la que se Modifica la Directiva 2010/31/UE Relativa a la Eficiencia Energética de los Edificios y la Directiva 2012/27/UE Relativa a la Eficiencia Energética*; European Union: Maastricht, The Netherlands, 2018.
9. Agencia Estatal Boletín Oficial del Estado. *Real Decreto 732/2019, de 20 de diciembre, por el Que se Modifica el Código Técnico de la Edificación, Aprobado por el Real Decreto 314/2006, de 17 de marzo*; Agencia Estatal Boletín Oficial del Estado: Madrid, Spain, 2019.
10. Agencia Estatal Boletín Oficial del Estado. *Real Decreto 390/2021, de 1 de junio, por el que se Aprueba el Procedimiento Básico para la Certificación de la Eficiencia Energética de los Edificios*; Agencia Estatal Boletín Oficial del Estado: Madrid, Spain, 2021.
11. López-Ochoa, L.M.; Las-Heras-Casas, J.; Olasolo-Alonso, P.; López-González, L.M. Towards nearly zero-energy buildings in Mediterranean countries: Fifteen years of implementing the Energy Performance of Buildings Directive in Spain (2006–2020). *J. Build. Eng.* **2021**, *44*, 102962. [CrossRef]
12. López-Ochoa, L.M.; Las-Heras-Casas, J.; Olasolo-Alonso, P.; López-González, L.M. Environmental and energy impact of the EPBD in residential buildings in hot and temperate Mediterranean zones: The case of Spain. *Energy* **2018**, *161*, 618–634. [CrossRef]
13. Olcina, J.; Serrano-Notivolit, R.; Miró, J.; Meseguer-Ruiz, O. Tropical nights in the Spanish Mediterranean coast: Recent evolution (1950–2014). *Clim. Res.* **2019**, *78*, 225–236. [CrossRef]
14. Miró, J.; Olcina, J. Cambio climático y confort térmico. Efectos en el turismo de la Comunidad Valenciana. *Rev. Investig. Turísticas* **2020**, *20*, 1–30.
15. Kottek, M.; Grieser, J.; Beck, C.; Rudolfm, B.; Rubel, F. World Map of the Köppen-Geiger climate classification updated. *Meteorol. Z.* **2006**, *15*, 259–263. [CrossRef]
16. Agencia Estatal Boletín Oficial del Estado. *Orden FOM/588/2017, de 15 de Junio, por la que se Modifican el Documento Básico DB-HE "Ahorro de Energía" y el Documento Básico DB-HS "Salubridad", del Código Técnico de la Edificación, Aprobado por Real Decreto 314/2006, de 17 de Marzo*; Agencia Estatal Boletín Oficial del Estado: Madrid, Spain, 2017.
17. Lucon, O.; Ürge-Vorsatz, D.; Ahmed, A.Z.; Akbari, H.; Bertoldi, P.; Cabeza, L.F.; Eyre, N.; Gadgil, A.; Harvey, L.D.; Jiang, Y.; et al. Climate Change 2014: Mitigation of Climate Change. In *IPCC Working Group III Contribution to AR5: Chapter 9—Buildings*; Cambridge University Press: Cambridge, UK, 2015.
18. Georgiou, G.; Eftekhari, M.; Lupton, T. Investigating the effect of tightening residential envelopes in the Mediterranean region. In Proceedings of the 14th International Conference on Sustainable Energy Technologies, Nottingham, UK, 25–27 August 2015.

19. Asdrubali, F.; Baldinelli, G.; Bianchi, F. A quantitative methodology to evaluate thermal bridges in Buildings. *Appl. Energy* **2012**, *97*, 365–373. [CrossRef]
20. Evola, G.; Margani, G.; Marletta, G. Energy and cost evaluation of thermal bridge correction in Mediterranean climate. *Energy Build.* **2011**, *43*, 2385–2393. [CrossRef]
21. Arias, N.; Bobadilla, A. Evaluación experimental y análisis de la mejora con aislamiento para el caso del puente térmico en el frente de forjado. *Inf. Constr.* **2017**, *69*, e188. [CrossRef]
22. Pérez-Carramiñana, C.; González-Avilés, Á.B.; Galiano-Garrigós, A.; Lozoya-Peral, A. Optimization of Architectural Thermal Envelope Parameters in Modern Single-Family House Typologies in Southeastern Spain to Improve Energy Efficiency in a Dry Mediterranean Climate. *Sustainability* **2022**, *14*, 3910. [CrossRef]
23. Figueiredo, A.; Kämpfb, J.; Vicente, R. Passive house optimization for Portugal: Overheating evaluation and energy performance. *Energy Build.* **2016**, *118*, 181–196. [CrossRef]
24. Pastor, R. Un edificio patrimonial de la huerta valenciana, la barraca, Construcción con Tierra, Investigación y Documentación. In Proceedings of the XI CIATTI 2014 Congresos de Arquitectura de Tierra en Cuenca de Campos, Valladolid, Spain, 26–28 September 2014.
25. Givoni, B. Comfort, climate analysis and building design guidelines. *Energy Build.* **1992**, *18*, 11–23. [CrossRef]
26. Givoni, B. *Man, Climate, and Architecture*; Elsevier: Amsterdam, The Netherlands, 1969.
27. Olgyay, V.; Olgyay, A.; Lyndon, D.; Olgyay, V.W.; Reynolds, J.; Yeang, K. *Design with Climate, Design with Climate*; Princeton University Press: Princeton, NJ, USA, 1963.
28. Bluyssen, P.M. Towards new methods and ways to create healthy and comfortable buildings. *Build. Environ.* **2010**, *45*, 808–818. [CrossRef]
29. Brager, G.S.; De Dear, R.J. Thermal adaptation in the built environment: A literature review. *Energy Build.* **1998**, *27*, 83–96. [CrossRef]
30. Cuce, E.; Sher, F.; Sadiq, H.; Cuce, P.M.; Guclu, T.; Besir, A.B. Sustainable ventilation strategies in buildings: CFD research. *Sustain. Energy Technol. Assess.* **2019**, *36*, 100540. [CrossRef]
31. Fabbri, K. Indoor thermal comfort perception: A questionnaire approach focusing on children. In *Indoor Thermal Comfort Perception: A Questionnaire Approach Focusing on Children*; Springer: Berlin/Heidelberg, Germany, 2015; pp. 1–302.
32. Galiano-Garrigós, A.; Marcos, C.L.; Kouider, T.; Gutiérrez, P.J.J. Juan Gutiérrez. Reassessing thermal comfort in modern architecture: E.1027 as a case study. *Build. Res. Inf.* **2022**, *50*, 230–254. [CrossRef]
33. Meir, I.A.; Roaf, S.C. The future of the vernacular towards new methodologies. In *Vernacular Architecture in the Twenty-First Century: Theory, Education and Practice*; Taylor & Francis: Abingdon, UK, 2014; pp. 215–230.
34. Nicol, F.; Roaf, S. Post-occupancy evaluation and field studies of thermal comfort. *Build. Res. Inf.* **2005**, *33*, 338–346. [CrossRef]
35. Nicol, J.F.; Humphreys, M.A. Adaptive thermal comfort and sustainable thermal standards for buildings. *Energy Build.* **2002**, *34*, 563–572. [CrossRef]
36. Tzempelikos, A.; Athienitis, A.K. The impact of shading design and control on building cooling and lighting demand. *Sol. Energy* **2007**, *81*, 369–382. [CrossRef]
37. Santamouris, M.; Sfakianaki, A.; Pavlou, K. On the efficiency of night ventilation techniques applied to residential buildings. *Energy Build.* **2010**, *42*, 1309–1313. [CrossRef]
38. Michael, A.; Demosthenous, D.; Philokyprou, M. Natural ventilation for cooling in mediterranean climate: A case study in vernacular architecture of Cyprus. *Energy Build.* **2017**, *144*, 333–345. [CrossRef]
39. Jalaei, F.; Jrade, A. Integrating Building Information Modeling (BIM) and energy analysis tools with green building certification system to conceptually design sustainable buildings. *J. Inf. Technol. Constr.* **2014**, *19*, 494–519.
40. Golneshan, A.A.; Yaghoubi, M.A. Simulation of ventilation strategies of a residential building in hot arid regions of Iran. *Energy Build.* **1990**, *14*, 201–205. [CrossRef]
41. Agencia Estatal Boletín Oficial del Estado. *Real Decreto 178/2021, de 23 de Marzo, por el que se Modifica el Real Decreto 1027/2007, de 20 de Julio, por el que se Aprueba el Reglamento de Instalaciones Térmicas en los Edificios*; Agencia Estatal Boletín Oficial del Estado: Madrid, Spain, 2021.
42. *Líder-Calener Unified Tool Software*; v2.0.2253.1167; Dirección General de Arquitectura, Vivienda y Suelo del Ministerio de Transportes, Movilidad y Agenda Urbana y por el Instituto para la Diversificación y Ahorro de la Energía (IDEA): Madrid, Spain, 2021.
43. Grupo de Termotecnia de la Asociación de Investigación y Cooperación Industrial de Andalucía (AICIA) en la Escuela Técnica Superior de Ingenieros de la Universidad de Sevilla. *HU CTE-HE 2019 y CEE (Version 2.0.2203.1160 de 26 de Abril de 2021) [Computer software]*; Dirección General de Arquitectura, Vivienda y Suelo del Ministerio de Transportes, Movilidad y Agenda Urbana e Instituto para la Diversificación y Ahorro de la Energía (IDEA): Madrid, Spain, 2020.
44. Aldersoni, A.A.; Chow, D.H.C. Adapting Traditional Passive Strategies within Contemporary House to Decrease High energy consumption Impact in Nejd Region, Saudi Arabia. *Earth Environ. Sci.* **2019**, *329*, 12007–12015.
45. Schulze, T.; Eicker, U. Controlled natural ventilation for energy efficient buildings. *Energy Build.* **2013**, *56*, 221–232. [CrossRef]
46. Alhamad, I.M.; AlSaleem, M.; Taleb, H. Passive heating and cooling potential strategies: A comparison between moderate summers and warm winters climate zones. *J. Phys. Conf. Ser.* **2019**, *1276*, 12059–12067. [CrossRef]

47. Pérez-Carramiñana, C.; Maciá-Mateu, A.; Sirvent-García, G.; Lledó-Llorca, I. Study of Natural Ventilation and Solar Control Strategies to Improve Energy Efficiency and Environmental Quality in Glazed Heated Swimming Pools in a Dry Mediterranean Climate. *Sustainability* **2022**, *14*, 8243. [CrossRef]
48. Albatayneh, A. Optimisation of building envelope parameters in a semi-arid and warm Mediterranean climate zone. *Energy Rep.* **2021**, *7*, 2081–2093. [CrossRef]
49. Albatayneh, A.; Alterman, D.; Page, A.; Moghtaderi, B. The significance of building design for the climate. *Environ. Clim. Technol.* **2018**, *22*, 165–178. [CrossRef]
50. Prieto, A.; Knaack, U.; Auer, T.; Klein, T. Passive cooling & climate responsive facade design exploring the limits of passive cooling strategies to improve the performance of commercial buildings in warm climates. *Energy Build.* **2018**, *175*, 30–47.
51. Ferrara, M.; Virgone, J.; Fabrizio, E.; Kuznik, F.; Filippi, M. Modelling Zero Energy Buildings: Parametric study for the technical optimization. *Energy Procedia.* **2014**, *62*, 200–209. [CrossRef]
52. Simon, F.; Ordoñez, J.; Girard, A.; Parrado, C. Modelling energy use in residential buildings: How design decisions influence final energy performance in various Chilean climates. *Indoor Built Environ.* **2019**, *28*, 533–551. [CrossRef]
53. Harkouss, F.; Fardoun, F.; Biwole, P.H. Passive design optimization of low energy buildings in different climates. *Energy* **2018**, *165*, 591–613. [CrossRef]
54. Causone, F.; Pietrobon, M.; Pagliano, L.; Erba, S. A high performance home in the Mediterranean climate: From the design principle to actual measurements. *Energy Procedia.* **2017**, *140*, 67–79. [CrossRef]
55. Ozarisoy, B.; Altan, H. Systematic literature review of bioclimatic design elements: Theories, methodologies and cases in the South-eastern Mediterranean climate. *Energy Build.* **2021**, *250*, 111281. [CrossRef]
56. Pesic, N.; Roset, J.; Muros, A. Natural ventilation potential of the Mediterranean coastal region of Catalonia. *Energy Build.* **2018**, *169*, 236–244. [CrossRef]
57. Ozarisoy, B.; Altan, H. A novel methodological framework for the optimisation of post-war social housing developments in the South-eastern Mediterranean climate: Policy design and lifecycle cost impact analysis of retrofitting strategies. *Sol. Energy.* **2021**, *225*, 517–560. [CrossRef]
58. O'Donovan, A.; O'Sullivan, P.D.; Murphy, M.D. Predicting air temperatures in a naturally ventilated nearly zero energy building: Calibration, validation, analysis and approaches. *Appl. Energy.* **2019**, *250*, 991–1010. [CrossRef]
59. Dervishi, S.; Pashako, F.; Dushaj, X.; Dervishi, I.O. Energy performance optimization of traditional housing in Mediterranean climate. *J. Build. Eng.* **2022**, *45*, 103423. [CrossRef]
60. Stazi, F.; Bonfigli, C.; Tomassoni, E.; Di Perna, C.; Munafò, P. The effect of high thermal insulation on high thermal mass:Is the dynamic behaviour of traditional envelopes inMediterranean climates still possible? *Energy Build.* **2015**, *88*, 367–383. [CrossRef]
61. Panchenko, V.A. Solar Roof Panels for Electric and Thermal Generation. *Appl. Sol. Energy* **2018**, *54*, 350–353. [CrossRef]
62. Heiselberg, P.K.; O'Donnavan, A.; Belleri, A.; Flourentzou, F.; Zhang, G.Q.; da Graca, G.C.; Breesch, H.; Justo-Alonso, M.; Kolokotroni, M.; Pomianowski, M.Z.; et al. *Ventilative Cooling Design Guide: Energy in Buildings and Communities Programme*; Department of Civil Engineering, Aalborg University: Aalborg, Denmark, 2018; p. 122.
63. Diakaki, C.; Grigoroudis, E.; Kolokotsa, D. Towards a multi-objective optimization approach for improving energy efficiency in buildings. *Energy Build.* **2008**, *40*, 1747–1754. [CrossRef]
64. Alahmer, A.; Mayyas, A.; Mayyas, A.A.; Omar, M.A.; Shan, D. Vehicular thermal comfort models; a comprehensive review. *Appl. Therm. Eng.* **2011**, *31*, 995–1002. [CrossRef]
65. Alahmer, A.; Abdelhamid, M.; Omar, M. Design for thermal sensation and comfort states in vehicles cabins. *Appl. Therm. Eng.* **2012**, *36*, 126–140. [CrossRef]
66. Pfafferott, J.; Herkel, S.; Jäschke, M. Design of passive cooling by night ventilation: Evaluation of a parametric model and building simulation with measurements. *Energy Build.* **2003**, *35*, 1129–1143. [CrossRef]
67. Rosaleny, M. La barraca valenciana. *Història i recuperació d'una arquitectura resilient Anu. D'arquitectura I Soc.* **2021**, *1*, 124–144.
68. *Design Builder*; v.7.0.1.006; DesignBuilder Software Ltd.: Stroud, Reino Unido, 2021.
69. Hajdukiewicz, M.; Geron, M.; Keane, M.M. Formal calibration methodology for CFD models of naturally ventilated indoor environments. *Build. Environ.* **2013**, *59*, 290–302. [CrossRef]
70. Zhai, Z. Application of Computational Fluid Dynamics in Building Design: Aspects and Trends. *Indoor Built Environ.* **2006**, *15*, 305–313. [CrossRef]
71. Ayad, S.S. Computational study of natural ventilation. *J. Wind. Eng. Ind. Aerodyn.* **1999**, *82*, 49–68. [CrossRef]
72. Ministerio de Fomento. *Documento Descriptivo Climas de Referencia*; Ministerio de Fomento: Madrid, Spain, 2017.
73. Chazarra, A.; Flórez, E.; Peraza, B.; Tohá, T.; Lorenzo, B.; Criado, E.; Moreno, J.V.; Romero, R.; Botey, R. *Mapas Climáticos de España (1981–2010) y ETo (1996–2016)*; Ministerio para la Transición Ecológica, Agencia Estatal de Meteorología: Madrid, Spain, 2018.
74. Melikov, A.K.; Kaczmarczyk, J. Air movement and perceived air quality. *Build Environ.* **2012**, *47*, 400–409. [CrossRef]
75. Birru, D.; Wen, Y.-J.; Rubinstein, F.M.; Clear, R.D. Ceiling fans as extenders of the summer comfort envelope. *ASHRAE Trans United States* **1983**, *89*. Available online: https://www.osti.gov/biblio/5241401 (accessed on 1 March 2023).
76. Zhai, Y.; Zhang, Y.; Zhang, H.; Pasut, W.; Arens, E.; Meng, Q. Human comfort and perceived air quality in warm and humid environments with ceiling fans. *Build Environ.* **2015**, *90*, 178–185. [CrossRef]
77. Arens, E.; Xu, T.; Miura, K.; Hui, Z.; Fountain, M.; Bauman, F. A study of occupant cooling by personally controlled air movement. *Energy Build.* **1998**, *27*, 45–59. [CrossRef]

78. Rohles, F.H.; Konz, S.A.; Jones, B.W. Enhancing Thermal Comfort with Ceiling Fans. In *Proceedings of the Human Factors Society Annual Meeting*; SAGE Publications: Los Angeles, CA, USA, 1982; pp. 118–122. [CrossRef]
79. Agencia Estatal Boletín Oficial del Estado. *Real Decreto Ley 14/2022, de 1 de Agosto, de Medidas de Sostenibilidad Económica en el ámbito del Transporte, en Materia de Becas y Ayudas al Estudio, así Como de Medidas de Ahorro, Eficiencia Energética y Reducción de la Dependencia Energética del Gas Natural*; Agencia Estatal Boletín Oficial del Estado: Madrid, Spain, 2022.
80. European Parlament and Council of the European Comunity. *Directive 2010/31/EU of the European Parliament and of the Council of 19 May 2010 on the Energy Performance of Buildings*; European Parlament and Council of the European Comunity: Brussels, Belgium, 2010.
81. EU Commission. *Commission Delegated Regulation (EU) No 244/2012 of 16 January 2012 Supplementing Directive 2010/31/EU of the European Parliament and of the Council on the Energy Performance of Buildings by Establishing a Comparative Methodology Framework for Calculating*; EU Commission: Brussels, Belgium, 2012.
82. Galiano-Garrigós, A.; González-Avilés, Á.; Rizo-Maestre, C.; Andújar-Montoya, M. Energy Efficiency and Economic Viability as Decision Factors in the Rehabilitation of Historic Buildings. *Sustainability* **2019**, *11*, 4946. [CrossRef]
83. CEN. *European Committee for Standardization—UNE EN 15459-1:2018 Energy Performance of Buildings—Economic Evaluation Procedure for Energy Systems in Buildings—Part 1: Calculation Procedures, Module M1-14*; European Committee for Standardization: Brussels, Belgium, 2018.

buildings

MDPI

Article

Vulnerability of Affordable Housing to Global Warming in South Africa: Case Study of a Masonry House in Johannesburg

Ryan Bradley

School of Civil and Environmental Engineering, University of the Witwatersrand, Johannesburg 2000, South Africa; ryan.bradley@wits.ac.za; Tel.: +27-11-717-7127

Abstract: Global warming is expected to lead to longer and more intense heatwaves, which will have negative environmental and socioeconomic impacts around the world. South Africa is projected to experience significant warming, with surface temperatures possibly increasing by up to 3 °C by mid-century. This warming trend has implications for architecture, as the demand for cooling in buildings could rise dramatically. However, socioeconomic conditions in developing countries may limit the use of air conditioning to mitigate indoor overheating. In South Africa, research has shown that government provided low-cost housing structures are thermally inefficient, with temperatures occasionally exceeding outdoor levels. Residents often rely on natural ventilation and personal actions to cope with heat. However, the effects of climate change may render these strategies insufficient if energy poverty and housing improvement are not addressed. This study aims to examine the impact of global warming on a high mass, naturally ventilated, affordable housing structure in Johannesburg, South Africa. Measured operative temperature data from a long-term experimental study, alongside adaptive temperature limits to evaluate overheating, highlight the vulnerability of indoor spaces without adequate insulation and/or thermal mass. The results underscore concerns about the performance of low-cost and affordable housing in warmer future climates in the South African interior.

Keywords: climate change; heatwaves; overheating; naturally ventilated; affordable housing

Citation: Bradley, R. Vulnerability of Affordable Housing to Global Warming in South Africa: Case Study of a Masonry House in Johannesburg. *Buildings* **2023**, *13*, 1494. https://doi.org/10.3390/buildings13061494

Academic Editors: Yingdong He and Nianping Li

Received: 19 April 2023
Revised: 2 June 2023
Accepted: 5 June 2023
Published: 9 June 2023

1. Introduction

The Intergovernmental Panel on Climate Change (IPCC) has reported that human activities have caused global warming, with global surface temperature reaching 1.1 °C above pre-industrial levels [1]. Climate change has also led to an increase in the frequency of heat waves (HWs), resulting in negative environmental and socioeconomic impacts in numerous regions worldwide [2]. As climate models predict rising surface temperatures in many regions across the world, the implications of climate change for architecture have become increasingly important. According to projections obtained from a dynamic regional climate model by the authors of the South African Risk and Vulnerability Atlas [3], South Africa is expected to experience the greatest future warming over the interior of the country, with temperatures projected to increase more than the global mean temperature [4]. Studies indicate that daily maximum temperatures may increase by up to 3 °C by mid-century based on high greenhouse gas (GHG) emission scenarios [2,5]. It is important to note that the predicted increase/decrease rates assigned to these climate models have some uncertainty, largely related to global emission scenarios. Nonetheless, even moderate GHG concentrations in the models indicate that HWs are expected to occur more frequently, last longer, and become more intense in South Africa [2].

Kruger and Mbatha (2021) presented evidence of gradual warming over the interior of South Africa, showing yearly average temperature deviation from the base period (Figure 1) for OR Tambo International Airport, Johannesburg, which has the most reliable long-term record in the region [6]. The presented data show that all years beyond 2001 have had average surface temperatures higher than the long-term average, with increased (1) hot extremes, (2) number of hot days and nights, and (3) duration of warm spells in the province (Gauteng) [6]. The observed rise in temperatures over the past two decades in Johannesburg may partially be ascribed to the amplification of the urban heat island (UHI) effect, and studies have highlighted the increased tendency of overheating, and associated health risks are associated with the UHI effect in large metropolitan areas [7,8]. The distinct trend in Figure 1 is, however, not unique to Johannesburg; similar patterns have been recorded across long-standing climate monitoring stations distributed throughout South Africa [6]. This widespread phenomenon implies that multiple environmental and human-induced factors may be contributing to the escalating temperatures. Many stations across South Africa, including Johannesburg International Airport, have also recorded their two hottest years on record in the last decade, with 2015 being the hottest (in many cases) due to the strongest El Niño event on record (Figure 2) [9,10]. In the South African interior during the summer, the occurrence of El Niño is typically associated with warmer and drier conditions, whereas the occurrence of La Niña is linked with cooler and wetter conditions [10].

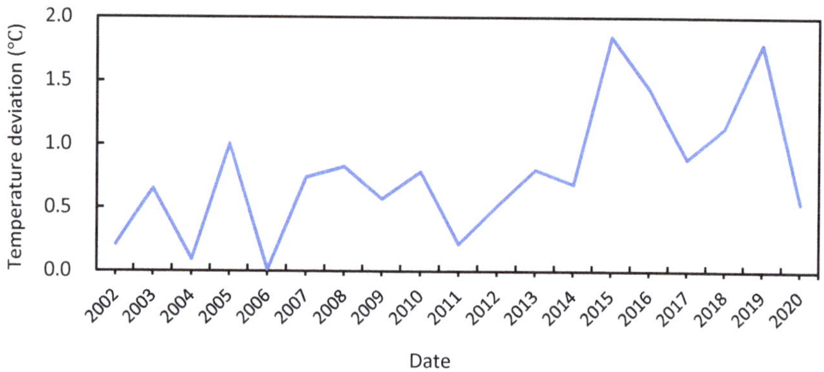

Figure 1. Annual average surface temperature deviation from the base period (1981–2010) at OR Tambo International Airport (Johannesburg): 2002–2020 (Data from [6]).

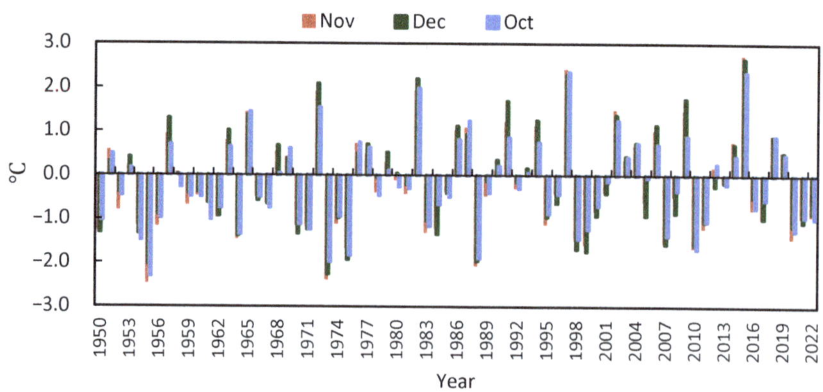

Figure 2. Nino3.4 SST (sea surface temperatures) anomalies averaged over the NINO34 region 5° N–5° S; 170–120° W: 1950–2022 (Data from US National Oceanic and Atmospheric Administration (NOAA) website: https://psl.noaa.gov/data/correlation/nina34.anom.data (accessed on 3 March 2023)).

Global warming poses a significant threat to the future performance of existing residential buildings, particularly in terms of energy demand and overheating [11–13]. In response, there has been a growing body of research dedicated to the topic, which has involved either full-scale testing or, most commonly, thermal and energy simulation studies [11–23]. The predominant approach typically involves either the exploration of extreme historical data or the forecast of future scenarios. The existing literature indicates that climate change-induced overheating could be a significant problem in many regions across the world [14–16]. However, the impact has large dependency on geographical location (climate) and building type [12]. Critical areas of concern include future deterioration of indoor comfort, at-risk populations, and upgrading regulations for future warmer climates. Moreover, studies have identified an anticipated increase in cooling loads and a need for the adoption of alternate cooling strategies [17,18]. For instance, a recent study by Andrić et al. (2021) reviewed numerous numerical studies on buildings in various climates. The findings suggest that cooling demand in 2050 could increase substantially, i.e., by up to 1050% compared to reference levels, contingent on specific building properties and prevailing climate conditions [19]. Several adaptation strategies have been examined to mitigate predicted degradation in future indoor thermal comfort. These include passive cooling strategies [17] and improvements in the thermal characteristics of building materials—insulation, glazing, and solar protection—to alleviate the risk of overheating [16,21–23].

In many developing countries, however, socioeconomic conditions may inhibit people's ability to mitigate overheating and indoor discomfort through adaptive retrofitting or artificial means such as air conditioning (AC) [24]. For instance, a 2015 study [25] reported that the penetration rate of AC in residential households in South Africa is just 4%, but it is as high as 40% in the highest income households. Wright et al. (2022) examined perceptions of thermal comfort among people in a structurally disadvantaged community in Limpopo province, South Africa, and found that only 17 out of 406 respondents (<4%) selected "Use an air conditioner" as a mechanism to cope with thermal discomfort [26]. Instead, people largely relied on adaptive actions such as sitting outdoors in the shade or opening windows and doors. Other studies have also shown that lower-income households depend on employing personal action to improve their thermal situations, such as natural ventilation [23,27].

However, climate change may negate the impact of these actions in future climate years if energy poverty and infrastructure improvement (retrofitting) are not considered alongside personal actions. Thapa (2022) conducted a simulation study, which revealed that naturally ventilated low-rise concrete dwellings in northeast India would be unable to suppress the effects of global warming, and the indoor environment would deteriorate in future conditions [23]. In situ studies conducted in the South African interior have also highlighted deficiencies with some types of low-cost residential buildings in the present climate [26–28]. For instance, Naicker et al. (2017) measured indoor air temperature and relative humidity in 59 households across Johannesburg, South Africa, and found that temperatures in low-cost government houses exceeded outdoor temperatures by as much as 5 °C. The reported deficiencies with most of these formal housing structures were that (1) the walls were only 150 mm thick, (2) most lacked ceilings, and (3) none had any type of roof insulation [27]. In a similar study, Mabuya and Scholes (2020) reported that most post-apartheid era houses examined in their study were thermally inefficient [28]. Wright et al. (2022) also examined indoor dwelling and outdoor temperatures in Giyani, Limpopo province, and found that the maximum measured indoor temperature was approximately 36 °C, while the maximum outdoor temperature was 40 °C. The outdoor and indoor temperatures were comparable on several hot summer days during the study period, which meant that natural ventilation and sitting outdoors offered little to no reprieve from the heat [26].

The primary objective of this study is to investigate the potential influence of global warming on the performance of high-mass residential buildings in South Africa. To achieve this, data from a long-term experimental study conducted on an earth masonry house in Johannesburg were analyzed [29,30]. The house was monitored from 2018 to 2021, during which various passive cooling techniques were implemented to improve its thermal performance. Notably, the study period encompassed the second and fourth warmest years ever recorded in Johannesburg, as illustrated in Figure 1. This provided an opportunity for a practical investigation of global warming's potential effects on a high-mass structure within the South African context. Although the summer of the second warmest year (2019) coincided with a neutral ENSO (El Niño Southern Oscillation) phase [31], the annual average temperature exceeded Johannesburg's long-term average by more than 1.75 °C, making it second only to the hottest year ever recorded in 2015.

Similar trends were observed for annual average maximum temperatures, as displayed in Table 1 [31–33]. In 2020, the average maximum temperature was marginally lower than those in 2018 and 2019 but still comparable to the 30-year base period. This can be primarily attributed to a moderate La Niña phase during the latter half of the year [34]. Table 2 presents the average maximum and daily mean temperatures for the early summer months (October and November), which correspond to the data and analysis discussed in this paper. The temperatures recorded at OR Tambo Airport during October and November 2019 significantly exceeded the historical long-term means for the same period, as evidenced in Table 2.

Table 1. Average daily maximum surface temperatures, OR Tambo Airport (Johannesburg) [1].

Year	Average Maximum Temp. (°C)	Rank	ENSO:
1981–2010 [2]	22.3	-	-
2015	24.2	1	El Niño (very strong)
2018	23.2	4	El Niño (weak)
2019	23.9	2	Neutral/El Niño (very weak)
2020	22.4	37	La Niña (moderate)

[1] Temperature data obtained from SAWS publications [31–33]. [2] Normal (30-year base period).

Table 2. Temperature data for OR Tambo Airport: Early summer (1 October–30 November).

Year	Average Maximum Temp. (°C)	Daily Mean Temp. (°C)
1961–1990 [1]	23.8	17.5
2015 [2]	27.8	20.5
2018 [2]	25.8	18.9
2019 [2]	27.7	20.8
2020 [2]	25.4	19.7

[1] Data for 1961–1990 obtained from Meteonorm v7.3 software. [2] Data obtained from SAWS.

The extraordinary heat of the 2019 summer prompted a more in-depth comparative analysis with other years monitored during the testing period. This analysis exposed the vulnerability of rooms without sufficient insulation, thermal mass, and ventilation to overheating during periods of very high temperatures. The first-floor spaces share some deficiencies with much of the formal low-cost housing infrastructure in South Africa, and the results presented herein support concerns raised in recent studies [26,28] about the performance of low-cost and affordable housing in the face of warmer future climates in the South African interior.

2. Materials and Methods

2.1. House Layout and Characteristics

The two-story house assessed in the study was located on the University of the Witwatersrand campus in Johannesburg, South Africa (coordinates: 26°11′09″ S 28°01′30″ E). The building layout is shown in Figure 3a,b. The first-floor spaces were designated as bedrooms and assumed to be occupied between 8:00 p.m. and 6:00 a.m. (i.e., a night-time space). The roof of the house consisted of two steep catenary vaults, enclosing the first-floor spaces, as shown in Figure 4a. The roof structure is somewhat unique: pitched roofs with wooden trusses topped with tiles or metal sheeting are the most common roofing arrangement across South Africa; the steep curved profile of the roof was chosen for its architectural performance, offering more usable floor space than pitched roofs and complying with dimensional and space requirements more effectively. All spaces met the dimensional requirements specified in the National Building Regulations [35].

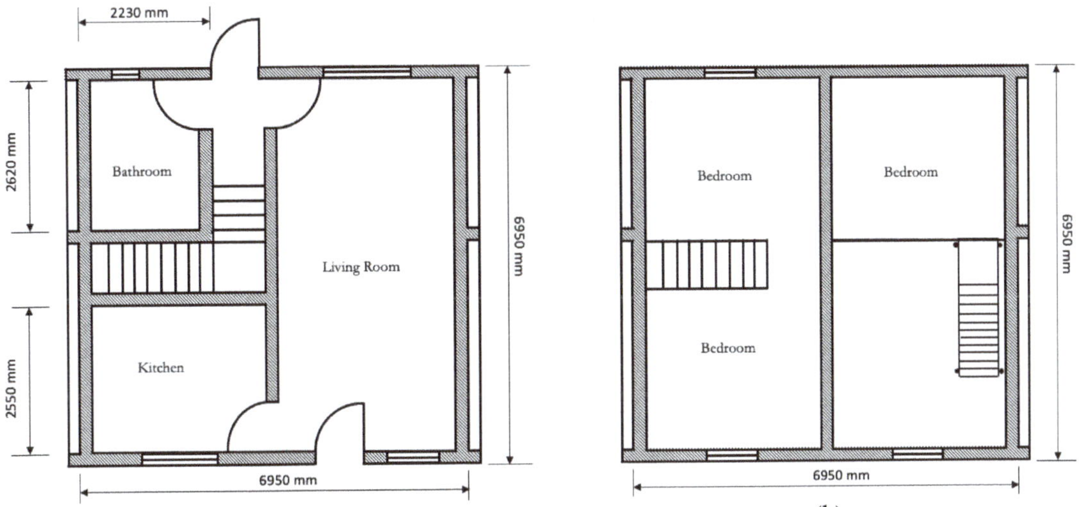

(a) (b)

Figure 3. Floorplan: (**a**) ground floor; (**b**) first floor.

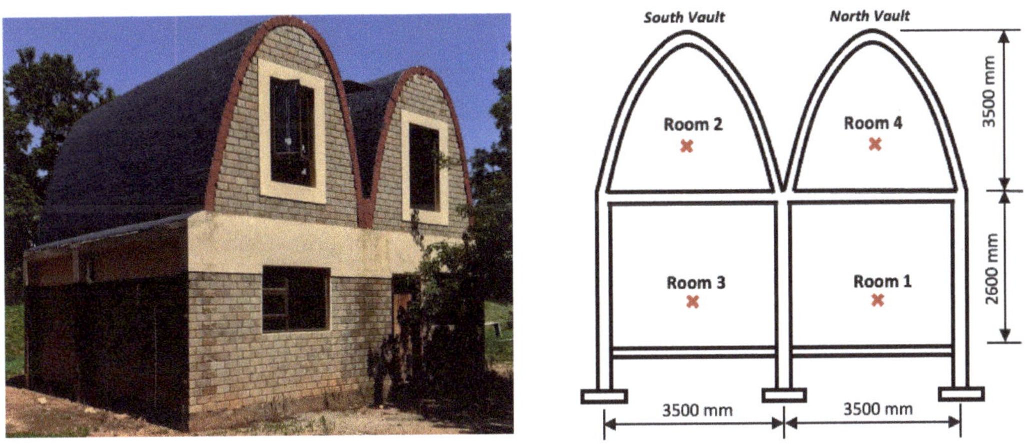

(a) (b)

Figure 4. Housing structure: (**a**) building from the south-east; (**b**) spaces investigated.

The load and non-load bearing ground floor walls were constructed using interlocking blocks, made from recycled building rubble, while the vaults were made of compressed stabilized earth blocks (CSEBs) using traditional mortar. The shell, including internal plaster, was 170 mm thick, and the vertical walls were 220 mm thick. The first-floor slab was 170 mm deep and consisted of a rib and block system. The windows are comprised of aluminum framing and single-glazed low-E glass. No overhangs or shading elements were considered in the present study. The estimated and manufacturers'-quoted thermal transmittance (U-value) for the main elements of the building envelope are given in Table 3. Values for the wall and floor slabs were calculated using the approaches outlined in

CIBSE Guide A [36]. The maximum U-value for masonry walling given in the deemed-to-satisfy approach in NBR [37] is 2.86 W/m^2K (given in the code by the minimum thermal resistance of R = 0.35 m^2K/W). Fenestration and openings also comply with the deemed to-satisfy requirements for ventilation and the minimum energy performance requirements (fenestration area < 15 % of net floor area per story).

Table 3. Estimated U-values.

Materials Description	U-Value (W/m^2K)
Solid dry-stack (mortarless) block masonry (220 mm)	2.11 [1]
CSEB block masonry (150 mm) with cement-sand plaster (20 mm)	2.38 [1]
Window glazing (6.38 mm) (single-glazed low-E glass: 83% transmission of visible light and 70% solar energy)	3.57
Solid concrete floor slabs (150 mm) on sand (uninsulated)	1.01 [2]

[1] R-values (m^2K/W) for internal and external surfaces assumed 0.13 and 0.04, respectively. [2] R-values (m^2K/W) for internal and external surfaces assumed 0.17 and 0.04, respectively.

2.2. Experimental Setup

Operative temperature (T_{op}) was measured 1.1 m above the floor using 40 mm black globe thermometers positioned in the center of each space/room. The thermometers recorded temperatures every 15 min using an Agilent data-acquisition unit (Agilent Technologies Inc., Santa Clara, CA, USA). The evaluation encompassed four distinct spaces, which were designated as Room 1, Room 2, Room 3, and Room 4, as depicted in Figure 4b. Outdoor air temperature was measured using a thermo-hygrometer wirelessly connected to an Oregon Scientific Weather Station (Oregon Scientific Inc., Tualatin, OR, USA). The accuracy of the thermocouples in the black globe thermometers was around ±0.5 °C, and the manufacturer's quoted accuracy for the outdoor thermometer was ±1 °C. More details on the instrumentation and setup can be found in [30].

2.3. Overheating Criteria

The assessment for indoor overheating was undertaken with the adaptive temperature limits specified by BS EN 15251 [38] in combination with the guidelines developed by the Chartered Institution of Building Services Engineers (CIBSE) [39]. Adaptive comfort models and accompanying temperature limits consider the occupants' ability to adapt to their environment by adjusting their behavior, clothing, and surroundings to achieve comfort under a range of indoor conditions. This contrasts with traditional, prescriptive models, based on fixed indoor temperature ranges.

The adaptive comfort model used in BS EN 15251 expresses the comfort temperature through its relationship with the outdoor temperature, and the following equation is used to estimate the comfortable temperature (T_{comf}) in naturally ventilated buildings:

$$T_{comf} = 0.33\ T_{rm} + 18.8\ (\text{where } T_{rm} > 10\ °C). \tag{1}$$

T_{rm} is the exponentially weighted running mean temperature (°C) for the day under consideration. This calculation puts higher importance on the most recent days. The equation for the determination of the running mean temperature is given by Equation (2):

$$T_{rm} = [1 - \alpha]\{T_{od\text{-}1} + \alpha T_{od\text{-}2} + \alpha^2 T_{od\text{-}3}\ldots\}. \tag{2}$$

α is a constant and Tod-1, Tod-2, Tod-3, etc., are the daily mean temperatures for yesterday, the day before yesterday, and so on. A value of 0.8 for the parameter 'α' provides the best correlation between T_{rm} and T_{comf} based on comfort surveys conducted throughout Europe [39,40]. Category II (Normal expectation) limits prescribed in BS EN 15251 were assumed in the present study, i.e., an acceptable temperature range of ±3 °C, about the comfort temperature (T_{comf}) for naturally ventilated buildings. The maximum acceptable temperature (T_{max}) is subsequently given by Equation (3):

$$T_{max} = T_{comf} + 3\ ^{\circ}C. \quad (3)$$

The CIBSE guidelines use three criteria to evaluate the risk of overheating [39]. These criteria account for the severity and frequency of overheating and are defined based on the difference (ΔT) between the indoor operative temperature (T_{op}) and the maximum acceptable temperature (T_{max}), as given by Equation (4). Overheating in a building or internal space is identified if it violates any two of the three criteria given in Table 4.

$$\Delta T = T_{op} - T_{max} \text{ (rounded to the nearest whole degree)} \quad (4)$$

Table 4. Overheating assessment criteria.

Criterion	Assessment Criteria	Limit
1	Percentage of occupied hours during which $\Delta T \geq 1\ ^{\circ}C$	Up to 3% of occupied hours
2	Daily weighted exceedance (We) in any one day > 6 °Ch (degreeHours)	0 days
3	Maximum temperature level (T_{upp}) $\Delta T > 4\ ^{\circ}C$	0 h

The CIBSE guidelines set an absolute upper limit on indoor operative temperature, T_{upp}, which is 4 °C above the maximum temperature, as given in Equation (5).

$$T_{upp} = T_{max} + 4\ ^{\circ}C \quad (5)$$

3. Results

From January 2018 to March 2021, the building's thermal comfort was assessed through periodic monitoring. Various strategies were employed during this period to enhance the comfort of specific internal spaces, such as external shading, reflective roof paint, and insulation of the ceiling. However, this study does not primarily focus on the effectiveness of these alterations. Instead, it selectively presents results demonstrating the influence of extreme high temperatures on building performance. For comparative purposes, the analysis concentrates on the early summer months of October and November in 2018, 2019, and 2020, when darker roof paint was applied.

During the evaluation, certain spaces experienced retrofitting or substantial changes in roof color, and the data collected during these periods have been excluded; specifically, data for Room 2 from the last few days of October and all of November in 2018, as well as all days in October and November 2020, have been disregarded. It is worth noting that the roofs were maroon colored in 2018 and grey in 2019 and 2020, which could have influenced indoor temperature levels to some extent.

The subsequent sub-sections will examine the performance of all monitored spaces and provide a detailed analysis of overheating in the more heat-vulnerable first-floor rooms.

3.1. Ground versus First Floor Spaces

It is widely acknowledged that high-mass buildings, built using materials such as concrete and brickwork, typically exhibit a more significant temperature lag (the delay between outdoor temperatures attaining their highest or lowest points and the subsequent indoor temperature response) in comparison to buildings composed of lightweight materials. Figure 5a,b shows the indoor operative temperature plotted alongside outdoor air temperature for a ground and first-floor space during the early summer period of 2018. A clear lag and attenuation of peak indoor temperatures is apparent for the ground floor space, which highlights the benefit of high thermal mass in maintaining more stable indoor temperatures throughout the day (despite the large fluctuations in outdoor temperatures, as shown in Figure 5). In contrast, the first-floor space did not moderate high temperatures as effectively (Figure 5b) and other means (beyond thermal mass) would need to be implemented to improve the overall thermal performance of the space. The higher recorded temperatures in the first-floor space can be ascribed to the increased exposure of the of the roof envelope and its lower surface albedo than the ground floor walls, as well as the buoyant warm air rising from the ground floor spaces (passages and doors were open during the assessments). Surface albedo and solar exposure are, however, the dominant aspects affecting heat gain in the first-floor spaces [26].

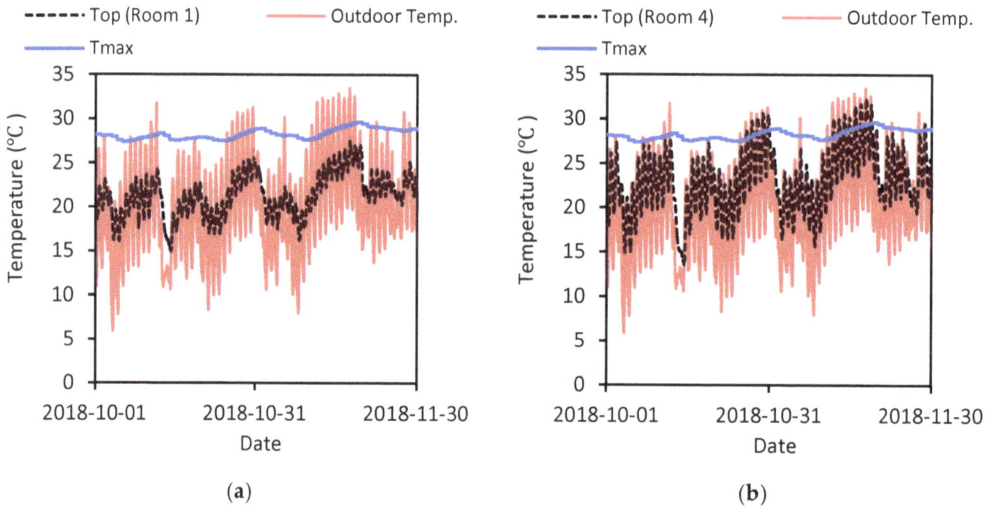

Figure 5. Indoor operative temperature and outside air temperature: (**a**) T_{op} Room 1 (2018); (**b**) T_{op} Room 4 (2018).

Figure 6a–d display the hourly indoor operative temperature plotted against the outdoor air temperature for all evaluated spaces during the study period. The steep positive slopes and high correlation coefficients (R) of the linear trendlines in Figure 6c,d suggest a strong positive correlation between outdoor and indoor temperatures, whereas the more gradual slopes in Figure 6a,b imply a weaker relationship for first-floor spaces. Although natural ventilation typically results in greater temperature fluctuations compared to artificially heated and cooled spaces, the first-floor spaces exhibit a notably significant range in temperature through the study period. Furthermore, indoor operative temperatures frequently exceeded T_{max} in the first-floor spaces, whereas temperatures in the ground-floor spaces did not exceed this limit (Figure 5). Based on the data presented in Figures 5 and 6, only the first-floor spaces were assessed for overheating during the early summer period of 2018, 2019, and 2020 in the subsequent sections.

3.2. Exceedance of T_{max}

Figures 7–9 illustrate the outdoor air temperature and indoor operative temperatures for the first-floor spaces during the assessment period corresponding to the years 2018, 2019, and 2020, respectively. Indoor temperatures frequently surpassed the maximum threshold temperature (T_{max}) during the 2nd (2019), 4th (2018), and 37th (2020) warmest years on record. It is important to mention that some data from 2019 are unavailable due to backup power supply issues with the data logger.

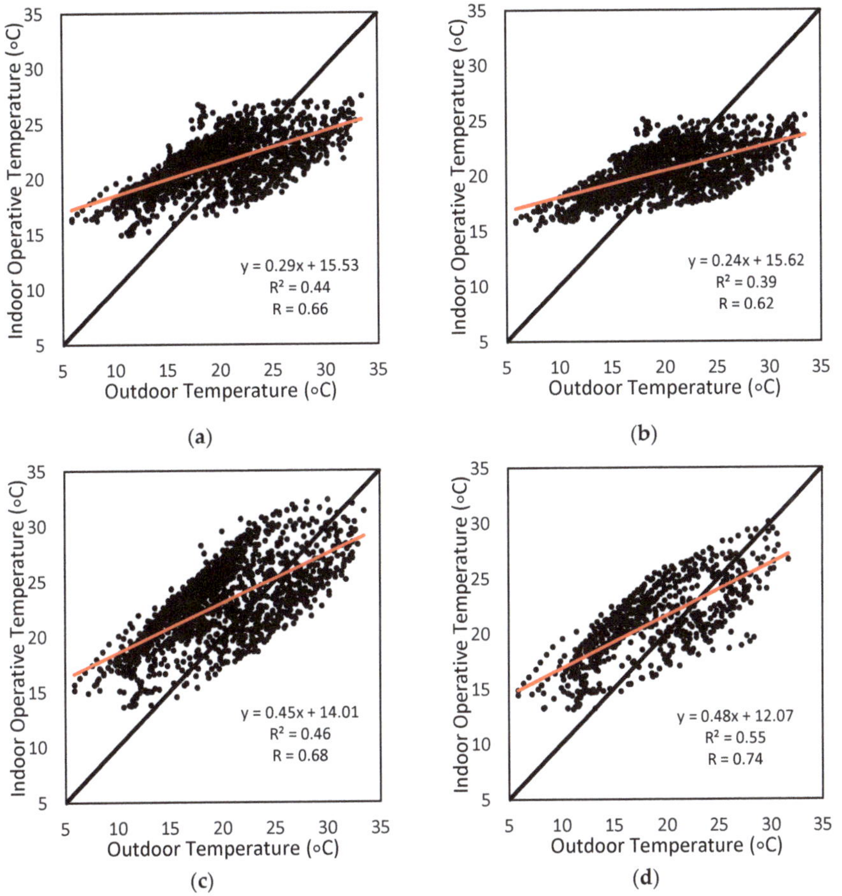

Figure 6. Indoor operative temperature versus outside air temperature (2018): (**a**) Room 1; (**b**) Room 3; (**c**) Room 4; (**d**) Room 2 (1 to 27 October only).

The most significant exceedance of T_{max} was observed in Room 4, as depicted in Figures 7d, 8d and 9d. On hot summer days, particularly during heatwaves, the operative temperature routinely surpassed the daily maximum temperature (T_{max}), beginning in the early afternoon, around 2:00 p.m., and persisting until around midnight. In contrast, Room 2 featured large windows in each gable wall, which were mostly open during late afternoons and early evenings. This facilitated natural ventilation and contributed to the cooler temperatures and fewer instances of T_{max} exceedance in this space [29]. However, regardless of the presence or absence of cross-ventilation, the first-floor spaces experienced numerous days with operative temperatures (T_{op}) exceeding T_{max}.

Upon comparing the plots for 2018 (Figure 7), 2019 (Figure 8), and 2020 (Figure 9), it becomes evident that the maximum threshold temperature (T_{max}) was surpassed most frequently during the 2019 assessment period, followed by 2020, and then 2018. Focusing on the worst-performing space, Room 4, the number of days with operative temperatures (T_{op}) exceeding T_{max} were 14, 35, and 19 for 2018, 2019, and 2020, respectively. The days of exceedance during the October–November period in 2019 considerably outpaced those in the same periods for 2018 and 2020. The lower number of days exceeding T_{max} for the 2018 period may be attributed to factors such as a different roof color, as well as lower minimum outdoor temperatures during October and November that year.

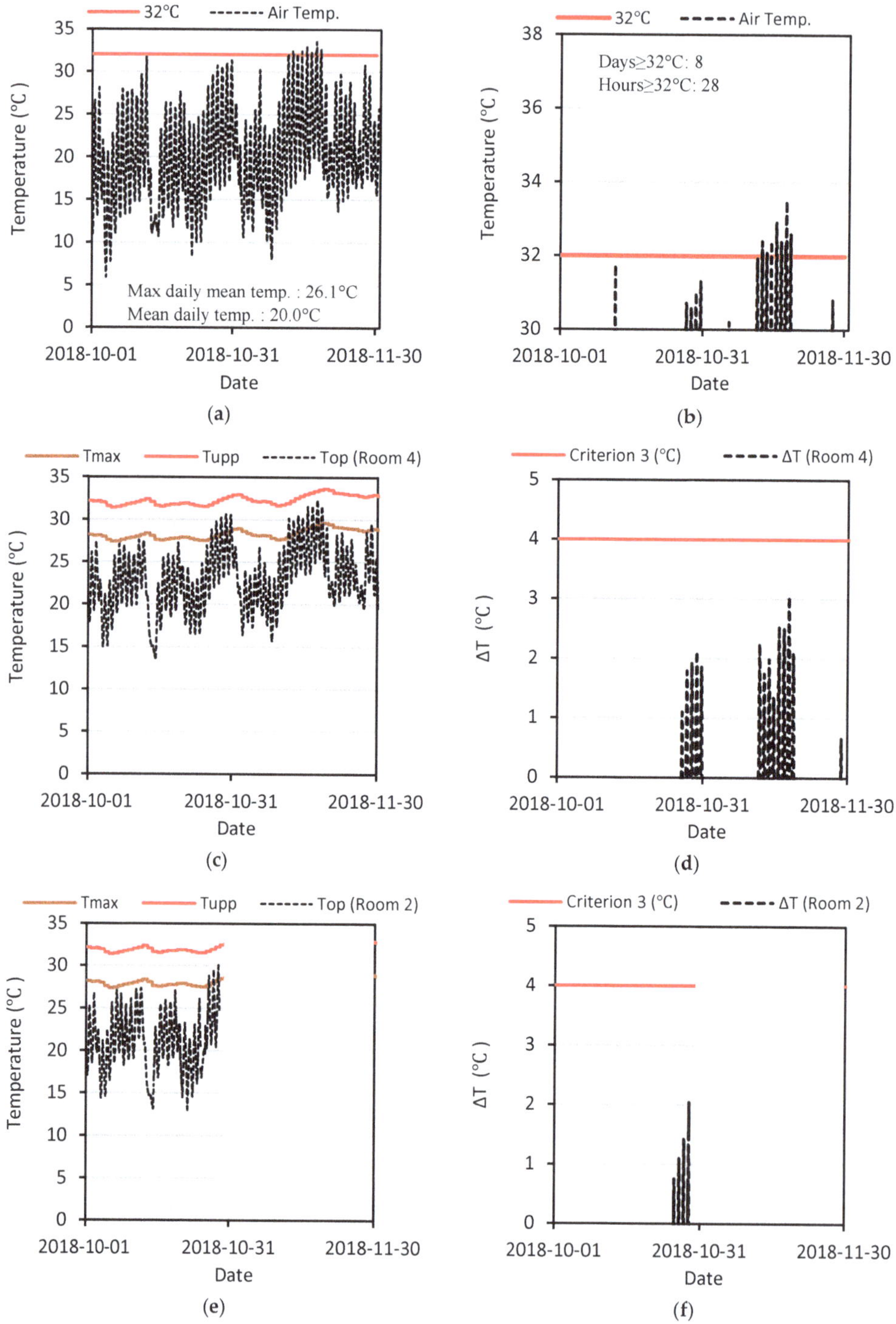

Figure 7. Temperatures, October through November 2018: (**a**) outdoor air; (**b**) outdoor air (max); (**c**) T_{op} Room 4; (**d**) ΔT Room 4; (**e**) T_{op} Room 2; (**f**) ΔT Room 2.

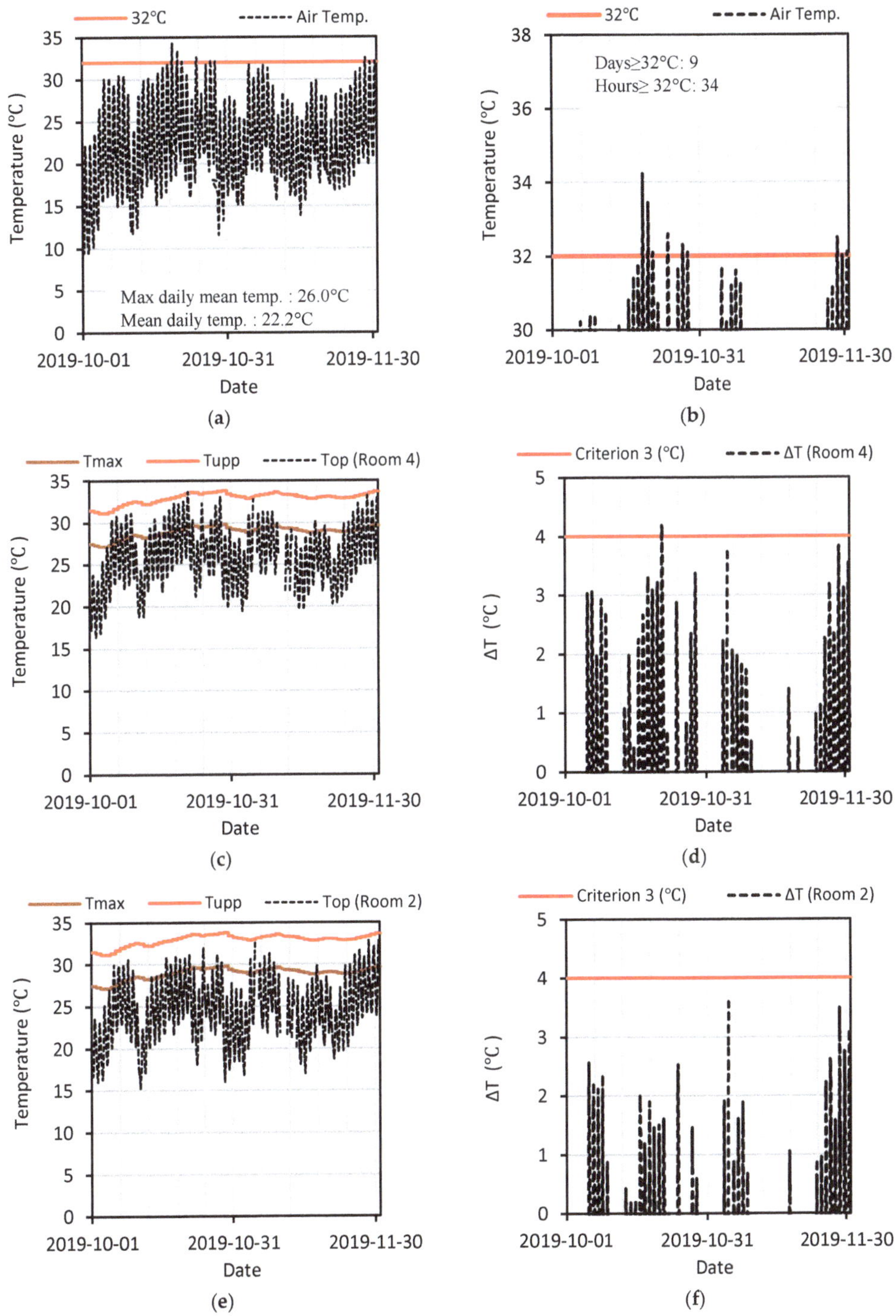

Figure 8. Temperatures, October through November 2019: (**a**) outdoor air; (**b**) outdoor air (max); (**c**) T_{op} Room 4; (**d**) ΔT Room 4; (**e**) T_{op} Room 2; (**f**) ΔT Room 2.

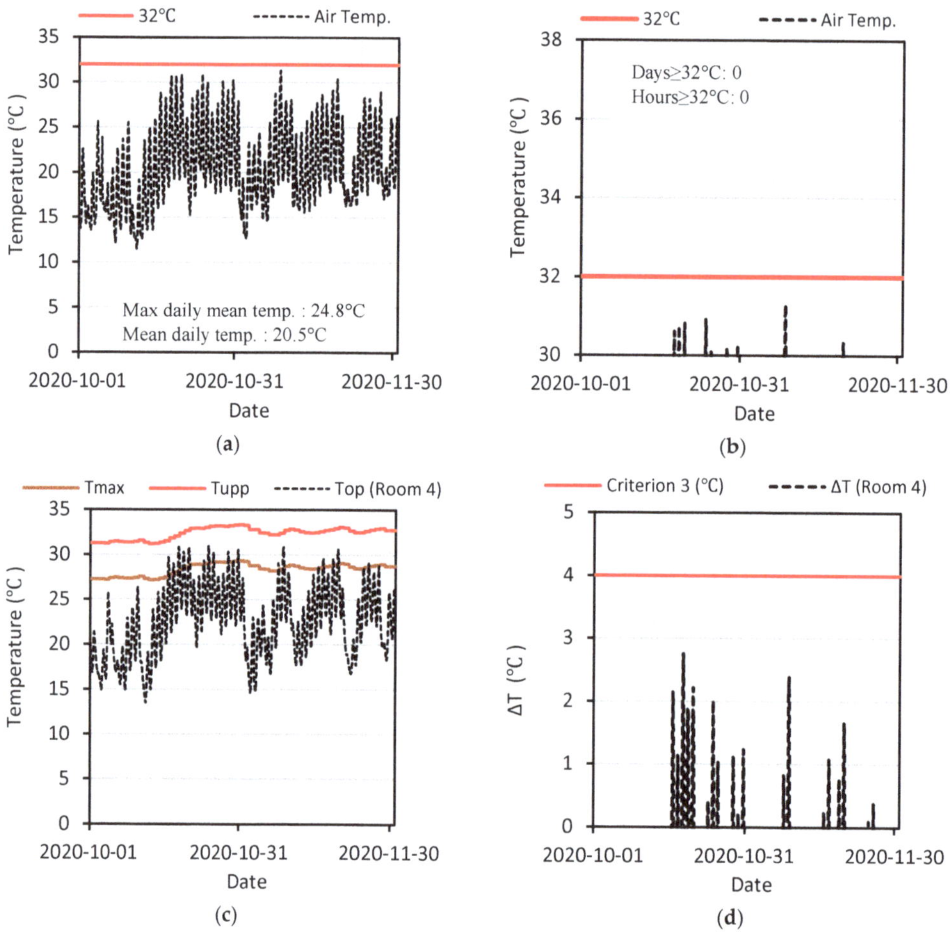

Figure 9. Temperatures, October through November 2020: (**a**) outdoor air; (**b**) outdoor air (max); (**c**) T_{op} Room 4; (**d**) ΔT Room 4.

3.3. Results of the Overheating Assessment: Room 4

Table 5 presents the findings from the overheating assessment for Room 4 corresponding to the early summer period. The room was considered a nighttime space (bedroom), occupied between 8:00 p.m. and 6:00 a.m. The following observations can be made:

1. In 2018 and 2019, the space violated criteria 1 and 2, resulting in a failure of the CIBSE assessment (i.e., the space overheated); however, in 2020, the space successfully met all three overheating criteria;
2. All criteria were most severely violated in 2019; however, the performance against criterion 1 was the most striking, with approximately three to five times more hours of exceedance compared to 2018 and 2020, respectively;
3. High outdoor temperatures recorded during October and November 2019 (Table 6) led to higher comfortable (T_{comf}) and maximum (T_{max}) temperatures during the period; nonetheless, the increase in these parameters was offset by the corresponding rise in indoor operative temperature (T_{op}), which meant that T_{max} was frequently exceeded during October and November 2019;
4. The first-floor spaces were significantly affected by extreme temperatures, leading to a notable increase in both the duration and intensity of overheating. During the months

of October and November 2019, the space experienced 12.5% exceedance hours, which was severe enough to suggest that the space failed to meet the requirements for the entire summer period (October–March), without requiring any further assessment (i.e., assuming all hours are below T_{max} for the remaining summer months, the % hours above T_{max} + 1 °C would still be approximately 4%).

Table 5. Overheating assessment of Room 4: October and November (only hours between 8:00 p.m.–6:00 a.m.).

Year	Exceed T_{max} (Y/N) *	Criterion 1 (<3%)	Criterion 2 (<6 °Ch)	Criterion 3 (<4 °C)	Overheat (Y/N)
2018	Y	Fail (3.4%)	Fail (We = 6.25 °Ch)	Pass (T = 3 °C)	Y
2019	Y	Fail (12.5%)	Fail (We = 7.75 °Ch)	Pass (T = 3 °C)	Y
2020	Y	Pass (2.3%)	Pass (We = 4.75 °Ch)	Pass (T = 2 °C)	N

* Room 2 is assumed occupied between 8 p.m. and 6 a.m.

Table 6. Outdoor temperatures recorded at the house: 1 October–30 November.

Year	Maximum Temp. (°C)	Mean Daily Maximum Temp. (°C)	Daily Mean Temp. (°C)
2018	33.5	27.0	20.0
2019	34.2	29.1	22.2
2020	31.3	25.6	20.5

3.4. Results of the Overheating Assessment: Room 2

In the evaluation of Room 2 as a nighttime space, it demonstrated adherence to all overheating criteria during October and November 2019, in contrast to Room 4. The significantly improved performance of Room 2, in comparison to Room 4, can be directly ascribed to the efficacy of cross-ventilation within the space. Several key attributes contributed to the successful implementation of nighttime cross-ventilation in Room 2, including: (1) a long, narrow floor plan; (2) an elevated location on the first-floor; (3) the absence of internal partitions or obstructions; (4) a large diurnal outdoor temperature variation; and (5) the presence of two large windows aligned on opposite sides of the space (Figure 10a). Figure 10b highlights the effectiveness of the cross-ventilation in question during a heatwave on 21 October 2019, with the windows remaining open during the late afternoon and night once the outdoor temperature decreased below the indoor temperature. It is important to highlight that both first-floor rooms, Rooms 2 and 4, did not meet the overheating criteria when assumed continually occupied (24/7).

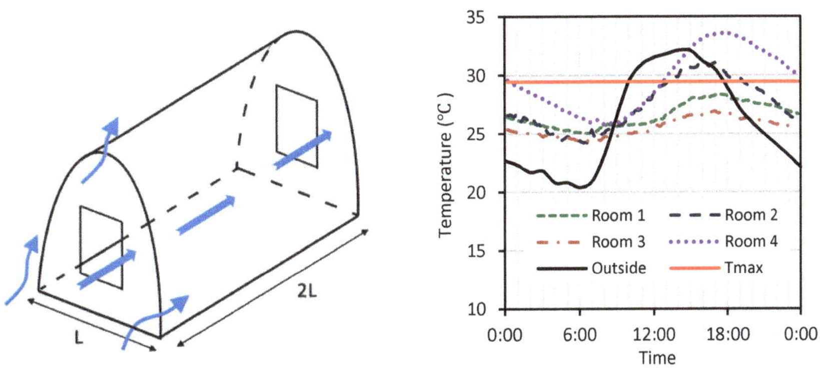

(**a**) (**b**)

Figure 10. Cross-ventilation in Room 2: (**a**) illustration of cross-ventilation (Room 2); (**b**) temperatures, 21 October 2019.

4. Discussion

The evaluation of overheating unveiled an underperformance in the first-floor spaces, during notably hot summer days and months. The persistent, intense heat of heatwaves had a significant impact, particularly on the first overheating criterion assessing the frequency and duration of overheating. Room 2, with efficient cross-ventilation, emerged as the most effective first-floor space during the evening, highlighting the essential role of nighttime ventilation in mitigating overheating under severe heat conditions. Past research using building simulation tools has shown that dwellings lacking adequate ventilation are particularly vulnerable to overheating problems, a situation likely to worsen with the advent of warmer climates [15,21]. In this context, Room 4 failed the overheating assessment, while Room 2 met all criteria during the especially hot summer of 2019 when designated as a bedroom (nighttime space).

However, despite the improvement in indoor comfort due to nighttime ventilation, Room 2 would not meet the criteria of the Chartered Institution of Building Services Engineers (CIBSE) overheating assessment if occupied continuously (24/7). The role of natural ventilation in combating overheating was limited by the assumed occupancy of the space, and it alone may not sufficiently mitigate overheating in future climates. Some studies indicate that climate change will reduce the effectiveness of natural ventilation in certain regions [12,16]. Current experimental and simulation studies further suggest the efficacy of natural ventilation as a cooling strategy is already limited in South Africa's hottest areas [26,41].

Room 4 and Room 2 on the first floor frequently recorded indoor temperatures that surpassed the maximum acceptable levels as stipulated by BS EN 15251 and the World Health Organization (WHO). Notably, this occurrence was more prevalent during the late afternoon and early evening during heatwaves. Room 4 reached a high of approximately 34 °C, while Room 2 reached a maximum of 33 °C, both in the late afternoon. Other researchers have also reported similar observations in simulation studies, e.g., [17] they have examined energy-efficient terraced houses and reported that operating temperatures in indoor rooms remained high throughout the day and night during heatwaves, ranging from a minimum of 24.7 °C to a maximum of 32.1 °C. The WHO recommends keeping daytime room temperatures below 32 °C and nighttime temperatures under 24 °C [42]. This is especially crucial in preventing heat-related health issues among vulnerable groups, such as infants, adults over 60, and individuals with chronic health conditions.

In stark contrast, ground floor spaces showed good performance, maintaining indoor temperatures below the maximum permissible limit even during the 2019 summer peak. This success is largely attributed to reduced solar exposure, lower thermal transmittance, and the favorable surface albedo. The effect of these features on building performance and thermal comfort is well-documented in free-standing residential buildings. A comparative performance assessment between ground and first-floor spaces pinpoints inadequate building fabric as the primary cause for extremely high bedroom temperatures. Several studies have presented similar observations with existing buildings, highlighting issues with building envelopes and the need for climate proofing of residential building stock [20,22,27]. Consequently, attention has shifted towards retrofitting or upgrading existing building envelopes for future climates, spanning a range of energy-efficient practices such as cool roofs, insulated wall systems, and green roofs [43–47].

In the context of low-cost and affordable housing in South Africa, this study supports recent research [26–28] reporting deficiencies in some types of formal low-cost housing. It necessitates the reevaluation of local building regulations, especially the deemed-to-satisfy approach (in which minima wall thickness, glazing areas, etc., are specified), and the introduction of adaptive comfort criteria to improve housing performance and thermal comfort in current and future climates.

5. Conclusions

In South Africa, surface temperatures are projected to increase at a rate surpassing the global mean, with the central interior region experiencing longer and more intense heatwave events. The escalating frequency, intensity, and duration of heatwaves associated with climate change highlight the crucial impact of thermal comfort on human health. The ability to alleviate extreme heat will disproportionately affect underprivileged communities, which lack the necessary resources to mitigate or address these challenges.

The residential building evaluated in this study demonstrated satisfactory performance during typical summer months. However, first-floor areas were vulnerable to elevated indoor temperatures under typical summer conditions. In contrast, the extreme summer conditions from late 2019 to early 2020 resulted in substantial indoor temperatures (peaking at 34 °C), and the failure of the Chartered Institution of Building Services Engineers' (CIBSE) overheating criteria in a first-floor space without cross-ventilation.

The adaptive strategy of opening windows to enhance thermal comfort proved to be highly effective in mitigating overheating in well-ventilated sections of the house. However, as climates grow warmer, these personal actions may lose effectiveness, necessitating additional cooling strategies. Occupants might need to consider retrofitting the building fabric, particularly if low-cost construction practices continue. This study supports recent observations that climate change, especially prolonged periods of extreme heat (heatwaves), may pose a threat to the thermal comfort and health of inhabitants in affordable and low-cost housing structures in South Africa's future climates. The findings emphasize the importance of addressing these challenges through improved construction practices and building regulations to protect vulnerable populations from the adverse health effects of extreme heat.

The study's in situ experimental nature fixed certain parameters, namely, building layout, orientation, wall thickness, fenestration location, and climatic zone, thus limiting the investigation's scope. Therefore, these factors and others should be incorporated in future building simulation modeling. Beyond the building examined in this study, future climate scenario simulations should also consider more conventional housing types to quantify global warming's impact more accurately on South Africa's building stock.

Funding: This research received no external funding. The APC was funded by the University of the Witwatersrand.

Data Availability Statement: Data presented in this study are available on request from the author.

Acknowledgments: The South African Weather Service (SAWS) is acknowledged for providing historic temperature data utilized in several analyses presented in this paper.

Conflicts of Interest: The author declares no conflict of interest.

References

1. Urgent Climate Action Can Secure a Liveable Future for All. Available online: https://www.ipcc.ch/report/ar6/syr/downloads/press/IPCC_AR6_SYR_PressRelease_en.pdf (accessed on 21 March 2023).
2. Mbokodo, I.; Bopape, M.-J.; Chikoore, H.; Engelbrecht, F.; Nethengwe, N. Heatwaves in the Future Warmer Climate of South Africa. *Atmosphere* **2020**, *11*, 712. [CrossRef]
3. DST (Department of Science and Technology). *South African Risk and Vulnerability Atlas*; Department of Science and Technology: Pretoria, South Africa, 2010.
4. Engelbrecht, F.A.; Adegoke, J.; Bopape, M.-J.; Naidoo, M.; Garland, R.; Thatcher, M.; McGregor, M.; Katzfey, J.; Werner, M.; Ichoku, C.; et al. Projections of rapidly rising surface temperatures over Africa under low mitigation. *Environ. Res. Lett.* **2015**, *10*, 085004. [CrossRef]
5. Engelbrecht, F.A.; McGregor, J.L.; Engelbrecht, C.J. Dynamics of the Conformal-Cubic Atmospheric Model Projected Climate-Change Signal over Southern Africa. *Int. J. Climatol.* **2009**, *29*, 1013–1033. [CrossRef]
6. Kruger, A.C.; Mbatha, S. *Regional Weather and Climate of South Africa: Gauteng*; South African Weather Service: Pretoria, South Africa, 2021.

7. Taylor, J.; Wilkinson, P.; Davies, M.; Armstrong, B.; Chalabi, Z.; Mavrogianni, A.; Symonds, P.; Oikonomou, E.; Bohnenstengel, S.I. Mapping the Effects of Urban Heat Island, Housing, and Age on Excess Heat-Related Mortality in London. *Urban Clim.* **2015**, *14*, 517–528. [CrossRef]
8. Santamouris, M. Recent Progress on Urban Overheating and Heat Island Research. Integrated Assessment of the Energy, Environmental, Vulnerability and Health Impact. Synergies with Global Climate Change. *Energy Build.* **2020**, *207*, 109482. [CrossRef]
9. Niño 3.4 SST Index. Available online: https://psl.noaa.gov/enso/dashboard.lanina.html (accessed on 3 March 2023).
10. El Niño Theme Page. Pacific Marine Environmental Laboratory. Available online: https://www.pmel.noaa.gov/elnino/impacts-of-el-nino (accessed on 13 March 2023).
11. Coley, D.; Kershaw, T. Changes in Internal Temperatures within the Built Environment as a Response to a Changing Climate. *Build. Environ.* **2010**, *45*, 89–93. [CrossRef]
12. Wang, H.; Chen, Q. Impact of Climate Change Heating and Cooling Energy Use in Buildings in the United States. *Energy Build.* **2014**, *82*, 428–436. [CrossRef]
13. Porritt, S.M.; Cropper, P.C.; Shao, L.; Goodier, C.I. Ranking of Interventions to Reduce Dwelling Overheating during Heat Waves. *Energy Build.* **2012**, *55*, 16–27. [CrossRef]
14. Escandón, R.; Suárez, R.; Sendra, J.J.; Ascione, F.; Bianco, N.; Mauro, G.M. Predicting the Impact of Climate Change on Thermal Comfort in A Building Category: The Case of Linear-Type Social Housing Stock in Southern Spain. *Energies* **2019**, *12*, 2238. [CrossRef]
15. Dino, I.G.; Meral Akgül, C. Impact of Climate Change on the Existing Residential Building Stock in Turkey: An Analysis on Energy Use, Greenhouse Gas Emissions and Occupant Comfort. *Renew Energy* **2019**, *141*, 828–846. [CrossRef]
16. Heracleous, C.; Michael, A. Assessment of Overheating Risk and the Impact of Natural Ventilation in Educational Buildings of Southern Europe under Current and Future Climatic Conditions. *Energy* **2018**, *165*, 1228–1239. [CrossRef]
17. Ozarisoy, B. Energy Effectiveness of Passive Cooling Design Strategies to Reduce the Impact of Long-Term Heatwaves on Occupants' Thermal Comfort in Europe: Climate Change and Mitigation. *J. Clean Prod.* **2022**, *330*, 129675. [CrossRef]
18. Roshan, G.; Oji, R.; Attia, S. Projecting the Impact of Climate Change on Design Recommendations for Residential Buildings in Iran. *Build. Environ* **2019**, *155*, 283–297. [CrossRef]
19. Andrić, I.; Corre, O.L.; Lacarrière, B.; Ferrão, P.; Al-Ghamdi, S.G. Initial Approximation of the Implications for Architecture due to Climate Change. *Adv. Build. Energy Res.* **2021**, *15*, 337–367. [CrossRef]
20. Gupta, R.; Gregg, M. Preventing the Overheating of English Suburban Homes in a Warming Climate. *Build. Res. Inf.* **2013**, *41*, 281–300. [CrossRef]
21. Hamdy, M.; Carlucci, S.; Hoes, P.-J.; Hensen, J.L.M. The Impact of Climate Change on the Overheating Risk in Dwellings—A Dutch Case Study. *Build. Environ.* **2017**, *122*, 307–323. [CrossRef]
22. Mavrogianni, A.; Wilkinson, P.; Davies, M.; Biddulph, P.; Oikonomou, E. Building Characteristics as Determinants of Propensity to High Indoor Summer Temperatures in London Dwellings. *Build. Environ.* **2012**, *55*, 117–130. [CrossRef]
23. Thapa, S. Risk of Overheating in Low-Rise Naturally Ventilated Residential Buildings of Northeast India—An Effect of Climate Change. *Archit. Sci. Rev.* **2022**, *65*, 14–41. [CrossRef]
24. Abbas, K.; Li, S.; Xu, D.; Baz, K.; Rakhmetova, A. Do Socioeconomic Factors Determine Household Multidimensional Energy Poverty? Empirical Evidence from South Asia. *Energy Policy* **2020**, *146*, 111754. [CrossRef]
25. Covary, T.; Du Preez, K.; Götz, T. Energy Efficient Air Conditioners. Wuppertal Institute for Climate, Environment and Energy. 2015. Available online: http://www.bigee.net/media/filer_public/2015/07/22/bigee_south_africa_ac_20150722.pdf (accessed on 7 March 2023).
26. Wright, C.Y.; Wernecke, B.; Kapwata, T.; Kunene, Z.; Mathee, A.; Hey, J.V.; Theron, L. Perceptions of Thermal Comfort and Coping Mechanisms Related to Indoor and Outdoor Temperatures among Participants Living in Rural Villages in Limpopo Province, South Africa. *South Afr. J. Psychol.* **2022**, *52*, 449–459. [CrossRef]
27. Naicker, N.; Teare, J.; Balakrishna, Y.; Wright, C.Y.; Mathee, A. Indoor Temperatures in Low Cost Housing in Johannesburg, South Africa. *Int. J. Environ. Res. Public Health* **2017**, *14*, 1410. [CrossRef] [PubMed]
28. Mabuya, B.; Scholes, M. The Three Little Houses: A Comparative Study of Indoor and Ambient Temperatures in Three Low-Cost Housing Types in Gauteng and Mpumalanga, South Africa. *Int. J. Environ. Res. Public Health* **2020**, *17*, 3524. [CrossRef]
29. Bradley, R.A. Assessing the Effectiveness of Several Passive Design Strategies Using the CIBSE Overheating Criteria: Case Study of an Earth Brick Shell House in Johannesburg, South Africa. *Archit. Sci. Rev.* **2022**, *65*, 232–246. [CrossRef]
30. Bradley, R.A.; Gohnert, M.; Fitchett, A. Long-Term Monitoring of an Earth-Masonry Shell House in Johannesburg, South Africa: Thermal Performance. *Proc. Inst. Civ. Eng.—Constr. Mater.* **2021**, *174*, 21–33. [CrossRef]
31. South African Weather Service. Annual State of the Climate 2019. Available online: https://www.weathersa.co.za/Documents/Corporate/Annual%20State%20of%20the%20Climate%202019.pdf (accessed on 4 March 2023).
32. South African Weather Service. Annual State of the Climate 2020. Available online: https://www.weathersa.co.za/Documents/Corporate/Annual%20State%20of%20the%20Climate%202020_19032021121122.pdf (accessed on 4 March 2023).
33. South African Weather Service. Annual State of the Climate 2021. Available online: https://www.weathersa.co.za/Documents/Corporate/Annual%20State%20of%20the%20Climate%202021_04042022114230.pdf (accessed on 4 March 2023).

34. El Niño/Southern Oscillation (ENSO). Available online: https://www.ncei.noaa.gov/access/monitoring/enso/sst (accessed on 13 March 2023).
35. *SANS 10400-C*; Application of the National Building Regulations—Part C: Dimensions. South African National Standards (SANS): Pretoria, South Africa, 2016.
36. CIBSE. *CIBSE Guide A: Environmental Design*, 8th ed.; Chartered Institution of Building Services Engineers: London, UK, 2019; ISBN 9781906846541.
37. *SANS 10400-XA*; Application of the National Building Regulations—Part X: Environmental Sustainability—Part XA: Energy Usage in Buildings. South African National Standards (SANS): Pretoria, South Africa, 2011.
38. *BS EN 15251*; Indoor Environmental Input Parameters for Design and Assessment of Energy Performance of Buildings Addressing Indoor Air Quality, Thermal Environment, Lighting and Acoustics. British Standards Institution (BSI): London, UK, 2007.
39. *CIBSE TM52*; The Limits of Thermal Comfort: Avoiding Overheating in European Buildings. Chartered Institution of Building Services Engineers: London, UK, 2013.
40. Tuohy, P.G.; Humphreys, M.A.; Nicol, F.; Rijal, H.B.; Clarke, J.A. Occupant Behaviour in Naturally Ventilated and Hybrid Buildings. *Am. Soc. Heat. Refrig. Air Cond. Eng. (ASHRAE) Trans.* **2009**, *115*, 16–27.
41. Conradie, D.C. South Africa's Climatic Zones: Today, Tomorrow. In Proceedings of the International Green Building Conference and Exhibition: Future Trends and Issues Impacting on the Built Environment, Sandton, South Africa, 25–26 July 2012.
42. World Health Organization. Climate Change, Heat and Health. 2022. Available online: https://www.who.int/news-room/factsheets/detail/climate-change-heat-and-health (accessed on 13 April 2023).
43. Karimpour, M.; Belusko, M.; Xing, K.; Boland, J.; Bruno, F. Impact of Climate Change on the Design of Energy Efficient Residential Building Envelopes. *Energy Build.* **2015**, *87*, 142–154. [CrossRef]
44. Dabaieh, M.; Wanas, O.; Hegazy, M.A.; Johansson, E. Reducing Cooling Demands in a Hot Dry Climate: A Simulation Study for Non-Insulated Passive Cool Roof Thermal Performance in Residential Buildings. *Energy Build.* **2015**, *89*, 142–152. [CrossRef]
45. Pisello, A.L.; Castaldo, V.L.; Piselli, C.; Fabiani, C.; Cotana, F. Thermal Performance of Coupled Cool Roof and Cool Façade: Experimental Monitoring and Analytical Optimization Procedure. *Energy Build.* **2017**, *157*, 35–52. [CrossRef]
46. Fitchett, A.; Govender, P.; Vallabh, P. An Exploration of Green Roofs for Indoor and Exterior Temperature Regulation in the South African Interior. *Environ. Dev. Sustain.* **2020**, *22*, 5025–5044. [CrossRef]
47. Bradley, R.A. Mitigating the Potential of Overheating in an Earth-masonry Shell House Situated in Johannesburg through Passive Design Strategies. In Proceedings of the 5th Biennial Residential Building Design and Construction Conference, State College, PA, USA, 5–6 March 2020; Memari, A., Klinetob-Lowe, S., Eds.; Pennsylvania Housing Research Centre: University Park, PA, USA, 2020.

buildings

MDPI

Article

Risk Control of Energy Performance Fluctuation in Multi-Unit Housing for Weather Uncertainty

Jiahe Wang *, Masayuki Mae and Keiichiro Taniguchi

Department of Architecture, Graduate School of Engineering, The University of Tokyo, 7-3-1 Hongo Bunkyo-ku, Tokyo 113-8656, Japan; mae.masayuki@gmail.com (M.M.); taniguchi@arch1.t.u-tokyo.ac.jp (K.T.)
* Correspondence: wjh15668495060@gmail.com

Abstract: With the acceleration of urban development, the population density of urban cities has increased. As the spatial characteristics of multi-unit housing (MUH) perfectly fit this developmental trend and, simultaneously, have high energy efficiency, the number of MUHs has increased rapidly in recent decades. Although many studies have proposed high energy efficiency strategies, weather uncertainty leads to errors between the operational performance of building energy and simulated values. This study introduces a robust optimization framework that incorporates uncertainty considerations into the optimization process to suppress energy consumption fluctuations and improve the average building energy consumption performance. Neural networks are used to model the uncertainty of multiple weather elements as normal distributions for each hour, and the accuracy of the uncertainty model is validated by calculating the mean absolute percentage error (MAPE) between the mean values of the distribution and the measurement values, which ranges from 3% to 13%. The clustering algorithm is proposed to replace the sampling method to complete the sampling work from the normal distribution space of the weather elements to serve the subsequent optimization process. Compared with the traditional method, the sampling results of the clustering algorithm show better representativeness in the sample space. The robust optimization results show that the average energy consumption of the optimal scheme decreases by 13.4%, and the standard deviation decreases by approximately 17.2%, which means that the optimal scheme, generated by the robust optimization framework proposed in this study, has lower average energy consumption results and a more stable energy consumption performance in the face of weather uncertainty.

Keywords: building performance gap; robust optimization; uncertain weather conditions; deep learning; clustering algorithm; multi-unit housing

Citation: Wang, J.; Mae, M.; Taniguchi, K. Risk Control of Energy Performance Fluctuation in Multi-Unit Housing for Weather Uncertainty. *Buildings* **2023**, *13*, 1616. https://doi.org/10.3390/buildings13071616

Academic Editors: Nianping Li and Yingdong He

Received: 31 May 2023
Revised: 11 June 2023
Accepted: 20 June 2023
Published: 26 June 2023

1. Introduction

Over the past few decades, global urbanization has been expedited by rapid industrial and economic development. According to data provided by the United Nations, the urban population is expected to account for 68% of the global population by 2050, resulting in increased energy consumption [1]. Moreover, the energy consumption of residential buildings in various countries accounts for 16–50% of the total national energy consumption, with a global average of approximately 31% [2].

With the acceleration of urban development, the population densities of modern cities have increased. As the spatial characteristics of multi-unit housing (MUH) perfectly fit this development trend and result in high energy efficiency, which is consistent with the design thinking for energy saving, the number of MUHs in many countries has increased rapidly in recent decades, the total number of completed multi-family housing units in the United States increased by 6% in 2020, reaching 375,000 units, which marks the highest annual number of completed multi-family housing units in the past thirty years [3]. In 2018, the number of newly registered residential buildings in British Columbia reached its peak (46,463) since 2002, with an increase of 75.5% MUH [4]. In Japan, the number of houses in MUHs reached 23.34 million, accounting for 43.5% of the overall residential sector, an increase of 2.5 times since 1998 [5].

Wu et al. [6] conducted a statistical analysis of the publication trends of articles related to NZEBs in the Web of Science database over the past 23 years. The results revealed a significant increase in NZEB research from 1994 to 2018. In terms of practical application, energy-saving buildings must achieve the energy-saving performance expected on the basis of the simulation results at the operational stage.

Measured historical weather data, such as typical meteorological year (TMY) weather data, are used as the input of the simulation software to calculate the energy consumption to predict building energy consumption in advance during the simulation stage. However, due to the uncertainty of weather conditions, a gap usually exists between the building-energy-consumption simulation results calculated by using historical weather data, which is a fixed value, and the actual building energy consumption, which is called the "energy performance gap" [7]. Shi et al. [8] investigated 21 energy-efficient buildings and showed that performance gaps are always present and unpredictable, regardless of the building type, location, climate, design, and construction method. Mahabir et al. [9] studied the differences in the energy consumption of highly efficient residential buildings in the same area under different weather conditions, and the results showed that the uncertainty of weather conditions led to a difference of −40% to +40% in the energy consumption results.

To compensate for the lack of fixed weather data to account for such uncertainties, several researchers have used historical weather data from years past to reproduce fluctuations in uncertainty. Sun et al. [10] used 32 years (1982–2013) of measured meteorological data to represent uncertainty, and Wang et al. [11] used weather data from 10–15 years in four cities to assess the energy variations in office buildings. However, such methods are usually limited by historical data; there is the possibility that extreme weather cannot be considered, and the probability of the prevalence of various weather conditions within the fluctuation range cannot be effectively calculated.

In recent years, a collection of probabilistic calculation methods for weather condition uncertainties has been developed. Wang et al. [11] determined the range and frequency of weather fluctuations by analyzing historically measured weather data over the past 15 years. Sun et al. [12] improved the variables in the calculation formula for weather elements from fixed values to probability distribution to calculate the probability model of weather elements using the statistics of measured data from hundreds of sites. Both statistical and probabilistic calculation methods often do not consider the correlation between weather elements, and it is difficult to reproduce time series of weather elements with high precision.

Thus, the fixed value of the historical weather data results in a deviation between the simulated building energy consumption and the actual data, which is the performance gap. At present, uncertainty is rarely included in the calculation of objective functions in energy consumption optimization studies. In fact, Luo et al. [13] and Imran et al. [14] used multi-objective optimization algorithms and artificial intelligence to optimize building energy consumption; however, the objective functions were established based on ideal fixed conditions. In the global optimization process, the deviation caused by a fixed weather value may lead to the final optimization result not being optimal, or even degraded. Thus, it is necessary to consider uncertainty as comprehensively as possible during the optimization phase to ensure the robustness of the optimization results.

Based on the literature review, it is evident that the lack of estimation of weather uncertainty conditions during the simulation stage has led to deviations between simulated results and actual operating conditions. This issue is widely present across various types of buildings, including MUH buildings with a significantly increasing number, so this issue cannot be ignored. In addition, the optimization framework at this stage rarely considers uncertainty, which further leads to the deviation between the simulation results and the actual fluctuations. To ensure the robustness of optimization results in the face of uncertainty, the methodology to reproduce weather uncertainty comprehensively and accurately and to combine uncertainty with the global optimization process is critical. This study proposes a highly efficient and robust optimization framework for calculating building energy fluctuations based on the reproduction of weather uncertainty during the simulation phase and optimizing the average value and standard deviation of energy consumption fluctuations in order to reduce the average energy consumption performance of MUHs and suppress the fluctuation range of energy consumption.

The robust optimization framework proposed in this study clarifies the method of using neural networks to predict meteorological uncertainty in the design phase and using clustering algorithms to improve the representativeness of samples in the optimized sampling process. Furthermore, it highlights the incorporation of uncertainty fluctuations during the optimization process, distinguishing it from traditional optimization procedures.

This article will consist of five chapters. Section 2 will introduce the establishment of meteorological uncertainty models and the use of clustering algorithms and the principles of robust optimization. Section 3 will primarily focus on introducing the objective models used in this study, while Section 4 will analyze the model's optimal results. Finally, Section 5 will provide a conclusion for the paper and suggest directions for future research.

2. Methodology

To account for the uncertainty fluctuations in weather conditions, during the optimization process, the robust optimization framework consists of three parts. First, probability distribution models are established for weather elements to describe uncertainties, and then the probability distribution model is sampled to establish weather uncertainty scenarios for the subsequent optimization process. Finally, building variables and uncertainty scenarios are input into the optimizer for robust optimization of building energy consumption performance.

Neural networks were used to construct distribution models of weather uncertainty fluctuations for subsequent optimization sampling, as described in Section 2.1. As this study focuses on the optimization of annual energy consumption fluctuations, the neural network produces a probabilistic model of the present weather elements instead of predicting the future in the context of the accumulation of forecast errors.

Sampling based on the aforementioned probability model was necessary to evaluate each building scheme in the optimization stage. However, multitarget optimization based on sampling usually leads to an unaffordable computational load. Moreover, traditional sampling methods are disadvantageous owing to their randomness. Thus, this study proposes the use of clustering algorithms instead of traditional sampling methods, as described in Section 2.2.

In the robust optimization stage, NSGA II was used as the optimizer to realize optimization. The optimization objectives are defined as the average energy consumption and the standard deviation of energy consumption of each building scheme in the face of various weather conditions from clustering to simultaneously optimize the stability of the average energy consumption and energy consumption performance in the face of weather fluctuations, as described in Section 2.3.

2.1. Uncertainty of Weather Conditions

The uncertainty of weather conditions is generated based on two neural networks. The dual-stage attention-based recurrent neural network (DARNN) [15] is used to predict the value of weather elements with weather elements as features, and the importance of weather elements to the air-conditioning load is explained through the prediction process based on the characteristic of attention mechanism. During the training process of the Bayesian recurrent neural network (Bayesian-RNN) [16], the weights and bias are established as a normal distribution based on the training set to realize the establishment of a normal distribution model of the forecast target. Therefore, weather elements are taken as prediction targets to establish their uncertainty models.

2.1.1. Dataset for Neural Networks

This section describes the two datasets used to generate the probabilistic model of weather uncertainty: a dataset of weather conditions and a dataset of building energy consumption. For the DARNN, due to the importance of calculating weather elements for computing air-conditioning energy consumption, the dataset is composed of weather data as a feature and the air-conditioning energy consumption as a label. For the Bayesian-RNN, the dataset only contains weather data, and, based on the importance calculation results from the DARNN, non-important weather elements are used as features, while important weather elements are used as labels.

The climate of Toyama Prefecture is characterized by cold and heavy snow in winter and heat and humidity in summer; it is a typical rainy and snowy area in Japan.

The weather condition dataset comes from the meteorology of Toyama Prefecture collected by the Japan Meteorological Agency [17], which is the hourly measurement data from 2019 and 2020, including elements of sea level pressure (hPa), station pressure (hPa), precipitation (mm), outdoor temperature (°C), global horizontal radiation (MJ/m^2), dew temperature (°C), vapor pressure (hPa), relative humidity (%), wind speed (m/s), and cloud cover.

The dataset of building air-conditioning energy consumption is the annual energy consumption calculated based on the weather data for 2019 and 2020 using EnergyPlus [18]. This is in reference to the simulation settings in Section 2.

Furthermore, the dataset summary of the two neural networks is presented in Table 1.

Table 1. Summary of training and testing datasets for neural networks.

	Training Data	Test Data
DARNN	2019/01–12 weather data and energy consumption	2020/07 and 12 energy consumption
Bayesian-RNN	2019/01–12 weather data	2020/01–12 weather data (solar radiation, relative humidity, outdoor temperature)

2.1.2. Importance Interpretation of Weather Elements

In this study, the dual-stage attention mechanism of the DARNN is employed to calculate the importance of weather elements for air-conditioning energy consumption. The main function of the attention mechanism [19] is to introduce the neural network to calculate the contribution weight of the encoder to the decoder. An ordinary attention usually requires three values, namely the query tensor Q, the key tensor K, and the value tensor V, as shown in Equation (1). The attention score is calculated as the importance of the input value to the prediction target, as shown in Equation (2).

$$\begin{aligned} Q &= W_q Q \\ K &= W_k K \\ V &= W_v V \end{aligned} \tag{1}$$

where W_q, W_k, W_v means weights matrix.

$$\begin{aligned} A &= K^T Q \\ A' &= \text{softmax}(A) \end{aligned} \tag{2}$$

The DARNN consists of two parts: the encoder and the decoder. Both stages use the attention mechanism and involve several steps, including calculating attention scores, computing attention weights, updating inputs or calculating CoVe (contextualized vector), and computing hidden states. Therefore, in the calculation process of the attention score and weight, the important results for the weather elements are obtained. The specific description and adoption of rationality for the importance interpretation of the DARNN were proven in a previous study by the author of [20]. The details of hyperparameters are shown in Table 2.

Figure 1 provides an overview of the process of using the DARNN to predict the air-conditioning energy consumption based on weather elements. The attention mechanism in the neural network calculates the weights for each weather element with respect to the target prediction. Since a total of 10 weather elements are considered, each with a time span of one year (8760 h), the resulting weight data structure is [10, 8760]. To facilitate statistical analysis, the weight data structure is transformed into [10, 1] by taking the average along the time dimension, serving as the final outcome of importance interpretation.

Taking three units on the middle floor of the building as representatives, namely, units 201, 202, and 203, and considering the importance of weather elements to the air-conditioning energy consumption in winter and summer, the importance interpretation results of the DARNN are shown in Figure 2. The 2019 dataset was used as the training set for the DARNN, and the July and December 2020 datasets were used as the test set. The results show that in both winter (December) and summer (August), the three most important factors were solar radiation, outdoor temperature, and relative humidity.

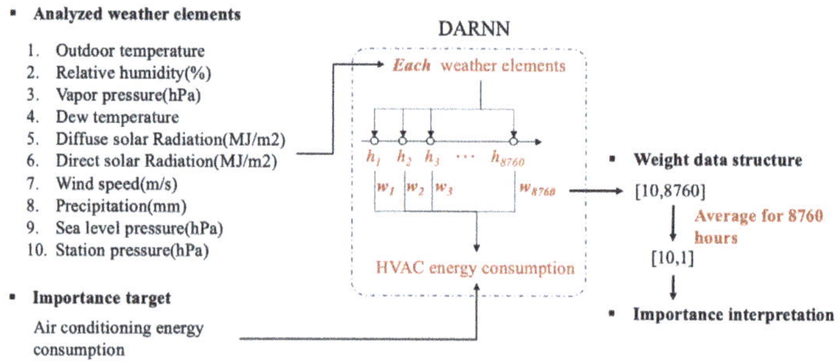

Figure 1. Framework of importance interpretation of weather elements.

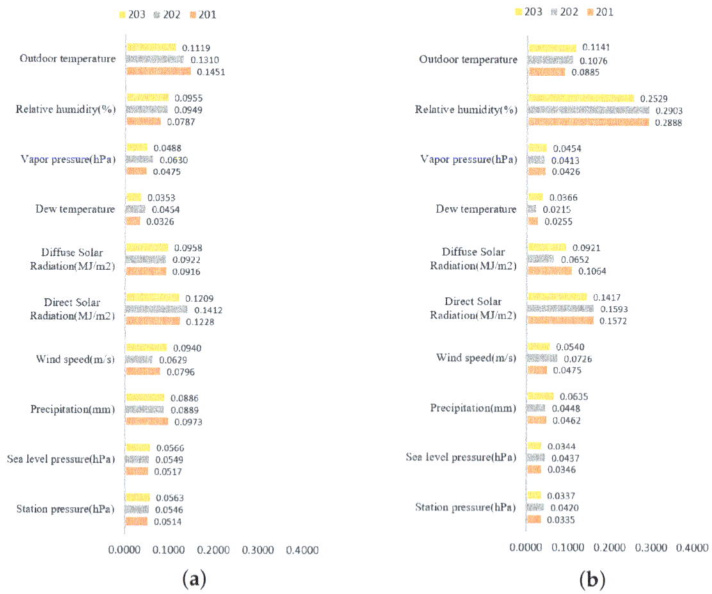

Figure 2. Importance ranking of weather elements to energy consumption. (**a**) Importance ranking of weather elements to energy consumption in winter; (**b**) importance ranking of weather elements to energy consumption in summer.

Table 2. Hyperparameter of DARNN.

Hyperparameter	Encoder	Decoder
Learning rate	0.001	0.001
Batch size	64	64
Hidden size	128	256
Num_layers	2	2
Sequence length	24	24

2.1.3. Uncertainty Modeling for Weather Elements

Based on the importance interpretation results, solar radiation, relative humidity, and outdoor temperature were used as uncertainty modeling objects. As there are latent correlations between weather elements, weather elements other than those mentioned above are used as input values for the Bayesian-RNN, and the outputs are the uncertainty models of the object weather elements, as shown in Figure 3, and the details of the hyperparameters are shown in Table 3. The result of uncertainty modeling was that 8760 normal distribution models were established for each element, that is, one normal distribution model at each hour of the year, to reflect the change in weather elements over time. The predictive accuracy and underlying principles of the Bayesian-RNN for weather uncertainty have been validated in the paper [20].

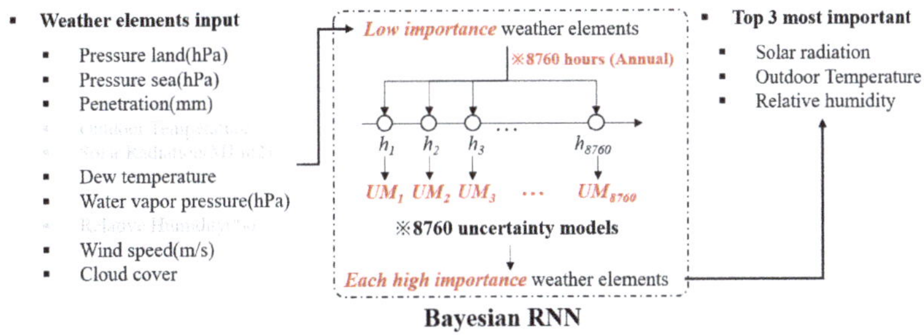

Figure 3. Framework of uncertainty modeling.

Where UM_n means uncertainty model.

Table 3. Hyperparameter of Bayesian-RNN.

Hyperparameter	Value
Learning rate	0.0001
Batch size	256
Hidden size	128
Num_layers	2 (BayesianLSTM layter, Linear layer)
Sequence length	24

To verify the reliability of the generated uncertainty model for three weather elements, this study sampled 3000 sets of data based on the normal distribution model of each hour and compared the mean value of the sampled data with the measurement data using the mean average percentage error (MAPE) as an indicator during winter (December) and summer (August), as shown in Table 4.

Taking the second week of December as an example, the line chart describes the uncertainty modeling results of the three weather elements, and the different red concentrations in the chart represent the mean plus or minus n times (n = 1, 2, 3) standard deviation, as shown in Figure 4.

Table 4. MAPE of uncertainty models for three weather elements.

	Winter (December)	Summer (August)
Outdoor temperature	9.50%	2.60%
Relative humidity	3.90%	4.20%
Solar radiation	13.00%	9.10%

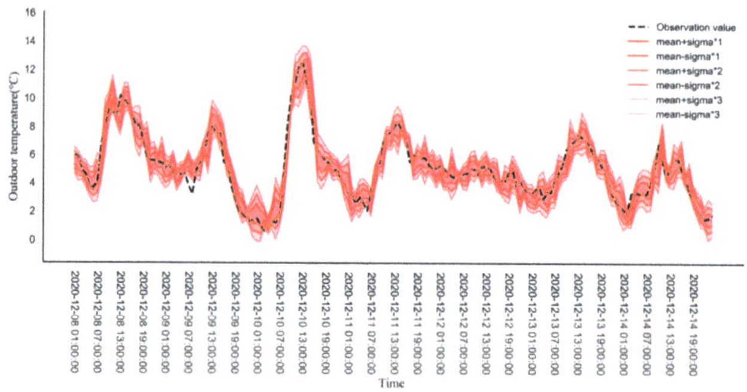

(**a**)

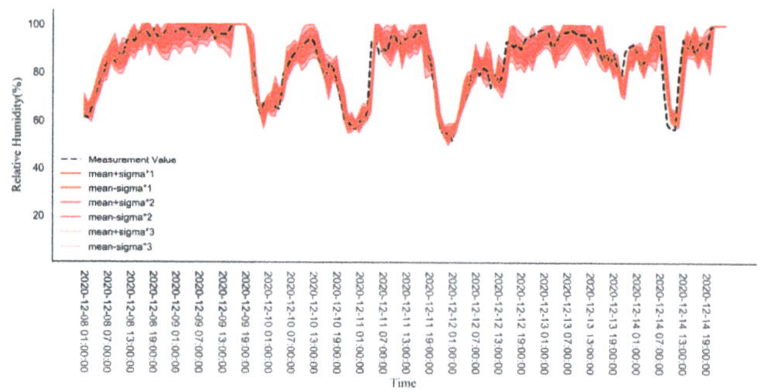

(**b**)

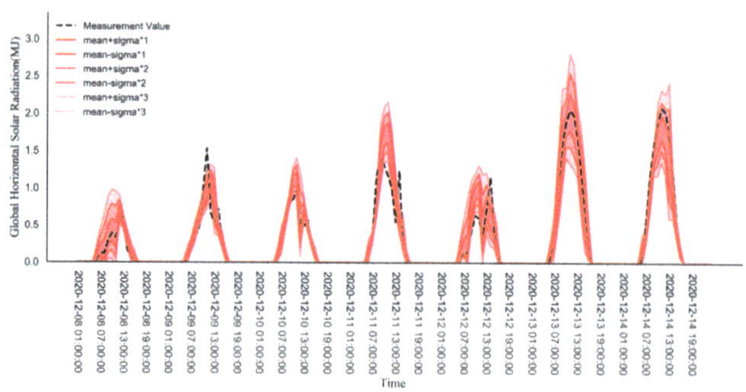

(**c**)

Figure 4. Description of the prediction distribution for three weather elements. (**a**) Description of the prediction distribution of outdoor tmperature in winter; (**b**) Description of the prediction distribution of relative humidity in winter; (**c**) Description of the prediction distribution of solar radiation in winter.

2.2. Sampling by Clustering Algorithm

Sampling based on the uncertainty model described above is required to account for uncertainty in the subsequent optimization process. However, traditional sampling methods, such as the Monte Carlo sampling method, usually have difficulty ensuring the representativeness of the sampling results in the sample space when extracting a small number of sample results. Although increasing the number of samples can solve this problem, the calculation load of the subsequent process increases significantly. Thus, this study used the k-means algorithm [21] to obtain the cluster center as the final sampling result. Although the clustering algorithm is a type of classification technique, its clustering principle ensures that the representativeness of the clustering is centered in the entire sample space; i.e., the clustering centers are included in the sampling space and evenly distributed. Thus, the cluster centers are used as the sampling results.

2.2.1. Comparison between Clustering Algorithm and Traditional Sampling Methods

The Monte Carlo and Latin hypercube samplings are commonly used traditional sampling methods. However, due to the reliance on randomness in the sampling process, there are instances where the representativeness of the sampling results cannot be guaranteed. In this study, these two methods are used as benchmarks for comparing the sampling effectiveness of clustering algorithms. As the sampling objects in this study are multiple normal distributions, to compare the representativeness of the sampling results in the sample space between the clustering algorithm and the traditional sampling method with a small number of samples, the standard normal distribution with a mean of 0 and a standard deviation of 1 was sampled 25 times using the Monte Carlo method and the Latin hypercube method. In addition, for the same normal distribution, 100,000 samples were first randomly selected, and, thereafter, the samples were clustered using the k-means algorithm to obtain 25 cluster centers as 25 samples, as shown in Figure 5.

The results show that the sampling results of both the Monte Carlo and Latin hypercube sampling methods have varying degrees of sample concentration, which leads to a lack of representativeness for some normal distributions, especially at the edges. In contrast, the clustering centers as the sampling results effectively solve the above problem, which is uniformly distributed in the normal distribution space based on probability, and the edges are effectively sampled.

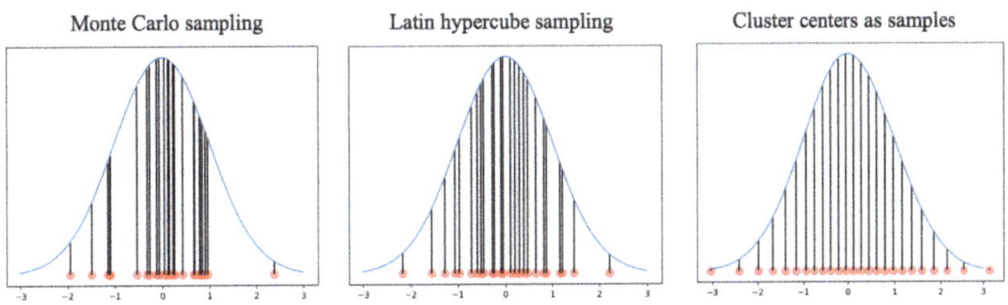

Figure 5. Comparison of clustering results with traditional sampling methods.

2.2.2. Clustering for Establishment of Uncertainty Scenarios for Weather Elements

Each weather element was sampled 100,000 times based on the normal distribution at each hour; thus, the sampling data results with the structure [8760, 100,000] were obtained. As there are three weather elements, the final data structure obtained through the sampling process is [3, 8760, 100,000]. The dimension of the number of samples was clustered to obtain 100 cluster centers due to which the data structure became [3, 8760, 100], as shown in Figure 6. Thus, there were 100 clustering centers as samples at each hour of the three weather elements.

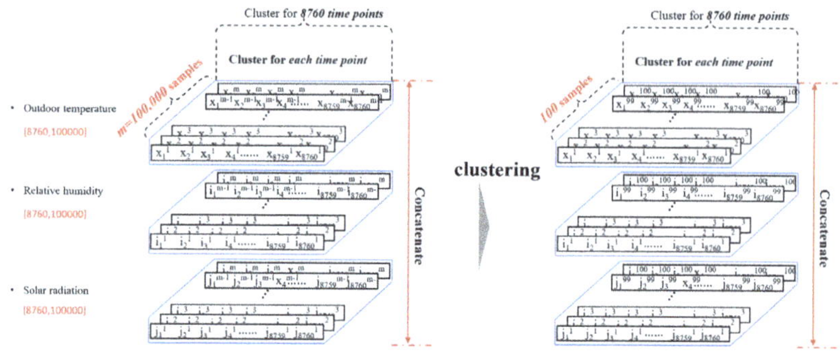

Figure 6. Date structure of clustering result.

2.3. Robust Optimization

The proposed robust optimization framework in this study differs from traditional optimization approaches in that it introduces uncertainty into the optimization process to realize that the results of each scheme are range values. The mean value and standard deviation of the range are used as the optimization objective functions. The inputs, outputs, and optimization flow of the robust optimization framework are described in this section.

The input side of robust optimization contains two components, variables, and uncertainty scenarios, both of which affect the building energy consumption, as described in Section 2.3.1.

Two optimization objectives are included: the average value and the standard deviation of building air-conditioning consumption for each building scheme in the face of various weather uncertainty scenarios to represent the average performance and fluctuation, as described in Section 2.3.2.

Section 2.3.3 describes the optimizer used in this study during the optimization process and the order in which the variables and uncertainties participate in the calculation of the optimization objectives.

2.3.1. Variables and Uncertainty of Robust Optimization

The optimization variables considered in this study are defined as building elements that can be determined by engineers and designers during the design phase. The variables came from three fields, the geometric information of windows and shading devices, building orientation, and thermal insulation performance, a total of 25 elements. Table 5 shows the range of variation in each element in the case building, and more detailed building information will be introduced in Section 3. These variables are discrete variables, and the value accuracy of each variable is indicated by "Accuracy" in the table. In addition, X9 through X25 are variables of thermal insulation performance, and the range of values is the four thermal insulation performance standards mentioned in Section 3.3, which do not involve value accuracy.

The clustering results in Section 2.2 are used as inputs for the robust optimization of weather uncertainty scenarios. During the optimization process, the optimizer randomly generates building schemes based on variables and, thereafter, uses simulation software to calculate the energy consumption performance of each scheme under various weather uncertainty scenarios.

2.3.2. Objectives of Robust Optimization

Building energy performance is affected by both building variables and weather uncertainties, as shown in Equation (3), so the energy consumption result of each building scheme generated from variable space is a dataset rather than a fixed value.

$$\text{Building energy performance} = f(x_1^m, x_2^m, \ldots x_{25}^m, \xi_{\text{weather}}) \quad (3)$$

where x denotes the optimization variables, x^m denotes the mth set of building schemes selected from the variable space, $x_{1...25}$ denotes each variable in the mth building scheme (such as the building orientation and window size), and ξ indicates the weather uncertainty scenarios generated in Section 3.2.

Table 5. Characterization of optimization variables.

NO.	Optimization Variables	Range/Value	Accuracy
X1	South window height [m]	[1.7, 2.2]	0.05
X2	South window width [m]	[1.0, 5.8]	0.05
X3	East window height [m]	[0.5, 2.0]	0.05
X4	East window width [m]	[0.9, 5.5]	0.05
X5	West window height [m]	[0.5, 2.0]	0.05
X6	West window width[m]	[0.9, 5.5]	0.05
X7	South shading device width [m]	[0.1, 3.0]	0.01
X8	Building orientation [°]	[−90, 90]	0.1
X9	External wall thermal insulation [-]	G2, G1, H25, H4	
X10	Internal wall thermal insulation [-]	G2, G1, H25, H4	
X11–X19	South window thermal insulation [-]	G2, G1, H25, H4	
X20–X22	East window thermal insulation [-]	G2, G1, H25, H4	
X23–X25	West window thermal insulation [-]	G2, G1, H25, H4	

To ensure the robustness of the optimization results, the optimization objective was divided into two parts: the average value of the results dataset was used as the average performance evaluation index in the face of uncertainty scenarios, and the standard deviation was the evaluation index of the fluctuation range, as shown in Equations (4) and (5).

$$\text{Objective function Avg} = \text{Avg.}\begin{pmatrix} f\left(x_1^m, x_2^m, \ldots x_n^m, \xi_{\text{weather } 1}\right) \\ f\left(x_1^m, x_2^m, \ldots x_n^m, \xi_{\text{weather } 2}\right) \\ \vdots \\ f\left(x_1^m, x_2^m, \ldots x_n^m, \xi_{\text{weather } j}\right) \end{pmatrix} \tag{4}$$

$$\text{Objective function Std Dev} = \text{Std Dev.}\begin{pmatrix} f\left(x_1^m, x_2^m, \ldots x_n^m, \xi_{\text{weather } 1}\right) \\ f\left(x_1^m, x_2^m, \ldots x_n^m, \xi_{\text{weather } 2}\right) \\ \vdots \\ f\left(x_1^m, x_2^m, \ldots x_n^m, \xi_{\text{weather } j}\right) \end{pmatrix} \tag{5}$$

where $\xi_{weather1...j}$ represents multiweather uncertainty scenarios.

Thus, considering that there are two air-conditioned rooms in each unit, namely, the LDK and bedroom, a situation in which the optimization is difficult to converge because of too many optimization objectives when each household is optimized individually has been avoided. Four optimization objectives are defined in terms of air-conditioning energy consumption in buildings:

1. The average of the total energy consumption in the LDK room facing uncertainty scenarios.
2. The standard deviation of the total energy consumption in the LDK room facing uncertainty scenarios.
3. The average of the total energy consumption in the bedroom facing uncertainty scenarios.
4. The standard deviation of the total energy consumption in the bedroom facing uncertainty scenarios.

2.3.3. Robust Optimization Flow

NSGA-II [22] was used as the optimizer; its configuration is shown in Table 6. In the optimization process, a set of variable values is extracted by the optimizer from the variable space to generate a design scheme, and the air-conditioning energy consumption of the scheme is calculated under various uncertainty scenarios based on EnergyPlus. The mean and standard deviation of the energy performance of each design scheme were returned to the optimizer to calculate the fitness, and the above process was repeated, as shown in Figure 7.

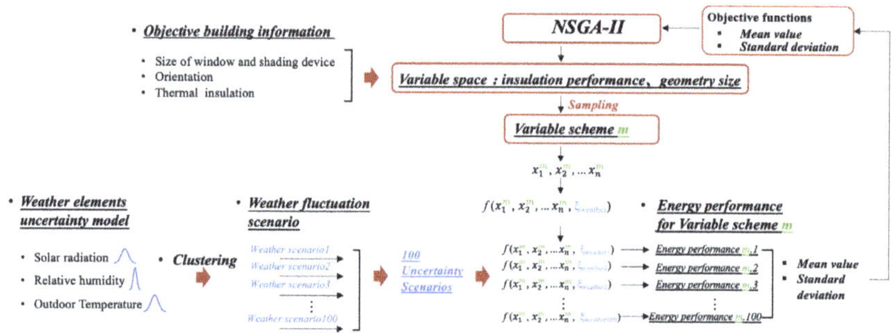

Figure 7. Robust optimization flow.

Table 6. Configurations of NSGA-II.

Settings and Operator	Method/Value
Max generation	100
Population size	40
Offspring size	10
Sampling	Lartin hypercube
Selection	Tournament
Mutation	Polynomial Mutation (0.1)
Crossover	Simulated Binary Crossover (0.5)

Compared with traditional optimization results, the optimization flow proposed in this study produces an optimal energy consumption distribution probability model rather than an optimal fixed energy consumption value. In the design stage, it can more comprehensively describe the fluctuation of building energy consumption in the face of weather uncertainty and help decision-makers make more accurate decisions.

3. Case Study and Simulation Settings

3.1. Geometry of Typical Residential Building

In this study, an LDK south-facing three-story MUH with three units on each floor has been considered. One LDK (living room, dining, and kitchen), bedroom, toilet, entrance with a corridor in each unit, plan of each floor, and size of each unit are shown in Figure 8. All units have exterior windows on the south wall of the LDK and on the bedroom walls of the east- and west-side units; the facade size of the unit is shown in Figure 9, with the west unit as an example.

3.2. Simulation Settings

In this study, the mathematical modeling of weather uncertainties, described in Section 2, and the resulting fluctuations in the energy consumption results were focused on during the design phase, and ideal fixed values were used for other simulation settings. The details of the fixed simulation settings are listed in Table 7 (refers to [23]). Only the LDK and bedroom were air-conditioned rooms, and the family composition was envisaged as a couple. Assuming that the infiltration frequency was 0.5/h and the enthalpy efficiency of the total heat exchanger was 70%, the ventilation frequency was set to 0.15 times.

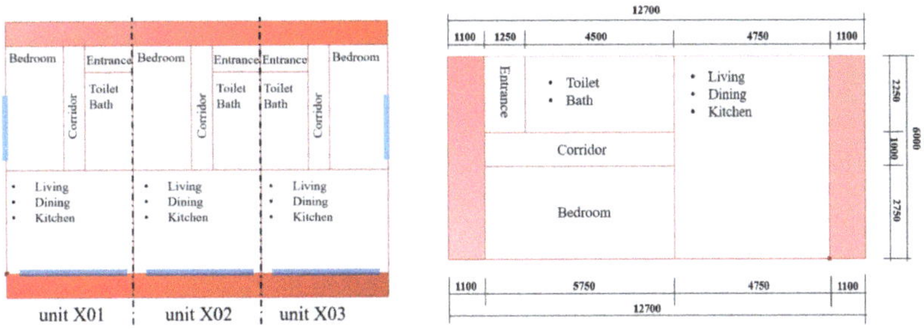

Figure 8. Plan of each floor and size of each unit.

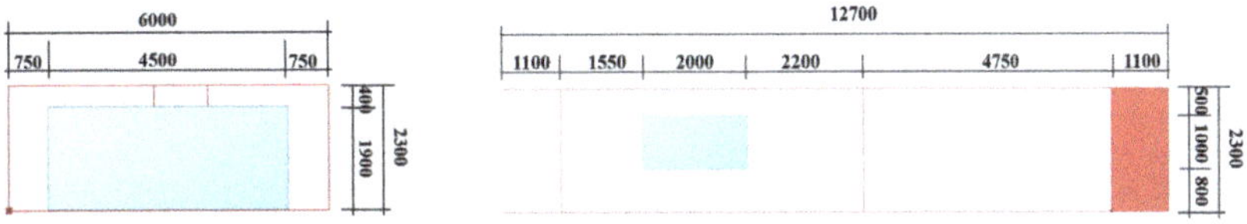

(**a**) (**b**)

Figure 9. Facade size of one unit. (**a**) South facade; (**b**) west facade.

Table 7. Simulation settings.

Air-Conditioning Room	LDK Bedroom
Cooling period	July-September
Heating period	December-March
Cooling/Heating Setpoint	27 °C (28 °C during night)/20 °C
Family Stucture	The couple(two people)
Infiltration rate	0.5 ACH (Energy recovery efficiency 70%)

3.3. Thermal Insulation Performance

Four thermal insulation performance standards were considered in this study: the Japan 1992 energy-saving standard (H4), the Japan 2013 energy-saving standard (H25), the HEAT20 G1 housing exodermis insulation standard (G1), and the HEAT20 G2 housing exodermis insulation standard (G2) [24]. These four thermal insulation performance standards also represent the thermal insulation performance of different grades of building envelopes in Japan and serve as a benchmark for comparing the optimization results of this study. The U values set for each thermal insulation standard for buildings in the seven climate zones of Japan are listed in Tables 8–11. The building in question in Toyama Prefecture is located in the fifth climate zone. Thus, the energy consumption of the building, along with the thermal insulation performance of the fifth climate zone, was used as a benchmark for the optimization results.

Table 8. Thermal insulation performance standard of H4.

	Area 1	Area 2	Area 3	Area 4	Area 5	Area 6	Area 7
External wall ($W/m^2{\cdot}K$)	0.9	0.9	2.4	2.7	3	3	3.2
External ceiling ($W/m^2{\cdot}K$)	0.42	0.42	1.05	1.12	1.55	1.55	2.3
Window ($W/m^2{\cdot}K$)	2.91	2.91	2.91	2.91	4.65	4.65	4.65
External floor ($W/m^2{\cdot}K$)	0.9	0.9	2.0	2.1	2.9	2.9	4
Internal wall ($W/m^2{\cdot}K$)	3.04	3.04	3.04	3.04	3.04	3.04	3.04
Internal ceiling/floor ($W/m^2{\cdot}K$)	2.79	2.79	2.79	2.79	2.79	2.79	2.79

Table 9. Thermal insulation performance standard of H25.

	Area 1	Area 2	Area 3	Area 4	Area 5	Area 6	Area 7
External wall (W/m^2·K)	0.7	0.7	0.9	0.9	1.3	1.3	1.3
External ceiling (W/m^2·K)	0.27	0.27	0.35	0.35	0.35	0.35	0.35
Window (W/m^2·K)	2.33	2.33	2.33	2.33	4.65	4.65	4.65
External floor (W/m^2·K)	0.65	0.65	0.8	0.8	1	1	1
Internal wall (W/m^2·K)	2.333	2.333	2.333	2.333	2.333	2.333	2.333
Internal ceiling/floor (W/m^2·K)	2.079	2.079	2.079	2.079	2.079	2.079	2.079

Table 10. Thermal insulation performance standard of G1.

	Area 1	Area 2	Area 3	Area 4	Area 5	Area 6	Area 7
External wall (W/m^2·K)	0.65	0.65	0.7	0.75	0.75	0.98	0.98
External ceiling (W/m^2·K)	0.23	0.23	0.25	0.25	0.25	0.3	0.3
Window (W/m^2·K)	1.6	1.6	1.9	1.9	1.9	2.33	2.33
External floor (W/m^2·K)	0.54	0.54	0.6	0.6	0.6	0.75	0.75
Internal wall (W/m^2·K)	1.406	1.406	2.333	2.333	2.333	2.333	2.333
Internal ceiling/floor (W/m^2·K)	2.079	2.079	2.079	2.079	2.079	2.079	2.079

Table 11. Thermal insulation performance standard of G2.

	Area 1	Area 2	Area 3	Area 4	Area 5	Area 6	Area 7
External wall (W/m^2·K)	0.5	0.5	0.5	0.35	0.35	0.7	0.7
External ceiling (W/m^2·K)	0.24	0.24	0.24	0.19	0.19	0.24	0.24
Window (W/m^2·K)	1.3	1.3	1.3	1.3	1.3	1.9	1.9
External floor (W/m^2·K)	0.5	0.5	0.5	0.26	0.26	0.58	0.58
Internal wall (W/m^2·K)	1.153	1.153	1.153	2.333	2.333	2.333	2.333
Internal ceiling/floor (W/m^2·K)	1.609	1.609	1.609	2.079	2.079	2.079	2.079

4. Results

4.1. Building Information after Optimization

Through optimization, a series of optimal solutions were selected as examples to illustrate the optimization results.

Before optimization, the thermal insulation performance of the fifth zone of the four thermal insulation standards described in Section 3 was used in the case study building. The optimal thermal insulation performance for each part of the building is listed in Table 12. As the excessive thermal insulation performance of the external wall leads to an additional cooling load, the thermal insulation of the external wall of the optimization result is the fifth area of the G2 standard. However, the optimization results emphasize the importance of the thermal insulation performance of internal walls to reduce the heating and cooling caused by the temperature difference between adjacent rooms or units due to the difference in unit location and ratio to exterior walls. The thermal insulation performance of the third area, which was selected as the thermal insulation of the internal walls, was superior to that of the fifth area.

Table 12. Thermal insulation of wall after optimization.

	Thermal Transmittance (W/m^2·K)	Thermal Insulation Standard and Climate Area
External wall	0.35	G2 5th zone
Roof	0.19	G2 5th zone
Floor	0.26	G2 5th zone
Internal floor/ceiling	2.709	G2 5th zone
Internal wall	1.153	G2 3th zone

The optimal window sizes are listed in Table 13. Compared to the original situation, the south window area becomes 45.3%, the west window area becomes 54%, and the east window area becomes 73.5%. As the thermal insulation performance of each window was optimized independently, the optimization results for the thermal insulation performance of the window are shown in Tables 14–16. Compared with the original thermal insulation performance of 1.3 W/m^2·K for all windows, some windows in the optimization results remained unchanged, and some windows declined to a certain extent.

Table 13. Size of window after optimization.

	Height (m)	**Width (m)**
South window (LDK)	1.75	1.97
West window (Bedroom)	0.64	0.9
East window (Bedroom)	0.67	2.34

Table 14. Thermal insulation of south window after optimization.

Thermal Transmittance (W/m^2·K)				
	Unit No.	×01	×02	×03
South window	10×	1.3	1.3	1.6
	20×	1.3	1.9	1.3
	30×	1.3	1.9	1.3

Table 15. Thermal insulation of east window after optimization.

Thermal Transmittance (W/m^2·K)			
	103	203	303
East window	2.33	1.9	1.3

Table 16. Thermal insulation of west window after optimization.

Thermal Transmittance (W/m^2·K)			
	101	201	301
West window	1.3	1.3	1.6

In addition, the building orientation was optimized from south to 18.5° southwest, and the depth of the southern shading devices was optimized from 1.1 m to 1.53 m.

4.2. Comparison of Energy Consumption before and after Robust Optimization

This section presents the improvement in the robustness of the building energy performance before and after optimization when considering weather uncertainty scenarios. The optimal scheme selected from the optimization results is the most effective for robust optimization. For the convenience of statistical analysis and visualization of optimization results for multiple objectives, the average energy consumption of the LDK and the bedroom are combined into the average energy consumption of the entire building, while the standard deviations of the LDK and bedroom are merged into the standard deviation of the overall energy consumption of the entire building.

One hundred weather uncertainty scenarios were generated as described in Section 2.2.2; thus, there were 100 energy performance results for each design scheme. Figure 10 shows the air-conditioning energy consumption results of an entire building for thermal insulation standards H4, H25, G1, and G2 (5th climate zone) and the optimal scheme, i.e., the probability distribution function (PDF) curve generated based on the mean and standard deviation of 100 results. For the four thermal insulation standards on the display, G2 had the ideal mean and standard deviation, that is, the lowest mean and smallest standard deviation.

Compared with the thermal insulation standard of G2, the thermal insulation performance of the external wall in the optimal scheme of this study almost no change. However, the thermal insulation performance of the internal wall is improved, and the building geometry information (windows and shading device size, building orientation) is adjusted.

Usually, due to the influence of location and floor, there is a significant difference in the indoor environment among different units, which leads to heat transfer between adjacent units or rooms except for the outdoor environment. When considering the uncertainty of weather conditions, the heat transfer between adjacent units becomes even more complex, exacerbating fluctuations in energy consumption for units. The optimization results reduce the heat transfer between adjacent units by improving the insulation performance of the inner wall, thereby improving the average energy consumption performance and reducing the fluctuation of building energy consumption. In addition, by increasing the depth of shading objects, reducing the size of windows, and adjusting the orientation of buildings, the impact of solar radiation on buildings is effectively reduced. Therefore, even in the face of the same uncertainty of solar radiation, the optimization results have a more stable performance of energy consumption.

For the optimal scheme, the average energy consumption decreased by 13.4% compared with that of G2, and the standard deviation decreased by approximately 17.2%. In other words, in the face of uncertainty, the optimal scheme proposed in this study has lower average energy consumption results and a more stable energy consumption performance than most energy-saving standards in Japan at the current stage.

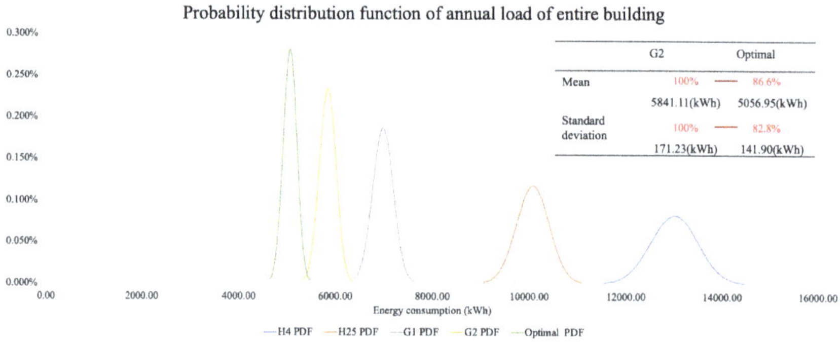

Figure 10. Probability distribution function of annual load of entire building.

The optimization results verified that the robust optimization framework proposed in this study can effectively optimize the energy performance stability and average energy consumption of the scheme under uncertainty, thereby alleviating the gap between simulation and measurement values, and ensuring a good performance of the building during the actual use stage. At present, the majority of environmental performance evaluation systems, such as LEED [25], primarily rely on referencing the actual operational performance when assessing the energy consumption performance level of buildings, which emphasizes the importance of building energy performance in actual use. The results of this study can assist decision-makers in designing high-performance buildings that have a practical significance in actual usage, rather than solely relying on fixed simulation results.

5. Conclusions

This study combines a weather uncertainty scenario modeling method using deep learning and a highly representative sampling method using a clustering algorithm and proposes a robust optimization framework to achieve the optimization of building energy performance considering uncertainty. The optimization framework is based on the high-precision reproduction of weather element uncertainty in the simulation stage to achieve the modeling of fluctuations in air-conditioning energy consumption and improve the evaluation indices to the mean and standard deviation to ensure the ability to evaluate building energy fluctuations during the simulation stage.

The weather uncertainty scenario modeling method establishes the uncertainty fluctuations of solar radiation, relative humidity, and temperature as normal distributions for each hour of the year. The reliability of this method is demonstrated by calculating the mean absolute percentage error (MAPE) between the mean values of the normal distribution and the measurement values, which ranges from 3% to 13%.

Compared to traditional sampling methods, using the cluster center of a clustering algorithm as sampling results leads to a more uniform sample results distribution, which results in better representativeness of the overall sampling space. Moreover, for a normal distribution, the clustering algorithm does not overlook regions with lower probabilities.

Furthermore, taking the mean value and standard deviation of energy fluctuation as the optimization objectives, the optimization results show that in the face of weather uncertainty, the average energy consumption of buildings has decreased by 13.4%, and the standard deviation of energy consumption fluctuations has decreased by 17.2%. The goals of low average energy consumption and low energy fluctuation of the building were achieved, namely, the risk from uncertainty fluctuation was controlled by robust optimization.

The scientific contribution of this study is to validate the feasibility of utilizing deep learning to establish a normal distribution model for certain meteorological elements, proposing and demonstrating the feasibility and advantages of using cluster centers as sampling points and providing other researchers with a sampling alternative with low random interference. In addition, the feasibility of combining a clustering algorithm, deep learning, and robust optimization to achieve optimization considering uncertainty is also demonstrated by the uncertainty modeling results and optimization results of this study. This demonstrates that further optimization of building solutions under the premise of considering uncertainty is necessary even in the context of the current high insulation performance standards.

In the robust optimization process of this study, the standard deviation is selected as one of the optimization objectives, which effectively reduces the fluctuation range. However, there are obvious conservative phenomena in the optimization results. After optimization, although the maximum energy consumption in the fluctuation range has been significantly reduced, the minimum energy consumption value has also become larger, that is, the minimum energy consumption result has worsened. Therefore, as a future research direction, it is necessary to consider new objective functions to address the conservatism issue.

Author Contributions: Conceptualization, J.W.; Methodology, J.W.; Software, J.W.; Validation, J.W; Formal analysis, J.W.; Data curation, J.W.; Writing—original draft, J.W.; Writing—review & editing, J.W.; Supervision, M.M.; Project administration, K.T. All authors have read and agreed to the published version of the manuscript.

Funding: This research received no external funding.

Institutional Review Board Statement: Not applicable.

Informed Consent Statement: Not applicable.

Data Availability Statement: Not applicable.

Conflicts of Interest: The authors declare no conflict of interest.

References

1. Ritchie, H.; Roser, M. Urbanization. Our World in Data. 2018. Available online: https://ourworldindata.org/urbanization (accessed on 31 March 2023).
2. Saidur, R.; Masjuki, H.H.; Jamaluddin, M. An application of energy and exergy analysis in residential sector of Malaysia. *Energy Policy* **2007**, *35*, 1050–1063. [CrossRef]

3. National Association of Home Builders Discusses Economics and Housing Policy. 2020 Multifamily Completion Data: Property Size. Available online: https://eyeonhousing.org/2021/08/2020-multifamily-completion-data-property-size/ (accessed on 8 June 2023).
4. BC Housing. *2018 BC Residential Building Statistics and Trends Report*; BC Housing: Vancouver, BC, Canada, 2018; p. 61.
5. Statistics Bureau, Ministry of Internal Affairs and Communications. 2018 Housing and Land Statistics Survey. Available online: https://www.stat.go.jp/data/jyutaku/2018/pdf/g_gaiyou.pdf (accessed on 28 May 2023).
6. Wu, W.; Skye, H.M. Residential net-zero energy buildings: Review and perspective. *Renew. Sustain. Energy Rev.* **2021**, *142*, 110859.
7. De Wilde, P. The gap between predicted and measured energy performance of buildings: A framework for investigation. *Autom. Constr.* **2014**, *41*, 40–49. [CrossRef]
8. Shi, X.; Si, B.; Zhao, J.; Tian, Z.; Wang, C.; Jin, X.; Zhou, X. Magnitude, causes, and solutions of the performance gap of buildings: A review. *Sustainability* **2019**, *11*, 937. [CrossRef]
9. Bhandari, M.; Shrestha, S.; New, J. Evaluation of weather datasets for building energy simulation. *Energy Build.* **2012**, *49*, 109–118. [CrossRef]
10. Sun, Y.; Gu, L.; Wu, C.J.; Augenbroe, G. Exploring HVAC system sizing under uncertainty. *Energy Build.* **2014**, *81*, 243–252. [CrossRef]
11. Wang, L.; Mathew, P.; Pang, X. Uncertainties in energy consumption introduced by building operations and weather for a medium-size office building. *Energy Build.* **2012**, *53*, 152–158. [CrossRef]
12. Sun, Y.; Heo, Y.; Tan, M.; Xie, H.; Jeff Wu, C.; Augenbroe, G. Uncertainty quantification of microclimate variables in building energy models. *J. Build. Perform. Simul.* **2014**, *7*, 17–32. [CrossRef]
13. Luo, Z.; Lu, Y.; Cang, Y.; Yang, L. Study on dual-objective optimization method of life cycle energy consumption and economy of office building based on HypE genetic algorithm. *Energy Build.* **2022**, *256*, 111749. [CrossRef]
14. Imran; Iqbal, N.; Kim, D.H. IoT Task Management Mechanism Based on Predictive Optimization for Efficient Energy Consumption in Smart Residential Buildings. *Energy Build.* **2022**, *257*, 111762. [CrossRef]
15. Qin, Y.; Song, D.; Chen, H.; Cheng, W.; Jiang, G.; Cottrell, G. A dual-stage attention-based recurrent neural network for time series prediction. *arXiv* **2017**, arXiv:1704.02971.
16. Blundell, C.; Cornebise, J.; Kavukcuoglu, K.; Wierstra, D. Weight uncertainty in neural networks. In Proceedings of the 32nd International Conference on Machine Learning, Lille, France, 6–11 July 2015; Volume 2, pp. 1613–1622.
17. Japan Meterological Agency. Search Past Weather Data, Toyama Prefecture Toyama. 2021. Available online: https://www.data.jma.go.jp/obd/stats/etrn/index.php?prec_no=55&block_no=47607&year=2021&month=1&day=1&view= (accessed on 9 March 2020).
18. National Renewable Energy Laboratory. EnergyPlus. 2021. Available online: https://energyplus.net/ (accessed on 18 April 2022).
19. Bahdanau, D.; Cho, K.; Bengio, Y. Neural machine translation by jointly learning to align and translate. *arXiv* **2014**, arXiv:1409.0473.
20. Wang, J.; Mae, M.; Taniguchi, K. Uncertainty modeling method of weather elements based on deep learning for robust solar energy generation of building. *Energy Build.* **2022**, *266*, 112115. [CrossRef]
21. MacQueen, J. Some methods for classification and analysis of multivariate observations. In Proceedings of the Fifth Berkeley Symposium on Mathematical Statistics and Probability, Oakland, CA, USA, 1 January 1967 ; Volume 1, pp. 281–297.
22. Deb, K.; Pratap, A.; Agarwal, S.; Meyarivan, T. A fast and elitist multiobjective genetic algorithm: NSGA-II. *IEEE Trans. Evol. Comput.* **2002**, *6*, 182–197. [CrossRef]
23. Institute for Built Environment and Carbon Neutral for SDGs. Explanation of Energy Consumption Calculation Method in the Criteria of Judgment of Housing Business Builder. 2012. Available online: http://www.heat20.jp/grade/index.html (accessed on 31 March 2023).
24. Insulation Technology Development Committee for Housing with an Eye on 2020, H. Investigation Committee of Hyper Enhanced Insulation and Advanced Technique for 2020 Houses Design Guide Plus. 2020 ; pp. 124–125. Available online: https://www.ibec.or.jp/ee_standard/build_standard.html (accessed on 31 March 2023).
25. U.S. Green Building Council. LEED Rating System. 2023. Available online: https://www.usgbc.org/leed (accessed on 9 June 2023).

buildings

MDPI

Article

Evaluating Savings Potentials Using Energy Retrofitting Measures for a Residential Building in Jeddah, KSA

Ahmed Felimban [1,2,*], Ulrich Knaack [1] and Thaleia Konstantinou [1]

1 Architectural Facades & Products Research Group, Department of Architectural Engineering + Technology, Faculty of Architecture and the Built Environment, Delft University of Technology, 2628 BL Delft, The Netherlands; u.knaack@tudelft.nl (U.K.); t.konstantinou@tudelft.nl (T.K.)
2 Faculty of Environmental Designs, King Abdulaziz University, Jeddah 21589, Saudi Arabia
* Correspondence: a.a.m.felimban@tudelft.nl or afelimban@kau.edu.sa; Tel.: +966-(0)-55770-0232 or +31-(0)-6-8587-5615

Abstract: Residential buildings in the Kingdom of Saudi Arabia (KSA) contribute to nearly half of the overall electricity consumption in the building stock, highlighting their significant role in energy consumption. While an upgraded energy code has been established and enforced for new buildings, existing buildings continue to operate at the same level of energy consumption. Therefore, there is a need for further energy upgrades in existing buildings. This study evaluates the energy savings potential of various energy retrofitting measures for a case study in Jeddah, KSA. Data from previous studies and current practices were collected and analyzed. Different energy upgrade measures, such as windows replacement, wall insulation upgrade, roof insulation upgrade, and air conditioning unit replacement, were selected and evaluated using a digital simulation tool called Design-Builder. The simulation results were compared to understand the potential percentage of energy savings. The average annual energy consumption (AAEC) was used as the primary performance indicator to compare the energy savings among the scenarios. The results demonstrate significant reductions in energy consumption for the proposed scenarios. Furthermore, the study examined the significant impact of uncertainties, specifically, the infiltration rate and AC setback temperature, on AAEC. In conclusion, the proposed scenarios have the potential to achieve substantial energy savings, ranging from 25% to 66%, depending on the number of energy retrofitting interventions employed. The findings of this study can serve as a useful reference for similar energy retrofitting projects.

Keywords: building energy performance; Jeddah; residential buildings; energy efficiency; average annual energy consumption (AAEC)

Citation: Felimban, A.; Knaack, U.; Konstantinou, T. Evaluating Savings Potentials Using Energy Retrofitting Measures for a Residential Building in Jeddah, KSA. *Buildings* **2023**, *13*, 1645. https://doi.org/10.3390/buildings13071645

Academic Editors: Yingdong He and Nianping Li

Received: 17 April 2023
Revised: 21 June 2023
Accepted: 24 June 2023
Published: 28 June 2023

1. Introduction

In the Kingdom of Saudi Arabia (KSA), building energy consumption is a significant contributor to oil consumption and a major expense for building users [1]. The government has introduced numerous initiatives to promote energy efficiency and renewable energy, recognizing the importance of energy efficiency in buildings [2]. While the Saudi building code (SBC) committee has endorsed the upgraded energy code and required it for new residential buildings in 2021, existing buildings continue to consume high levels of energy. Furthermore, the Saudi Energy Efficiency Center (SEEC) has initiated efforts to improve energy consumption in existing buildings, including enhancing the energy efficiency of household appliances and lighting products [2]. However, further investigation is needed to understand the energy-saving potential of energy upgrade measures for existing buildings. Upgrading the building envelopes is essential for reducing energy consumption, which can reach up to 68% [3]. However, retrofitting existing buildings for energy efficiency can be challenging due to design limitations, climate considerations, and occupant behavior [4]. Unfortunately, the retrofitting progress of existing buildings has been slow due to insufficient awareness, funding, and technical expertise [4].

This paper aims to evaluate the energy-saving potential of building envelope upgrades through energy retrofitting measures for a case study of a residential building in Jeddah, Saudi Arabia. The study compares various strategies for eight apartments to illustrate the potential energy savings achievable for each unit. The building under investigation is a typical low-rise residential structure, consisting of five floors and eight apartments. Key performance indicators, such as the average annual energy consumption (AAEC), are used to compare the energy consumption of the building before and after the energy retrofit scenarios. The proposed scenarios encompass improvements to the building envelope and HVAC systems.

The study explores the challenges and opportunities associated with retrofitting existing buildings in Jeddah city, emphasizing the importance of a holistic approach that considers multiple factors when retrofitting existing buildings. Additionally, the potential for energy savings resulting from retrofitting existing buildings is examined. The paper also discusses uncertainties and their effects on the AAEC, including the infiltration rate (ACH50) (Air changes per hour at 50 Pascal) and the desired user thermal comfort temperature. The simulation of energy from the basic model employed an infiltration rate of 20 ACH50, with attempts made to generate scenarios targeting a rate of 4 ACH50, as required by SBC standards. Outdoor scenarios were applicable to the entire building, rather than individual apartment upgrades. The simulation results reveal a significant reduction in the average annual energy consumption (AAEC) when a deep energy retrofit scenario was implemented.

The paper's conclusion compares the energy-saving potential among different scenarios and highlights certain uncertain factors that impact the AAEC, such as infiltration rates and setback temperatures. The study's results contribute significantly to the existing literature on energy retrofitting of existing buildings in Jeddah.

2. Methodology

A mixed-methods approach, encompassing qualitative and quantitative techniques, was employed in this study to provide comprehensive results contributing to the existing literature on energy retrofitting of existing buildings in Jeddah. Qualitative data were collected from the relevant literature and related studies, forming a solid foundation for exploring the energy-saving potential of retrofitting measures using the DesignBuilder (Version 7) digital software (EnergyPlus).

To examine the energy-saving possibilities and enhance the energy efficiency of existing apartments in Jeddah, a series of energy retrofitting scenarios were analyzed for eight apartments within the same building. The simulation tool was utilized to assess the current energy consumption and determine the potential energy savings for each scenario.

The objective of this paper was to evaluate and compare the energy-saving possibilities across different energy retrofitting scenarios, utilizing simulation software. To establish the simulation parameters, several steps were undertaken. First, a comprehensive review of the literature and related studies was conducted to summarize the essential design parameters and energy benchmark levels, which play a crucial role in highlighting specific variables. Second, the case study was described, as it was necessary for digital modeling. Data were collected from various sources, including floor plans, apartment orientation, component materials and U-values, user activities, and mechanical AC systems. Third, specific uncertainties that affect energy savings results, such as the infiltration rate and user thermal comfort (setback temperature), were identified. Fourth, an overview of the energy upgrade scenarios and interventions was formulated. Finally, the simulation results were analyzed to evaluate and compare the energy savings achieved by each energy upgrade scenario.

Further illustrations will be presented in subsequent sections, and Figure 1 provides a visual representation of the workflow, facilitating readers' understanding of the sequential steps in this study.

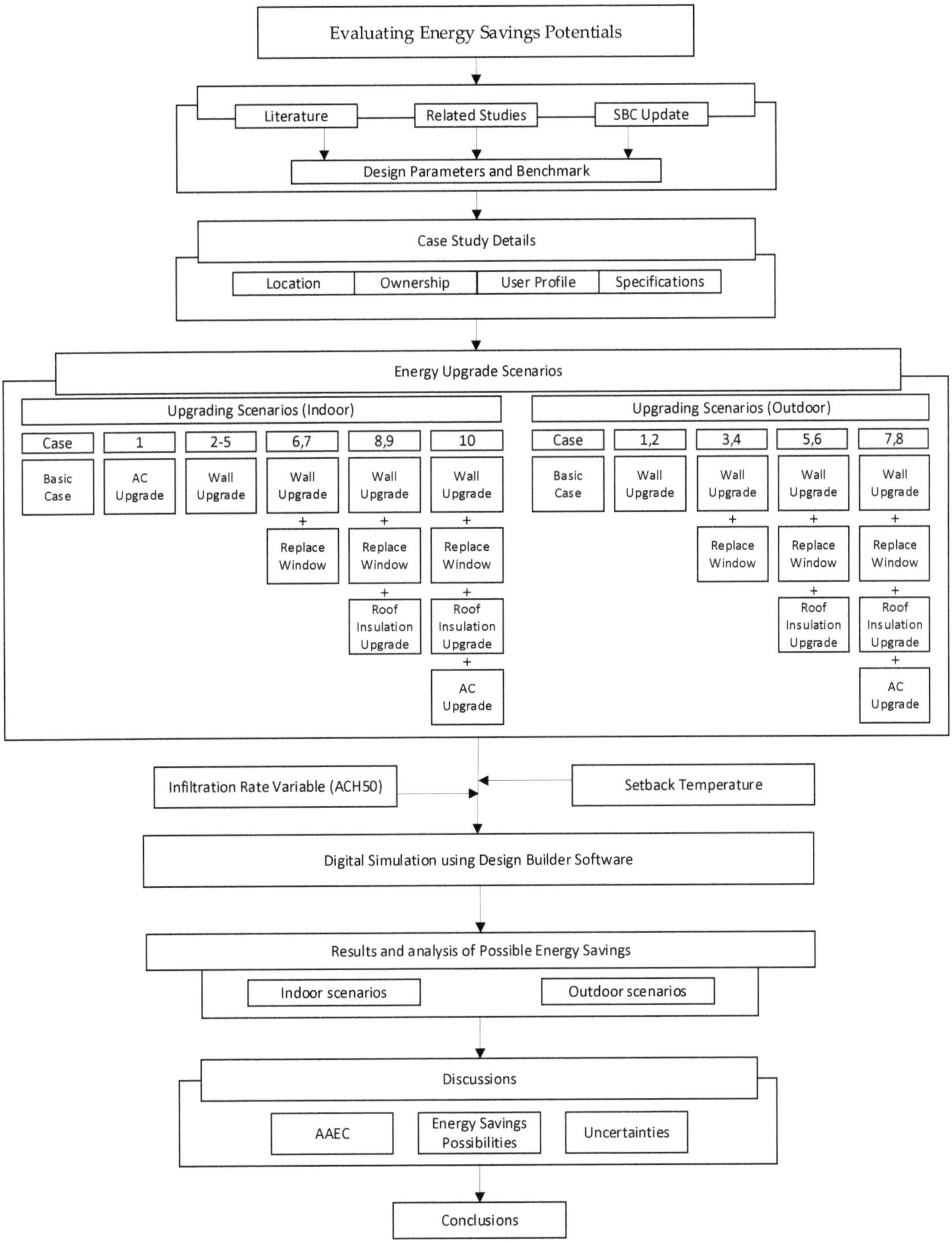

Figure 1. Research structure of the paper.

3. Literature Review and Related Studies

This section addresses the knowledge gap in previous studies on energy retrofitting upgrades and highlights the latest methods used in the field for energy retrofit measures. It also presents related studies and existing practices.

Researchers have extensively investigated the advantages of energy conservation measures for residential buildings in the Gulf Cooperation Council (GCC) countries. However, there has been a lack of investigation into the recent cost of living increases and the government's vision for energy conservation measures, particularly in the context of the Kingdom of Saudi Arabia (KSA). KSA citizens have only recently gained knowledge and basic information about energy savings and conservation measures. Some studies have demonstrated the potential for significant energy consumption savings through the application of various energy-saving measures. However, many of these studies did not explicitly consider the associated costs.

Several researchers, such as [5–7], have reported varying energy savings, ranging from 15% to 72%, when implementing different energy-saving measures, including insulation upgrading, U-value upgrading, window glazing upgrading, electrical devices upgrading, shading devices installation, and on-site energy generation units. Konstantinou presented five specific strategies (replace, add-in, wrap-it, add-on, and cover-it) that could be incorporated into refurbishment designs for aging residential buildings, providing a toolbox of refurbishment strategy possibilities to aid decision-making processes [8]. The main outcome of Konstantinou's study was the provision of a façade refurbishment toolbox to support the decision-making process in design. It is important to note that all the strategies mentioned in the study aimed to improve all building envelope components where heat loss occurs, both inside and outside the buildings. In contrast, in the KSA context, energy retrofitting strategies aim to improve all building envelope components where heat penetrates, specifically, from the outside of the building to the inside, due to the KSA context.

Several researchers have investigated different energy-saving measures, considering climate and economic conditions in GCC countries. These studies have provided fundamental knowledge on applicable and feasible strategies that could be implemented within the KSA context. However, the recent cost of living increases, the addition of value-added tax (VAT), and the introduction of new service fees require further investigation.

For example, a research study conducted in Kuwait in 2003 demonstrated savings of 3.25 million MWh over 10 years by implementing energy-saving measures, such as insulation, glazing upgrades, and reducing the window area in 42,403 retrofitted old residential buildings [9]. The main challenges in this project were the government's role as the main financer of the initial retrofitting costs and the long payback period of over 30 years, during which electricity tariff prices continued to be subsidized. Additionally, Krarti presented the economic and environmental benefits of improving energy efficiency in new and retrofitted buildings in Kuwait [10]. The study proposed three levels of retrofitting proposals for better energy efficiency. Furthermore, the research showed that implementing various energy-saving measures could result in savings of 8%, 23%, or more than 50% of annual energy usage.

Similarly, Krarti recommended similar implementation measures for buildings in the KSA to achieve comparable energy savings on a larger scale. Ameer suggested that doubling electricity prices (electricity tariffs) in Kuwait would incentivize the implementation of energy efficiency measures in the residential buildings sector, which would ultimately benefit the Kuwaiti government [11]. Currently, building users in Kuwait, as in other GCC countries, rely on government subsidies for their energy bills, although the amount of subsidies varies across countries.

In the UAE, Taleb tested the improvement of building thermal performance using 8 passive cooling strategies, achieving energy consumption reduction of up to 23.6% in residential buildings in Dubai [12]. Further, Alfaris observed a remarkable increase in energy performance efficiency of 25% by implementing low and medium energy conservation measures [6]. This led to energy consumption savings ranging from 14.4% to 47.6%, depending on individual operating conditions and occupants' behaviors. Rakhshan demonstrated a 40% reduction in summer peak demand and a 32% decrease in CO_2 emissions by improving wall insulation to a U-value of 0.3 W/(m^2K) and upgrading AC systems to a Coefficient of Performance (COP) of 2.7 [13]. Giusti and Almoosawi examined the impact of occupants' behaviors on electricity consumption, including raising the AC thermostat temperature to 24 °C, switching off domestic water heating when not needed, and adding roof insulation, all of which contributed to varying percentages of energy consumption savings [14].

Friess reviewed several passive measures, such as building orientation, thermal insulation, appropriate glazing types and orientation, excessive light levels and glare, and natural ventilation. These measures were found to achieve energy consumption savings of 30% in villas and up to 79% in high-rise office buildings [15]. Studies on energy-saving measures have been more extensively explored in the United Arab Emirates (UAE) compared to other GCC countries, yielding diverse results that have been recently implemented.

On the other hand, researchers have highlighted the significant impact of building user behaviors on energy savings. Al-Saadi demonstrated a substantial reduction of up to 42.5% in annual energy consumption through the implementation of various saving measures in a calibrated model of a typical house in Oman [16]. Alalouch emphasized the urgent need for large-scale retrofitting programs, which could effectively reduce energy consumption through suitable energy-saving measures [17].

Aldossary proposed various management and technical upgrades in KSA that could serve as benchmarks for enhancing energy efficiency in the country's residential stock. The author identified three prototype houses exhibiting maximum energy efficiency, surpassing international standards. These recommendations serve as exemplary standards for future implementation of retrofitting measures in KSA's residential buildings. Similarly, Krarti investigated optimal energy savings for residential buildings across 5 sub-climate zones in KSA, ranging from 26% to 47.3%. These savings were achieved by applying energy conservation measures to building envelope elements, such as wall insulation, roof insulation, window area, window glazing, window shading, and thermal mass, considering life cycle cost and energy savings [18]. Additionally, Alaidroos highlighted that the application of energy conservation measures within the KSA region could result in significant annual savings on energy cost subsidies, national oil consumption, and investment in new power plants [18].

Furthermore, Alaboud noted that implementing necessary measures, such as reducing the Window-to-Wall Ratio (WWR) from 15.3% to 5%, adding insulation to the roof and external walls, and increasing the thermostat temperature in houses by 1 °C, could lead to a substantial 35% decrease in cooling load [19]. The study suggested that if these measures were implemented, there would be a 35% reduction in cooling demand, which could be further increased with additional retrofitting plans, considering cost-effectiveness. In 2019, Krarti demonstrated that retrofitting residential buildings could reduce energy consumption by up to 60%. Moreover, by utilizing solar panels (PVs) on building roofs in KSA, surplus energy could be generated for internal use or returned to the grid [20]. Previous studies have explored different strategies for energy-saving measures focused on building energy performance.

In summary, the studies have shown significant energy savings, ranging from 15% to 72%, when implementing various energy retrofitting measures using appropriate combinations of interventions, as illustrated in Table 1. However, these studies have not taken into account recent changes in the KSA context, such as the recent increase in energy costs, updates in building codes, and the government's development towards the 2030 vision. Therefore, there are knowledge gaps that need to be addressed in current research.

Table 1. The Strategy Types and The Savings Interventions.

Strategy Type	Replace	Add (Inside)	Add (Outside)
Envelope energy-saving interventions	Wall filling (blocks) Insulation Air Sealant Window AC Systems	Wall filling (blocks) Insulation Air sealant Window	Shading Devices On-Site Energy

3.1. Retrofitting Strategies for Low-Rise Residential Buildings in Jeddah, KSA

Various factors influence the implementation of energy retrofit applications, including the micro-climate, thermal properties of building fabric, occupants' thermal comfort level, owners' acceptance of changes, and budget constraints [21]. However, a study by Ma emphasizes that sustainable energy retrofitting applications must follow a strategic design process, which involves careful decision making at different phases [22]. The same study proposes a systematic approach to achieving sustainable energy retrofit applications, which can be divided into three activities: pre-retrofit (identification of possible solutions), retrofit (testing of solutions), and post-retrofit (evaluation of the application) [22]. Similarly, energy retrofitting strategies require similar processes to ensure an effective application for maximizing energy efficiency.

Energy retrofits can be categorized in various ways [23–27]. Natural Resources Canada (NRCAN) classifies retrofit activities into three scales: minor retrofit, major retrofit, and deep retrofit. This study adopts the NRCAN categorization as a baseline [23]. The scales are classified based on the level of intervention required and the percentage of energy savings.

A—Minor Energy Retrofit

A minor energy retrofit involves easy upgrades that can be implemented with low-cost investments. This includes sealing gaps, upgrading lighting systems, improving electrical devices, implementing control systems, and conducting regular maintenance. These activities require minimal interventions and have little to no disturbance for building users.

B—Major Energy Retrofit

A major energy retrofit involves significant changes or upgrades aimed at reducing the building's energy consumption while causing minimal disruption for building users. Key activities include replacing, upgrading, or adding building elements, such as windows (frames, panes, glazing), wall thickness, insulation and shading systems and improving AC systems.

C—Deep Energy Retrofit

A Deep Energy Retrofit (DER) achieves substantial energy savings, potentially reducing energy costs by up to 60%. The International Energy Agency (IEA) defines Deep Energy Retrofit as follows:

> *"A major building renovation project in which site energy use intensity (including plug loads) has been cut by at least 50% compared to the baseline with a corresponding improvement in indoor environmental quality and comfort"* [28].

In the same study, Deep Energy Retrofit is described as a comprehensive approach that includes upgrades, additions, or changes to building systems, resulting in at least 50% energy consumption savings. These activities may involve major disruptions for building users, such as replacing the entire façade, adding a second-skin façade, or applying External Insulation Finishing System (EIFS) as insulation cover.

Table 2 illustrates the different levels of interventions and activities, providing an overview of the changes that can be implemented across various energy retrofit measures.

Table 2. Main differences between minor, major, and deep energy retrofits.

Minor	Major	Deep
Lighting upgrades AC systems upgrades Electrical devices upgrades Gaps fillings fixes Electrical devices maintenance Controlling systems	Windows (frames, pans, pans cavity, glazing) Wall (thickness, materials) Insulation Gaps filling Shading systems (fix, active)	Major energy retrofit activities Outdoor insulations (EIFS) Second-skin façades Replace the entire façade

In this study, energy retrofitting is defined as any changes (replacement, repairs, upgrades, or additions) that improve the building's energy performance efficiency and reduce energy consumption costs. The scope of the study focuses on exploring potential energy retrofitting strategies and their economic and environmental benefits.

Initially, in KSA, energy retrofitting strategies were not widely recognized among building users due to the low energy tariffs. However, attitudes and approaches changed when the tariffs increased significantly, as mentioned earlier. The Social Development Bank (SDB) in KSA defined the restoration loan/finance program as follows:

> *"A financing program designed for restoration, maintenance, repair of structural and emergency defects, for the purpose of additions or necessary modifications for private residential houses"* [29].

Until recently, energy retrofitting measures were not commonly practiced in most architectural firms in KSA. However, there has been a shift, and it is now becoming more standard practice as clients recognize its importance. The recent increase in energy prices and the definition of energy performance defects in buildings have contributed to this change, as they are now being closely monitored.

It is worth noting that energy prices in KSA remain competitive compared to average energy consumption prices in European countries, as shown in Figure 2. The SEEC initiatives have primarily focused on minor energy activities, such as promoting the labeling of electrical devices (including lighting and ACs) and providing discounts on efficient AC units. They have also increased public awareness of energy efficiency through the media. However, major and deep energy retrofits have not been comprehensively covered by the SEEC.

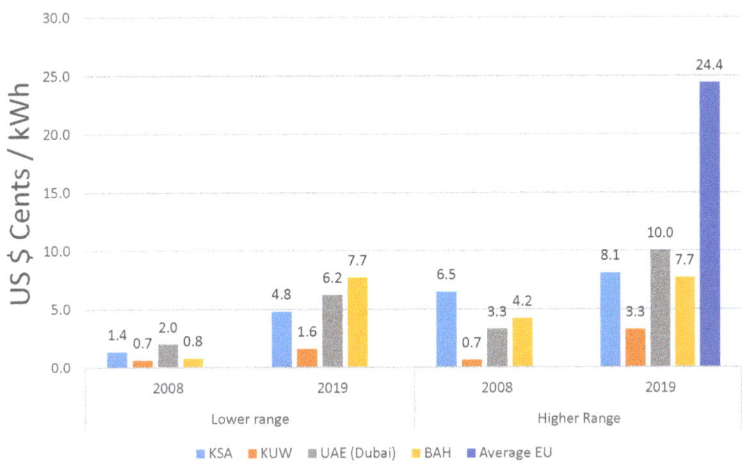

Figure 2. Electricity tariff prices (lower range (≤6000 kWh) and higher range (>6001 kWh)) comparing four GCC countries and the EU.

Energy retrofitting strategies in the KSA context have generally received low acceptance rates among homeowners due to various factors. In the past, this included high initial costs and limited interest in energy efficiency, primarily due to low energy tariffs, as shown in Figure 3.

Figure 3. Austah House's recent renovation project in Yanbu city, KSA.

Historically, energy retrofitting solutions needed to consider the initial costs of energy improvements compared to the current scenario, focusing on total energy savings and their potential impact on energy costs. Energy performance and cost are, therefore, primary indicators for evaluating energy consumption levels. The challenge lies in creating a comfortable indoor environment for building users in the harsh, outdoor, hot desert climate of KSA, where temperatures range between 32 and 49 °C (Jeddah), and indoor temperatures should ideally be maintained between 18 and 24 °C [1].

Despite the limited number of residential building renovation projects in KSA, energy upgrades have not been a significant consideration, even among existing projects that have not met upgraded SBC standards. Homeowners typically renovate their buildings for aesthetic or structural purposes, with little emphasis on energy efficiency. However, a recent documented renovation project of a residential building serves as an example of integrating energy upgrade measures alongside aesthetic improvements, resulting in substantial energy savings. This case highlights the common misconception that residential building renovation is solely for aesthetic purposes and underscores the importance of considering energy efficiency measures during renovation projects.

Austah House, located in Yanbu city in the western region of KSA, represents a recent renovation case that provides insight into available energy enhancement possibilities (see Figure 3). The following information is based on an interview with the owner and architect, Moaad Austah, and supporting pictures and information obtained through Twitter, a social media application [30].

The renovation of the Austah House was motivated by cultural aspects, building quality, and economic considerations, highlighting how building owners approach renovation activities, as shown in Figure 3. The construction of the building took 2 years and was completed in 2020, which was attributed to the fact that the building was occupied during the construction period.

The main energy-related changes incorporated into the house included upgrades to the walls, windows, lights, and air conditioning (AC) systems. The east and west façade walls were upgraded to a 40 cm thickness, consisting of a 15 cm block, 5 cm insulation panels, another 15 cm block, and 2.5 cm of mortar on both sides of the wall. The windows were upgraded to double-glazed windows with thermal break frames (6 mm glass, 12 mm air vacuumed, and 6 mm glass). Additionally, 18 AC window units were replaced with split AC units with energy efficiency ratings of 5 or 6 stars. The owner reported a 30% energy saving after the building renovation.

To assess the actual energy consumption, the study illustrated the monthly electricity bills over a 2-year period from February 2019 to December 2020. The average energy consumption on the first floor was approximately 135 kWh/m^2 per year, with an energy cost of around 24 S.R. (USD 6.4), as detailed in Figure 4. The Austah house case exemplified a diligent renovation approach, including several energy retrofitting measures, such as wall insulation upgrades, window replacements, AC upgrades, and lighting improvements. Further energy savings measures could potentially enhance the energy performance efficiency of the building.

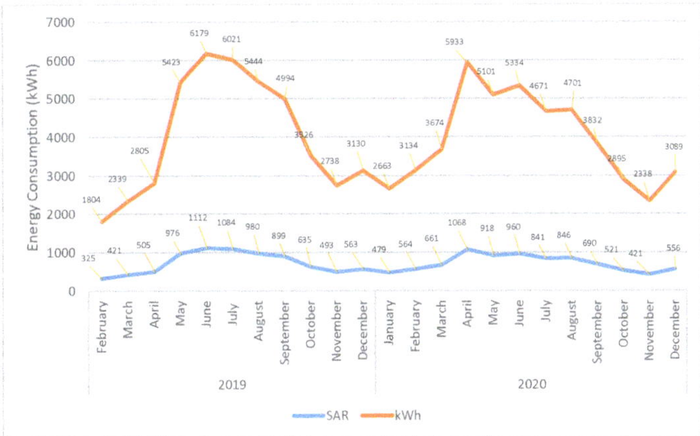

Figure 4. The 2019–2020 energy consumption (kWh) and energy cost (Saudi Riyal (SAR)) for 1st floor [31].

3.2. Energy Retrofitting Strategies

Existing retrofitting projects in KSA have primarily focused on either restructuring the building or enhancing its aesthetic appearance. Konstantinou [8] presented various refurbishment strategies used as a basis for the study, with an update for the Jeddah context. The strategies employed (replace, add-in, wrap-it, add-on, or cover-it) were specified in the specific context of KSA, excluding the cover-it strategy from the study scope. Table 3 illustrates the suitable strategy categories based on these approaches.

Table 3. List of energy upgrade strategies.

Strategies	Replace	Add-In	Wrap-It	Add-On
Description	Replace façade elements with better energy performance	Upgrade by adding from the inside of the building components (wall, window, insulation)	Wrap the building with a second layer	Adding a shading device or structure element to the outdoor façade
Interventions	Replace entirely, Replace partially (walls, windows, connections, insulation)	Increase wall thickness, internal insulation, cavity insulation, windows (panes, cavities, glazing), add sealant between components	External insulation (EIFS exterior insulation), second-skin façade	Adding (fixed, active) shading devices or adding balconies, merge balconies to indoor space if applicable
Benefits	New components with better performance, small disturbance to users	Appropriate for existing buildings, increase the thermal resistance, individual decision making	Increase the thermal resistance using external insulation, no thermal bridging	Better energy performance on the developed parts, heat prevention increase, increased indoor space in some cases
Limitations	Significant impact on the building users' activities, high initial costs	Thermal bridging needs attention, decrease in livable space	Not applicable for SBC limitations except for external insulation, high initial costs	Low WWR application limitations from the SBC

A—Replace Strategy

The replace strategy involves exchanging old building components (walls, windows, insulation, connections) with new ones, either separately or by altering the entire façade. The cost depends on the number of intervention activities and the energy efficiency level of the materials used. Fewer interventions result in lower costs, while replacing an entire façade significantly impacts the building's energy efficiency with higher costs. However, the level of disturbance to building users' activities should be considered early in the application process to minimize disruptions. The level of disturbance may vary depending on the extent of the replacement interventions, with greater replacement interventions resulting in more disturbances.

B—Add-In Strategy

The add-in strategy encompasses upgrading building components (walls, windows, insulation, connections) within the existing building. This strategy allows for maintaining the same outdoor façade, making it suitable for individual units of residential buildings. Attention needs to be given to addressing thermal bridging issues that arise from connections between building components. Increasing indoor wall thickness may reduce the livable space, which is crucial to consider during the design process.

C—Wrapping (Wrap-It) Strategy

The wrapping (wrap-it) strategy involves adding an extra layer to the building, such as a second-skin façade or external insulation (EIFS). The second-skin façade resolves thermal bridging issues and allows for aesthetic modifications to the building when outdoor wall thickness increase is not restricted by the Saudi Building Code (SBC). EIFS has great potential for improving thermal energy performance, eliminating thermal bridging. However, if the existing walls have stone finishing, additional investigation is required for the disassembly of the stone, considering time and cost implications. Although the EIFS option has been used recently, the costs have been higher compared to previous strategies.

D—Add-On Strategy

The add-on strategy involves adding shading devices or structural elements to the external façade layer, primarily targeting shaded parts of the building. Residential buildings in hot climate conditions generally have a low Window-to-Wall Ratio (WWR). Merged balconies can be expanded to provide additional indoor space, subject to merging restrictions imposed by the Saudi Building Code (SBC).

3.3. *Energy Performance Challenges of Residential Buildings in Jeddah*

Residential building envelopes in Jeddah were designed with insufficient thermal properties. Several studies have assessed the thermal performance of existing building envelopes, including walls, windows, floors, and roofs. Computational simulations were employed to analyze actual building cases [3,18,32]. These simulations have identified potential areas for energy savings [7,9,18]. Interestingly, 1 study demonstrated that with minimal interventions, energy savings ranging from around 15% to just below 50% could be achieved, depending on the specific energy-saving measures implemented [6]. Life cost analysis has also been utilized to determine the most energy-saving interventions in the long run.

The primary challenge observed in buildings requiring energy upgrades is the presence of defective thermal building designs, resulting in significant outdoor heat gains through different components of the building envelope, leading to thermal discomfort in indoor spaces (see Table 4). The study identified that high energy consumption is primarily influenced by infiltration rates and thermal bridges. In practice, mechanical cooling systems, such as air conditioning units (ACs), are commonly employed by building users to achieve thermal comfort. Therefore, the main issues that need to be addressed regarding current residential building energy performance are the lack of information regarding ACs at various levels, from building codes to construction, and the impact of low energy tariffs on the defective energy performance of building envelopes.

Table 4. Energy efficiency challenges, causes, and results.

	Challenges	Causes	Result
Energy Efficiency	Insufficient thermal performance of envelope (wall, windows, no insulation)	Old SBC standards, inappropriate design, no insulations with high U-value for walls	Outdoor heat gain, high energy demand, user discomfort
	Defective airtightness with a high infiltration rate	Inappropriate sealants or no sealants, poor components, materials	Outdoor heat gain, increased cooling load demand, high energy demand, user discomfort
	Thermal bridges	No insulation, poor design	Outdoor heat gain

When upgrading existing buildings to reduce high heat gain from the external environment, several factors need to be considered to improve indoor thermal comfort for occupants:

- Lack of knowledge on the cost benefits of essential technical solutions for energy retrofitting building envelopes. Defective thermal properties of walls, roofs, and windows must be addressed. Additionally, the old Saudi Building Code (SBC) did not require sufficient energy levels and allowed the use of low-thermal-resistance materials.
- High infiltration rates (air tightness) in indoor spaces were not included in the old SBC. Currently, there is a lack of knowledge regarding the most efficient infiltration rate for existing residential buildings.
- The presence of thermal bridges resulting from poor thermal designs requires consideration.

Although these constraints are interconnected, the previous discussion highlights that the energy requirements outlined in the old building code were the main driver of the current problem. Historically, low energy costs and occupant behavior also contributed to high energy consumption in residential buildings. When energy requirements and electricity tariffs were low, buildings were constructed with poor thermal properties, leading to increased cooling demands. Occupants responded to thermal discomfort in indoor spaces by using ACs, which were cost-effective at the time.

In recent years, GCC countries have implemented various policies aimed at reducing CO_2 emissions. These policies have necessitated changes at the governmental level to enhance building energy performance and promote the use of renewable energy sources, raising the overall energy efficiency standards in the region.

The Solar Decathlon Middle East (SDME) competition focused on the potential for net-zero buildings in hot, arid climate conditions, considering the large areas required for PV solar panels to achieve the net-zero concept [33]. The results emphasized the importance of efficient building designs combined with renewable energy sources. A project by Virginia Tech demonstrated the possibility of transforming a villa from an energy consumer to an energy producer, setting a benchmark for net-zero energy houses [34]. Although the SDME competition focused on a different building typology (villas), the results can provide valuable evidence of energy-efficient homes in hot, arid climate zones.

A recent local study by Aldossary proposed an energy benchmark range for apartments in low-rise buildings in KSA, ranging from 77 to 98 kWh/m^2 per year, resulting in lower carbon emissions [7]. Aldossary also provided specifications for building elements, including the layers, thickness, and U-Value in Table 5. The author suggested an optimal solution of a 35 cm external wall thickness with a 0.257 U-Value. Another study by Alaidroos extensively tested five energy efficiency measures, demonstrating significant energy-saving potential that could reduce energy cost subsidies, national oil consumption, and the need for new power plants [18]. The suggested energy range can serve as a benchmark and reference for future comparisons with the proposed scenarios.

Table 5. Building elements specifications [7].

Building Element	Specification	Thickness (cm)	U-Value (W/m^{2}k)
External wall	Mortar–red brick–mortar	24	2.7
Internal wall	Mortar–brick–mortar	24	3.38
Roof	Six layers (tiles, mortar, sand, insulation, and reinforced concrete)	40	2.8
Floor	Seven layers (ceramic, mortar, sandstone, concrete, insulation, basement concrete, and basement stone)	50	1.9
Windows	Single glazing	1	5.57
Doors	Wooden door	4	2.1

Furthermore, the study illustrates that current energy consumption in KSA ranges from 114 to 166 kWh/m^2 per year for apartments and 109 to 185.4 kWh/m^2 per year for villas [32]. The results of the simulation-based studies illustrated actual upgrading possibilities of energy performance for the existing residential buildings, especially after regulating the upgraded SBC. The detailed study will be used as a reference level while further simulation validation is needed. Simulation-based studies have demonstrated actual energy performance improvement possibilities for existing residential buildings, particularly after the introduction of upgraded SBC regulations. These detailed studies will be used as reference points, but further simulation validation is necessary.

The Saudi Building Code (SBC) underwent an upgrade in 2018, accompanied by an increase in energy tariffs. Subsequently, in January 2021, the SBC national committee endorsed an upgraded building code that applied to all new residential buildings constructed from 1 July 2021 onwards. Interestingly, after the SBC endorsement, there was a noticeable decrease in the issuance of construction permits in Jeddah for at least three months, until the committee revised some of the requirements [35]. This study specifically focuses on existing residential buildings in Jeddah categorized as Zone 1 according to the SBC classification, as depicted in Table 6 [36]. While the upgraded SBC assigns improved U-Values to optimize the energy performance of new residential buildings, the current performance levels of existing residential buildings fall significantly below these upgraded SBC standards. The upgraded SBC energy standards, considered the best minimum requirements for Zone 1, can serve as a baseline reference for any energy upgrade solutions. Consequently, energy upgrade measures through retrofitting strategies should align with the current SBC U-Values.

Table 6. The minimum U-Values and R-Values for air-conditioned spaces and non-air-conditioned spaces (C.I. (continuous insulation), C (C-Value (Thermal conductance)), F (F-Factor (heat transfer through the floor)) [36].

SBC Required U-Values and R-Values for Zone 1	With ACs		With No ACs	
	U-Value (W/m^{2}K)	R-Value (m^{2}K/W)	U-Value (W/m^{2}K)	R-Value (m^{2}K/W)
Ceiling	U-0.202	R-5.0 C.I	U-0.4	R-2.5 C.I
Wall				
Wall above ground	U-0.342	R-2.92 C.I	C-0.453	R-2.2 C.I
Wall under ground	C-6.473	R-2.92 C.I	C-6.473	None
Floor				
All	U-0.496	R-1.5 C.I	U-0.78	R-0.7 C.I
Steel beam	U-0.296	R-3.3	U-0.296	R-3.3
Other	U-0.188	R-5.3	U-0.288	R-3.3
Ground flooring	F-0.90	R-2.6 60cm	F-1.263	None
Doors	U-2.839		U-2.839	
Windows				
All connection	U-2.668	SHGC-0.25	U-3.695	None
Menwar (shaft)	U-4.259	SHGC-0.35	U-10.22	SHGC-0.35

Interestingly, the upgraded SBC allows for lower U-Values in non-air-conditioned spaces compared to air-conditioned spaces. This suggests that designing passive housing units may require higher U-Values if appropriately designed. The question arises as to who is responsible for verifying whether a housing unit is designed passively or not. However, the building code could define the minimum R-Values or U-Values requirements, while specifying whether the space is air-conditioned or not. Additionally, it needs to consider the possibilities of manipulation, especially given that most buildings in Jeddah require AC systems.

Cost plays a crucial role in the decision-making process when considering energy-saving measures. The ideal solutions are those that incur minimal costs, while offering efficient intervention measures.

The study's primary objective is to establish a framework of potential energy retrofitting interventions that meet, at the very least, the upgraded Saudi Building Code (SBC) energy standards for existing low-rise residential buildings in Jeddah.

3.4. *Jeddah's Climatic Challenges*

Jeddah experiences the highest number of Cooling Degree Days (CDDs) per year compared to other cities in Saudi Arabia, totaling 6587 CDDs, as shown in Table 7 [37].

Table 7. Cooling and heating degree-days for five cities in KSA.

City	Cooling Degree Days (CDDs) (°C-Days)	Heating Degree Days (HDDs) (°C-Days)
Jeddah	6587	0
Dhahran	5953	142
Riyadh	5688	291
Tabuk	4359	571
Abha	3132	486

Jeddah has a hot, dry climate with a maritime desert subzone [38]. The city's temperature ranges from a maximum of 48 °C to a minimum of 13 °C, with varying relative humidity levels, as explained in more detail in Table 8 [32].

Table 8. Temperatures and humidity levels in Jeddah city, KSA.

	Jan	Feb	Mar	Apr	May	June	July	Aug	Sep	Oct	Nov	Dec
Max. Temperature	32	35	39	42	42	48	45	41.5	42	43	38	36.5
Min. Temperature	13	15.4	18	19	20	23.4	24.8	25	23.8	20	20	17
Relative Humidity	59	56	60	58	56	58	49	52	66	61	65	51

Alaidroos and Al-Hadhrami highlighted that residential buildings in Jeddah exhibit exceptionally high cooling consumption, accounting for 71% of the total energy consumption, and have the highest number of Cooling Degree Days [18,39]. Felimban also emphasized that buildings without thermal insulation have a negative impact on occupants' energy behavior [1]. As a result, hot outdoor air easily penetrates indoor spaces, leading to increased indoor temperatures and prolonged usage of mechanical cooling systems.

3.5. *Overview of the Existing Residential Building Stock*

3.5.1. Residential Building Stock in KSA

The primary objective of this section is to identify representative building typologies that can serve as a foundation for defining design parameters for energy upgrade strategies for building envelopes. The aim is to gather relevant information to accurately assess the current energy performance of residential buildings in terms of common building types, the housing unit population, the Sakani (My House) financial support program, changes in low-rise building regulations, and ownership types.

In Saudi Arabia (KSA), low-rise residential buildings constitute approximately 50% of the overall building stock (including commercial, governmental, agricultural, and industrial buildings) [1]. Apartment units account for around 2.9 million units, representing approximately 53% of KSA's residential buildings [3]. In Makkah province, where Jeddah is located, there are approximately 1 million housing units.

The Sakani program aims to increase housing ownership from 47% to 70% by 2030. The program offers various housing solutions, including providing housing units or financial support for acquiring the first housing unit. Established in 2017, Sakani was launched to expedite homeownership among Saudi families. By August 2021, Sakani had successfully accommodated over 1.4 million households (see Figure 5), encompassing a range of housing products, such as residential free lands with loans, market unit loans, self-construction loans, under-construction unit loans, ready-made unit loans, subsidized mortgage transfers, loans for military members, loans for civilians, and loans for education members. To achieve the strategic goal, 40% more housing units are currently in the delivery process [40]. The majority of apartment buildings in KSA are low-rise structures, typically comprising 3–5 floors and classified as residential. Low-rise residential buildings in KSA are categorized into two types: purely residential or residential combined with commercial spaces, based on land use standards.

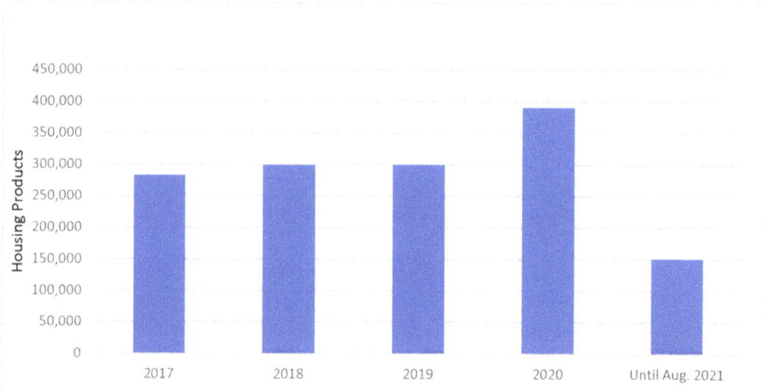

Figure 5. Comparison of Sakani Housing Products (1.4 Million) from 2017 to 2021.

The regulations for low-rise residential buildings have undergone two changes, resulting in different building use and ownership categories (old, recent, and new), as illustrated in Figure 6. The first upgrade mandated building owners to allocate parking spaces on the ground floor, while allowing for the construction of a villa on the rooftop, which this study categorizes as the recent type. This regulation change introduced multi-ownership to the building, although historically, potential homeowners were reluctant to purchase an apartment with shared ownership.

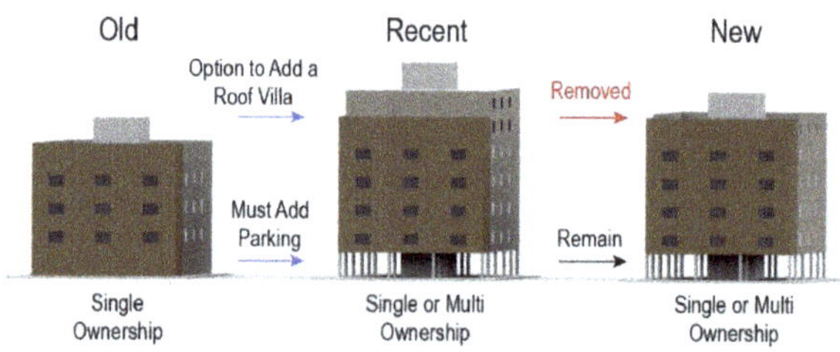

Figure 6. The regulation changes affect the building ownership types.

The regulation change offered additional yearly income for owners who constructed a villa on the roof and leased the remaining apartments. Consequently, during this period, many investors constructed low-rise residential buildings with villa roofs and sold them as separate entities (apartments and a roof villa), leading to the trend of multi-ownership within a single building.

Subsequently, low-rise building regulations were updated, and the construction of a villa on the roof was prohibited, although parking requirements remained. In summary, these regulation changes have impacted ownership types, shifting from single ownership to multi-ownership, which was unacceptable two decades ago.

The improvement in multi-ownership management can be attributed to the implementation of updated building regulations and the introduction of the Mullak program in February 2020. This study primarily focuses on low-rise residential buildings and investigates both multi-ownership and single-ownership scenarios to propose practical solutions. Figure 7 depicts the various types of low-rise residential buildings found in Jeddah, considering different construction ages. Despite variations in construction ages, these buildings exhibit comparable energy performance due to their similar construction approaches. However, discrepancies in ownership types may influence the proposed results, which will be discussed further in the subsequent text.

Figure 7. Different types of low-rise residential buildings illustrate the types of construction and ownership.

Several researchers have explored Saudi residential building characteristics from various perspectives, taking into account factors such as location and user profiles. Aldossary identified several prototypes based on official construction plans, while Alaidroos described construction methods and HVAC specifications for a base case villa in KSA. However, this study focuses specifically on 3–5 floor residential buildings with both multi-ownership and single ownership, excluding villas on the roof due to the latest SBC upgrade regulations. The primary emphasis is on multi-ownership buildings to accommodate more housing units, although single-ownership scenarios are also considered. The categorization of the selected residential buildings is based on construction methods and materials used, rather than their historical construction dates.

3.5.2. Common Construction Method

Usually, reinforced concrete is the most commonly used material in residential building construction due to its availability, affordability, and durability. Skeleton structure systems, also known as frame structures, are the predominant construction method employed in low-rise residential buildings in KSA [41]. The construction phases of a low-rise residential building generally follow typical steps, as depicted in Figure 8.

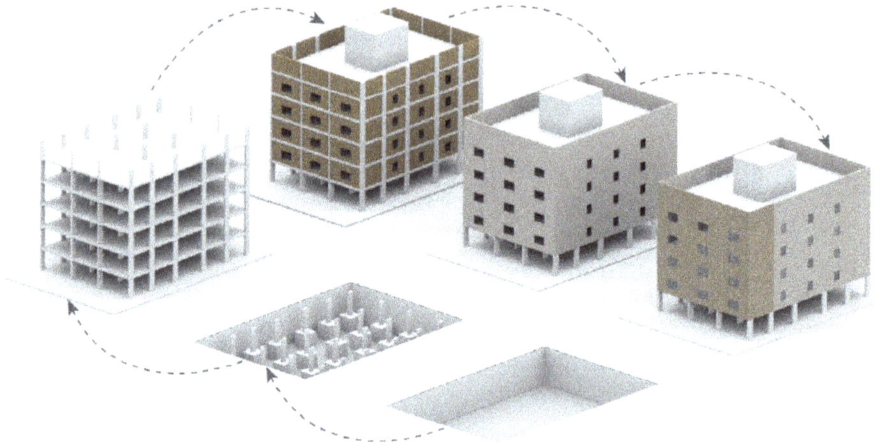

Figure 8. Typical construction method.

The construction process for low-rise residential buildings begins with the underground phase, which involves drilling and column foundation. The most common system used for low-rise buildings in KSA is the concrete skeleton structure, utilizing reinforced concrete for its affordability, availability, and durability. The construction of the skeleton structure follows specific procedures from the foundation to the roof slab.

The next phase is the block walls (wall-filling), where blocks are used to fill the spaces between the skeleton columns, as indicated by the red walls in Figure 9. The choice of material for wall-filling depends primarily on the allocated budget, with options ranging from cement blocks, red blocks, and Burkani blocks to the less common Siporex blocks. Each type of wall-filling material has different properties and thermal conductivity levels. Typically, walls are constructed using 20 cm blocks, as shown in Figures 9 and 10.

Figure 9. The 20 cm blocks for wall-filling within the skeleton structure.

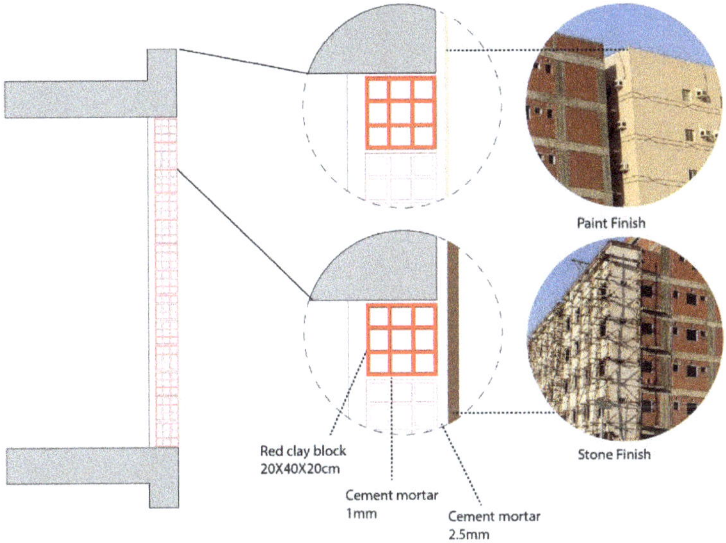

Figure 10. Typical wall section demonstrating materials used in construction.

The third phase involves cement finishing (mortar), which entails adding a 2 cm layer of mortar to both sides of the blocks, as depicted in Figure 9. The final phase involves the overall finishing of the building envelope through painting or the application of stone. The choice of painting or decorative elements depends on the budget of the building owner. Some owners with higher budgets may opt to add different types of stone to the front façade for aesthetic purposes, as shown in Figures 9 and 10.

Understanding the typical construction materials used in residential buildings helps define and clarify the available materials for each targeted envelope component within the Jeddah context [41]. These materials serve as a starting point for energy-upgrading applications, such as replacement or upgrading solutions for walls, insulation, windows (frames, glazing, and panes), and sealants, supporting the design possibilities. Additionally, this section discusses the materials related to existing residential buildings, as these materials play a significant role in outdoor heat gain. As mentioned earlier, the construction method encompasses four stages: skeleton structure, wall-filling, cement mortar, and finishing. Each stage utilizes various materials. The subsequent sub-section and Figure 10 explain the material variations for wall-filling (blocks), insulation, finishing, windows, and sealants.

A—Wall Fillings (Blocks)

According to Felimban, approximately 70% of residential buildings in Jeddah lack insulation [1,2]. In general, the walls of typical residential buildings consist of single bricks (red blocks, cement blocks, Burkani blocks, or Siporex blocks) covered with a 2 cm cement layer of mortar on both sides.

Walls with openings incorporate window frames with single glazed panes, as discussed in a later sub-section. The current thermal energy levels of the walls present an excellent opportunity for energy performance upgrades. Any energy upgrade scenario should consider the types of blocks commonly used in Jeddah's residential buildings. Wall upgrade interventions could involve increasing thickness by adding insulation layers and an air gap to the indoor space. All scenarios involve tightening the indoor space, while addressing thermal bridges resulting from connections, which is another challenge that needs to be considered. Figure 11 illustrates the commonly used wall materials, including cement blocks with holes, red blocks, Burkani blocks, and Siporex blocks. Since 2021, block factories have been required to meet specific thermal properties to comply with the upgraded SBC standards (see Table 9).

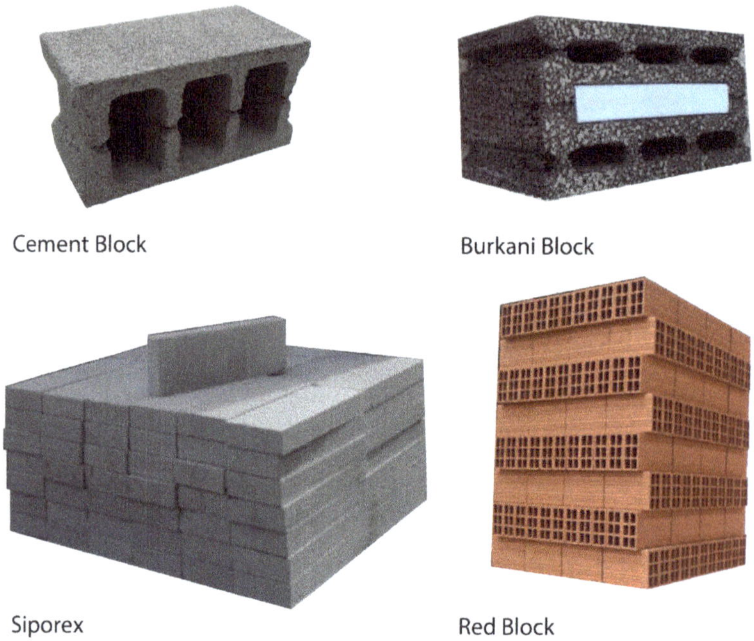

Figure 11. Various types of building blocks.

Table 9. Wall materials U-Values and R-Values [42].

Wall Materials (20 cm)	K-Value W/mK	R-Value m^2K/W	U-Value W/m^2K
Cement Block with Holes	0.976	0.204918	4.880
Red Block	0.382	0.52356	1.910
Burkani Block	0.36	0.555556	1.800
Autoclaved Aerated Concrete (Siporex)	0.156	1.282051	0.780

B—Insulation Materials

In hot climate regions, thermal insulation serves the purpose of resisting the penetration of hot air from the outside to the inside. Insulation layers are applied to the walls and roof surfaces, and they can be categorized as conventional (commercially available organic or inorganic products) or sustainable (natural or recycled) materials [43]. As mentioned earlier in this study, most residential buildings in KSA lack insulation, which significantly hinders energy efficiency by allowing outdoor heat transfer to indoor spaces. Commonly used and available insulation materials in KSA include polystyrene, polyurethane, mineral wool, glass wool, Perlite, and Siporex, as shown in Figure 12 [42,43]. Various factors, such as availability, cost, installation difficulty, soundproofing, and fire resistance, must be considered in the design process to select a suitable insulation material.

C—Finishing

The finishing layers typically involve a 2–2.5 cm cement layer of mortar on both sides of the wall, followed by painting or the application of stone on the main façade for aesthetic purposes. The thermal conductivity of mortar finish is 0.72 W/mK. When it comes to painting or stone selection, it largely depends on the building owner's budget. Some owners may have a higher budget and can incorporate different types of stone on the front façade for both aesthetic and thermal performance purposes. It is worth noting that certain painting companies now offer thermally resistant options, although these options come at a higher price compared to other materials. In energy upgrade scenarios, the finishing layer options, especially when a stone finish is involved, require careful consideration and planning in terms of the cost and time required for stone disassembly.

Figure 12. Various types of insulation materials.

D—Windows

Upgrading windows plays a significant role in enhancing energy efficiency, despite accounting for a maximum of 25% of the total façade area [44]. Window frames, glazing, and panes greatly influence the transfer of outdoor heat into indoor spaces through gaps between the frames and the wall, as well as within the window frames themselves. Additionally, consideration should be given to the Solar Heat Gain Coefficient (SHGC) through window panes. Aluminium frames are typically used for their durability and market availability. In recent years, the use of unplasticized Polyvinyl Chloride (uPVC) frames has increased due to their competitive prices and availability. Moreover, uPVC has demonstrated higher thermal resistance compared to other materials, such as aluminium or steel. Timber and steel frames are rarely used in window construction due to their high cost and limited availability, and thus, they are excluded from the scope of this study.

There are different types of window glazing and panes with varying U-Values and SHGC, as illustrated in Table 10. The thermal resistance levels of window glazing depend on factors such as the number of panes, color, and cavity size between the panes.

Table 10. Window glazing types and heat transfer level, LoE (Low Emissivity) [18].

SBC Requirement for Glazing	2.66	0.25
Glazing Type	U-value (W/m^{2}K)	SHGC
Single Clear	6.31	0.86
Double Clear Air	3.23	0.76
Double Clear Argon	2.61	0.76
Double LoE Clear Air	2.47	0.6
Double LoE Clear Argon	1.48	0.59
Double LoE Tint Air	2.43	0.39
Double LoE Tint Argon	1.46	0.37
Double LoE SEL Clear Air	2.32	0.42
Double LoE SEL Clear Argon	1.3	0.42
Double LoE SEL Tint Air	2.32	0.3
Double LoE SEL Tint Argon	1.3	0.28

E—Sealants

Sealant materials are used to fill the gaps between the wall (blocks) and the window frames, preventing the infiltration of air and water into indoor spaces [8]. In KSA, sealants play a significant role in mitigating outdoor heat transfer to indoor areas. Lower-quality sealants result in higher heat transfer to indoor spaces, leading to increased mechanical cooling demand. Sealant compositions are categorized as silicone sealants, hybrid polyurethane sealants, and polyurethane sealants.

4. Design Parameters and Energy Benchmark

This section aims to define the design parameters and energy consumption benchmark levels to establish the study scope and enable energy upgrade scenarios.

Key performance indicators (KPIs) are crucial for evaluating the scenarios. The primary KPI is the Annual Average Energy Consumption (AAEC) in kWh/m^2/year, which serves as a benchmark for comparing various upgrade scenarios to the original case. The evaluation of each apartment considered at least one of the following upgrades: U-values (W/m^{2}K), thickness (cm), infiltration rate (ACH50), Solar Heat Gain Coefficient (SHGC) and Window-to-Wall Ratio (WWR) for windows, and Coefficient of Performance (COP) for AC units.

The hot, arid, humid climate conditions in Jeddah necessitate mechanical systems in all indoor spaces, a finding supported by other researchers, such as Felimban and Alaidroos [1,18]. In addition, it is essential to consider the building location within the neighborhood, as this could also affect the AAEC for each apartment in the building. Additionally, the location of the building within the neighborhood must be taken into account, as it can also impact the AAEC for each apartment. The selection of building types was based on the number of housing units and the new buildings developed by the KSA Ministry of Housing. Apartments account for more than 50% of the total housing units in KSA [45].

This research focused primarily on buildings constructed using the concrete skeleton structure (CSS), as this is the predominant construction method in Jeddah. The historical aspect of the buildings was not considered. Moreover, the construction process of a CSS involves wall infilling with blocks, plaster/cement finishing, and aesthetical finishing. The Saudi Energy Efficiency Centre (SEEC) and Felimban highlight that over 70% of residential buildings in Jeddah lack thermal insulation, underscoring the need for energy retrofit upgrades of the existing building envelopes [1,46].

The Saudi Building Code (SBC) has upgraded its energy efficiency requirements and energy benchmarks. However, in February 2022, the SBC National Committee revised the energy efficiency requirements in response to feedback from construction companies and their reluctance to issue new construction permits for residential buildings. Table 11 illustrates the specific value changes in energy requirements resulting from the SBC upgrades by the National Committee.

Table 11. SBC energy requirements upgrades (Bold numbers indicate the changes) [36,44,47,48].

	Wall Construction (U-Value)	Roof Construction (U-Value)	Ground Floor Construction (U-Value)	Repeated Floor Construction (U-Value)	Window Glazing	Window Frame (U-Value)		(WWR) Max	Air Infiltration (ACH 50)	HVAC System Efficiency (COP)
Started to be applied by 1 July 2021	**0.342**	**0.202**	**0.49**	0.49	(U-value) **2.66** SHGC **= 0.25**	**2.66**		**25%**	**4**	**4**
Updated on 23 August 2021	**0.403**	**0.272**	0.49	0.49	(U-value) 2.66 SHGC = 0.25	2.66	Rate	25%	4	4
Updated 21 February 2022; ends by the end of 2023	**0.611**	0.272	0.49	0.49	(U-value) 2.66 SHGC = 0.25	2.66		25%	4	4

Aldossary's research established AAEC values for different residential buildings in KSA, albeit only covering the first two floors of low-rise residential building types [32].

Unfortunately, the top floors of buildings require more extensive energy upgrade interventions to achieve better performance, as they are more exposed to the sun's heat and radiation due to additional external surfaces. Researchers have observed AAEC values ranging from 116 to 165 kWh/m^2/year, which are predicted to be even higher for top floors. However, Aldossary proposed AAEC values in the range of 77–98 kWh/m^2 to achieve a low carbon energy consumption level [7].

Various researchers, including Aldossary, Alaidroos, Krarti, and Hijazi, have explored different sets of energy retrofit measures that can reduce energy consumption in the residential building sector by 37%, 41.5%, 50%, and up to 80% when applying hybrid systems [18,32,49,50]. The literature often presents optimistic predictions of energy savings for existing buildings when implementing different energy-saving measures. In this study, detailed energy retrofit scenarios have been defined to provide a more realistic estimation of energy-saving possibilities for specific units in Jeddah. Additionally, factors such as infiltration rate (ACH50) and user thermal comfort temperature (°C) have been included, as they impact the AAEC results. However, the calculation did not consider thermal bridges due to the complexity of methodologies required to obtain accurate results [51].

5. Case Study Descriptions and Simulation Process

5.1. Building Location and Position Selection

The selected case study was the residential building described in detail in Section 3.4. Jeddah's climate and location have been described in considerable detail in Section 3.4 and by Felimban, Talep, and Aldossary [1,32,52]. The building position that was eventually selected was based on simulation testing of six positions of a typical building in an urban setting. Then, the worst case was selected, where the average energy consumption was the highest. This will be further shown in the simulation progress section.

5.2. Building and Apartment Descriptions

Generally, the land area for a residential building varies among 20 m × 20 m, 20 m × 30 m, 25 m × 30 m, and 30 m × 30 m, with a built-up ratio maximum of 60% [36,53]. The building case was extracted from actual plans of a low-rise residential building (4 floors) provided by an architectural firm [54]. However, the case is based on a land size of 750 m^2 (25 m × 30 m), resulting in a built-up floor area of around 450 m^2. The selected building contains eight apartments (two per floor), and the first floor (ground floor) has parking spaces and other services, such as driver rooms and the main entrance. The apartments mainly face either west or east. However, the east and north sides face the neighboring buildings, while the west and south sides face the street. These factors have an effect on the AAEC for each apartment. Each apartment has three bedrooms, a living room, a kitchen, a dining room, a reception room, a maid room, and three bathrooms, as shown by the floor plans in Figures 13–15.

Figure 13. First floor plan (ground floor) (14 Parking spots, 6 Driver rooms, and 1 guest room).

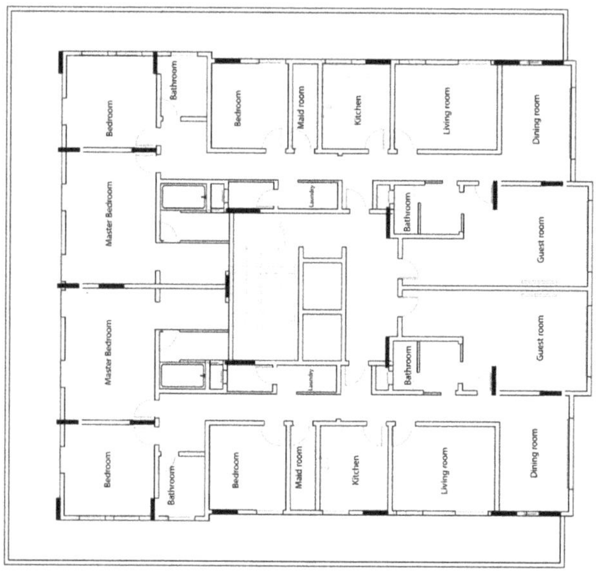

Figure 14. Repeated floor plan (3 bedrooms, 1 maid room, 1 kitchen, 1 living room, 3 bathrooms, 1 dining room, and 1 guest room).

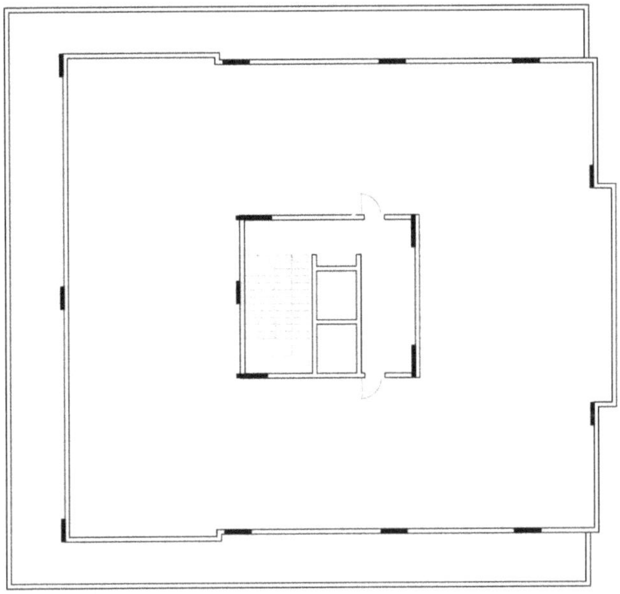

Figure 15. Roof floor plan.

The building fabric was defined and illustrated based on previous studies and material properties. Tables 12 and 13 demonstrate every component in respect to total U-values, component thickness, and other variables. The apartments on the east side of the building have the same floor area, which is around 215 m^2, while on the west side, the area is around 225 m^2.

Table 12. Building specifications.

	Number of Floors	Total Number of Apartments	Area of Apartments	Building Location	Total Number of Occupants in the Building	Cooling Set Point	Cooling Set Back
Description	4 floors + parking floor (parents (2) + kids (4) + a housemaid)	8 apartments (2 per floor)	West 215 m^2 East 225 m^2	Jeddah (South East)	56 occupants (parents (2) + kids (4) + a housemaid)	24 °C	26 °C

Table 13. Building fabric description and current energy values of building components.

Building Component	Detailed Description	Thickness (mm)	U-Value (W/m^2K)
Wall Construction	20 mm cement/plaster/mortar inside + 200 mm concrete block heavy weight + 20 mm cement/plaster/mortar outside	240	2.676
Roof Construction	20 mm ceramic/porcelain top side + 20 mm mortar + 80 mm sandstone, 1.83 W/mk + 5 mm asphalt1 + 200 mm concrete, reinforced with 1% steel + 20 mm plaster bottom	345	2.81
Ground Floor Construction	25 mm ceramic/porcelain top side + 25 mm mortar + 80 mm sandstone, 1.83 W/mk+ 100 mm concrete, reinforced with 1% steel + 5 mm asphalt1 + 50 mm cast concrete + 150 mm stone basalt + 2 mm soil–earth	437	2.269
Repeated Floor Construction	25 mm ceramic/porcelain top side + 25 mm mortar + 80 mm sandstone, 1.83 W/mk + 200 mm concrete, reinforced with 1% steel + 20 mm plaster bottom	350	2.403
Window Glazing	Single-clear (SHGC = 0.86)	3	5.894
Window Frame	Aluminum frames	5	5.881

Table 13. *Cont.*

Building Component	Detailed Description	Thickness (mm)	U-Value (W/m^{2}K)
	Details	Rate	
(WWR)	The percentage of the total window area to total wall area	10%	
Air Infiltration Rate	The assumed rate is based on the blower door test (BDT) rate, which assumes that the indoor area is pressurized under 50 PA	20 (ACH 50)	
HVAC System Efficiency	AC window type	1.8 COP	

5.3. User Profile

In the real world, every apartment has a different user profile, while in this example, specific information has been used to create a basis, against which other apartments can be compared. The typical number of users in an apartment is 7, including a housemaid; the average family size is 5.9 members [45,55]. The activity in the apartment varies depending on the parents' professions. However, in this study, it is assumed that user activities are based on a proposed schedule of activities and AC working duration hours, as demonstrated in Table 14. Furthermore, every room has a different number of hours during which the AC is used; the living room proved to be the most active room, with usage of 17.5 h per day, and the guest room was the least active room, using an average of 3 h per day, as Figure 16 illustrates.

Table 14. User activity schedule for a case model of a Saudi Family.

Activity	Sunday	Monday	Tuesday	Wednesday	Thursday	Friday	Saturday	Total Hours/ Room/Week	Total Hours/ Room/Month	Average Hours/ Room/Day
Master Bedroom	23:00– 06:00	23:00– 06:00	23:00– 06:00	23:00– 06:00	23:00– 06:00	23:00– 06:00	23:00– 06:00	49	210	7
Children's Bedroom 1	21:30– 06:00	21:30– 06:00	21:30– 06:00	21:30– 06:00	23:30– 08:00	23:30– 08:00	21:30– 06:00	59.5	255	8.5
Children's Bedroom 2	21:30– 06:00	21:30– 06:00	21:30– 06:00	21:30– 06:00	23:30– 08:00	23:30– 08:00	21:30– 06:00	59.5	255	8.5
Housemaid's Bedroom	23:30– 07:00	23:30– 07:00	23:30– 07:00	23:30– 07:00	23:30– 07:00	23:30– 07:00	23:30– 07:00	52.5	225	7.5
Dining Room	06:30– 07:30 16:00– 17:30 20:30– 22:00	06:30– 07:30 16:00– 17:30 20:30– 22:00	06:30– 07:30 16:00– 17:30 20:30– 22:00	06:30– 07:30 16:00– 17:30 20:30– 22:00	06:30– 07:30 16:00– 17:30 21:30– 23:00	06:30– 07:30 16:00– 17:30 21:30– 23:00	06:30– 07:30 16:00– 17:30 20:30– 22:00	7 10.5 10.5	120	4
Living Room	06:00– 23:30	06:00– 23:30	06:00– 23:30	06:00– 23:30	06:00– 23:30	06:00– 23:30	06:00– 23:30	122.5	525	17.5
Kitchen	06:00– 07:30 14:00– 18:00 20:00– 23:00	06:00– 07:30 14:00– 18:00 20:00– 23:00	06:00– 07:30 14:00– 18:00 20:00– 23:00	06:00– 07:30 14:00– 18:00 20:00– 23:00	06:00– 07:30 14:00– 18:00 20:00– 23:00	06:00– 07:30 14:00– 18:00 20:00– 23:00	06:00– 07:30 14:00– 18:00 20:00– 23:00	10.5 28 21.00	255	8.5
Guest Room	None	None	None	None	17:00– 24:00	17:00– 24:00	17:00– 24:00	21	90	3

The provided assumed activity hours were the minimum duration hours that varied among families. However, a compact schedule, i.e., a schedule where the people who lived in the house were there for the maximum number of hours, was taken as the basis to use later in the simulation program (DesignBuilder). The assumed schedule was applied to all the apartments to provide comparable numbers that could subsequently be validated. The occupancy percentage was 20% during the inactive hours (07:00–16:00).

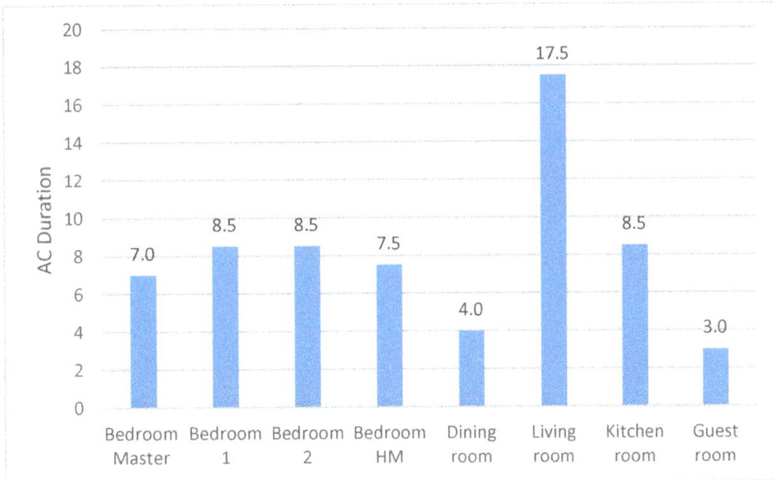

Figure 16. Comparison of the average AC duration for different rooms per day.

5.4. Building Ownership

The ownership of a residential building was primarily only single ownership until it developed into a multi-ownership model. In 2018, the "Mullak" ownership system was introduced to settle the required rules for single- and especially multi-ownership types of apartments [56]. In this study, the ownership of a building has a significant role in designing the energy retrofitting scenarios, which have been divided into single-ownership or multi-ownership types.

Typically, the construction of any residential building falls within three types of constructors: individual, private developer, or governmental. Each type has different business activities that fulfil the construction's primary goal. Therefore, the type of ownership falls under single- or multi-ownership, as Table 15 illustrates.

Table 15. Different business activities for several building contractors.

Contractor	Individual	Private Developer	Governmental	Ownership Type
Business Activities	Selling	Selling	Selling	Multi
	Living + Selling			
	Renting for Short Term	Renting for Short Term	Renting for Long Term	Single
	Living + Renting for Short Term			

The energy retrofitting scenarios have been divided into two primary types: indoor and outdoor. The indoor scenarios are possible for both ownership types, while the outdoor scenarios are only possible for the single-ownership type because of difficulties in the decision-making processes.

5.5. Simulation Description

The selected software was DesignBuilder [57], which allows engineer researchers to analyze the energy consumption of building energy. However, a comparative study of widely used dynamic simulation tools for buildings, such as EnergyPlus, TRNSYS, Simulink libraries CarnotUIBK and ALMABuild, IDA ICE, Modelica/Dymola, and DALEC, demonstrated a good consensus among these tools, despite the varying levels of input detail required by each tool [58]. The Design-Builder tool was chosen due to its availability in the market and its accessibility as a simulation software. It allows for the analysis and prediction of energy consumption in any structure using predefined datasets. The Design-Builder program is particularly user-friendly, making it suitable for educational purposes. It eliminates the need to extensively delve into software details and codes. The main features of using the Design-Builder software are its ability to simulate accurate environmental performance data, its fast simulation capabilities, and its ability to import various file types for 2D and 3D imaging. Additionally, one can save rendered images of any result at any stage [58,59].

The study modelled the case study in the DesignBuilder software using the collected actual floor plans from the Archteam firm. The data were entered based on previous studies described earlier in this section.

Initially, the floor plans were extracted from the provided documents, and a 3D model was constructed using the DesignBuilder software. The wall specifications were then added based on Table 13, which was derived from Table 5 and other relevant literature. Subsequently, the window and roof specifications were incorporated. Afterward, various datasets were inputted, including ACH50 (N50), setback temperature, climate data, and activity data. The simulation was then conducted to obtain annual energy consumption data, which were stored in an Excel file. The simulation covered 8 apartments, each with 17 scenarios (10 indoor and 7 outdoor), resulting in a total of 272 simulations per trial.

Due to various uncertainties, the simulation was repeated multiple times, accounting for factors such as the actual infiltration rate and the AC setback temperature, which are further elaborated upon in the subsequent sections. Each scenario's simulation time ranged up to 7 s. The primary objective of using AAEC (Annualized Average Energy Consumption) was to compare the energy consumption before and after implementing the upgrading measures for all eight apartments within a single building.

6. Energy Retrofitting Scenario Description

The available energy retrofit interventions were described in the previous study as a guideline for designing the energy retrofitting scenarios in this section. Table 16 illustrates every energy upgrade scenario, as it shows the interventions used. In addition, Tables 17 and 18 illustrate the scenarios designed to achieve the SBC requirements, where red colors mean that the value did not meet the SBC energy requirements, while green means that the value did meet the SBC energy requirements. The concept achieves high-resolution scenarios by starting with minimal changes and adding additional intervention to reach an efficient scenario that meets the SBC (green labels in Tables 17 and 18). The design was divided into two categories, indoor scenarios and outdoor scenarios, and these are described as follows.

Table 16. Overview of indoor and outdoor scenarios. Detail for the scenario construction in Tables 17 and 18.

Indoor Scenarios	
Base Case	Base Case Corner Face SW +SB + ACH50 4
Scenario 1	Mortar Finishing + Replace Windows (Creative Windows CO.)
Scenario 2	Wall (EPS 5 cm) + Cement Hollow Block (10 cm) + Mortar Finishing
Scenario 3	Wall (XPS 5 cm) (HFC) + Mortar Finishing
Scenario 4	Wall (XPS 7.5 cm) (HFC) + Mortar Finishing
Scenario 5	Wall (XPS 10 cm) (HFC) + Mortar Finishing
Scenario 6	Wall (XPS 10 cm) (HFC) + Mortar Finishing + Replace Windows (Wintek HD Plus Gray)
Scenario 7	Wall (XPS 10 cm) (HFC) + Mortar Finishing + Replace Windows (Creative Windows CO.)
Scenario 8	Wall (XPS 10 cm) (HFC) + Mortar Finishing + Replace Windows (Creative Windows CO.) + Roof XPS 10 cm
Scenario 9	Wall (XPS 10 cm) (HFC) + Mortar Finishing + Replace Windows (Creative Windows CO.) + Upgrade Roof with XPS 15 cm

Table 16. *Cont.*

Indoor Scenarios	
Scenario 10	Wall (XPS 10 cm) (HFC) + Mortar Finishing + Replace Windows (Creative Windows CO.) + Upgrade Roof with XPS 10 cm + Replace ACs with COP 4
Outdoor Scenarios	
Base Case	Base Case Corner Face SW + SB
Scenario 1	EIFS Wall (EPS 10 cm)
Scenario 2	EIFS Wall (XPS 10 cm)
Scenario 3	EIFS Wall (XPS 10 cm) + Replace Windows Wintek HD Plus Grey
Scenario 4	EIFS Wall (XPS 10 cm) + Replace Windows (Creative Windows CO.)
Scenario 5	EIFS Wall (XPS 10 cm) + Replace Windows (Creative Windows CO.) + Upgrade Roof with XPS 10 cm
Scenario 6	EIFS Wall (XPS 10 cm) + Replace Windows (Creative Windows CO.) + Upgrade Roof with XPS 15 cm
Scenario 7	EIFS Wall (XPS 10 cm) + Replace Windows (Creative Windows CO.) + Upgrade Roof XPS 10 cm+ Replace ACs with COP 4

Table 17. Indoor energy retrofit scenarios for a residential building in Jeddah (red color indicates did not meet the SBC and green color indicates the value meets the SBC).

	Base Case	Scenario 1	Scenario 2	Scenario 3	Scenario 4	Scenario 5	Scenario 6	Scenario 7	Scenario 8	Scenario 9	Scenario 10
N50 (ACH50)	20.00	4.00	4.00	4.00	4.00	4.00	4.00	4.00	4.00	4.00	4.00
Wall (U-Value) (W/m^{2}K)	2.68	2.68	0.53	0.49	0.35	0.27	0.27	0.27	0.27	0.27	0.27
Thickness (m)	0.24	0.24	0.39	0.29	0.32	0.34	0.34	0.34	0.34	0.34	0.34
Roof (U-Value) (W/m^{2}K)	2.81	2.81	2.81	2.81	2.81	2.81	2.81	2.81	0.27	0.19	0.27
Thickness (m)	0.345	0.345	0.345	0.345	0.345	0.345	0.345	0.345	0.445	0.495	0.445
G-Floor (U-Value) (W/m^{2}K)	2.27	2.27	2.27	2.27	2.27	2.27	2.27	2.27	2.27	2.27	2.27
Thickness m	0.44	0.44	0.44	0.44	0.44	0.44	0.44	0.44	0.44	0.44	0.44
R-Floor (U-Value) (W/m^{2}K)	2.40	2.40	2.40	2.40	2.40	2.40	2.40	2.40	2.40	2.40	2.40
Thickness (m)	0.35	0.35	0.35	0.35	0.35	0.35	0.35	0.35	0.35	0.35	0.35
Window Glazing Type	Single 3 mm	6 mm–12 mm air-6 mm	Single 3 mm	Single 3 mm	Single 3 mm	Single 3 mm	6 mm–12 mm air-6 mm	6 mm–12 mm air-6 mm	6 mm–12 mm air-6 mm	6 mm–12 mm air-6 mm	6 mm–12 mm air-6 mm
Window Glazing (U-Value) (W/m^{2}K)	5.89	2.13	5.89	5.89	5.89	5.89	2.69	2.13	2.13	2.13	2.13
Glazing Type	Clear	Gray reflective SHANG-HAI	Clear	Clear	Clear	Clear	Gray hd plus	Gray reflective SHANG-HAI	Gray reflective SHANG-HAI	Gray reflective SHANG-HAI	Gray reflective SHANG-HAI
Window Glazing (SHGC)	0.86	0.25	0.86	0.86	0.86	0.86	0.25	0.25	0.25	0.25	0.25
Window Frame Type	Aluminum	UPVC Creative Windows	Aluminum	Aluminum	Aluminum	Aluminum	UPVC wintek	UPVC Creative Windows	UPVC Creative Windows	UPVC Creative Windows	UPVC Creative Windows
(U-Value) (W/m^{2}K)	5.88	1.33	5.88	5.88	5.88	5.88	1.79	1.33	1.33	1.33	1.33
Window Ratio (WWR)	0.10	0.10	0.10	0.10	0.10	0.10	0.10	0.10	0.10	0.10	0.10
Lighting (W/m^2-100 lux)	5.00	5.00	5.00	5.00	5.00	5.00	5.00	5.00	5.00	5.00	5.00
AC Type	AC Window	AC Window	AC Window	AC Window	AC Window	AC Window	AC Window	AC Window	AC Window	AC Window	AC Split
(CoP)	1.80	1.80	1.80	1.80	1.80	1.80	1.80	1.80	1.80	1.80	4.00

Table 18. Outdoor energy retrofit scenarios for a residential building in Jeddah (red color indicates did not meet the SBC and green color indicates the value meets the SBC).

	Base Case	Scenario 1	Scenario 2	Scenario 3	Scenario 4	Scenario 5	Scenario 6	Scenario 7
N50 (ACH50)	20.00	4.00	4.00	4.00	4.00	4.00	4.00	4.00
Wall (U-Value) (W/m^{2}K)	2.68	0.31	0.27	0.27	0.27	0.27	0.27	0.27
Thickness (m)	0.24	0.36	0.36	0.36	0.36	0.36	0.36	0.36
Roof (U-Value) (W/m^{2}K)	2.81	2.81	2.81	2.81	2.81	0.27	0.19	0.27
Thickness (m)	0.35	0.35	0.35	0.35	0.35	0.45	0.50	0.45
G-Floor (U-Value) (W/m^{2}K)	2.27	2.27	2.27	2.27	2.27	2.27	2.27	2.27
Thickness (m)	0.44	0.44	0.44	0.44	0.44	0.44	0.44	0.44
R-Floor (U-Value) (W/m^{2}K)	2.40	2.40	2.40	2.40	2.40	2.40	2.40	2.40
Thickness (m)	0.35	0.35	0.35	0.35	0.35	0.35	0.35	0.35
Window Glazing Type	Single 3 mm	Single 3 mm	Single 3 mm	6 mm–12 mm air-6 mm	6 mm–12 mm air-6 mm	6 mm–12 mm air-6 mm	6 mm–12 mm air-6 mm	6 mm–12 mm air-6 mm
Window Glazing (U-Value) **(W/m^{2}K)**	5.89	5.89	5.89	2.69	2.13	2.13	2.13	2.13
Glazing Type	Clear	Clear	Clear	Gray hd plus	Gray reflective SHANGHAI	Gray reflective SHANGHAI	Gray reflective SHANGHAI	Gray reflective SHANGHAI
Window Glazing (SHGC)	0.86	0.86	0.86	0.25	0.25	0.25	0.25	0.25
Window Frame Type	Aluminum	Aluminum	Aluminum	UPVC wintek	UPVC Creative Windows	UPVC Creative Windows	UPVC Creative Windows	UPVC Creative Windows
Window Frame (U-Value) (W/m^{2}K)	5.88	5.88	5.88	1.79	1.33	1.33	1.33	1.33
Window Ratio (WWR)	0.10	0.10	0.10	0.10	0.10	0.10	0.10	0.10

Table 18. *Cont.*

	Base Case	Scenario 1	Scenario 2	Scenario 3	Scenario 4	Scenario 5	Scenario 6	Scenario 7
Lighting (W/m^2-100 lux)	5.00	5.00	5.00	5.00	5.00	5.00	5.00	5.00
AC Type	AC Window	AC Window	AC Window	AC Window	AC Window	AC Window	AC Window	AC Split
(CoP)	1.80	1.80	1.80	1.80	1.80	1.80	1.80	4.00

6.1. Indoor Scenarios

In Table 17, Scenario 1 involves the replacement of windows with an energy-efficient option. Scenarios 2–5 incorporate additional measures to enhance wall insulation with local materials to achieve the required SBC U-values. Scenarios 6 and 7 incorporate the wall upgrade aspect of Scenario 5, with the window replacement, while the only difference between Scenarios 6 and 7 is the type of windows used. Scenarios 8 and 9 follow the approach of Scenario 7 and upgrade the roof U-value with two distinct U-values. Finally, Scenario 10 builds upon Scenario 8 and replaces the air-conditioning systems with efficient alternatives.

6.2. Outdoor Scenarios

In Table 18, Scenarios 1 and 2 incorporate external insulation and finishing systems (EIFSs) as add-on measures to improve the U-value of the walls. Scenarios 3 and 4 build upon Scenario 2 and replace the windows. Scenarios 5 and 6 follow the approach of Scenario 4, including upgrading the roof U-values. Lastly, Scenario 7 incorporates the measures from Scenario 5, but also involves replacing the air-conditioning systems with energy-efficient alternatives.

Tables 17 and 18 demonstrate how and what the scenarios are. The central concept of designing the energy retrofit scenarios was to develop scenarios from a minor upgrade to a deeper upgrade using mixed energy retrofitting strategies (add-in, add-on, replace-it, and wrap-it) in order to reach the SBC energy requirements. The scenarios are intended to develop the targeted envelope component (wall, windows, and roof) to upgrade the heat-resistant value in order to achieve better performance. In both Tables 17 and 18, the green color indicates that the value reached the SBC minimum standards.

The infiltration rate was assumed as 20 ACH50, as recommended by Makawi, where higher results could be possible for the basic case [60,61]. The rationale for employing a value of 20 ACH50 to represent infiltration in simulation software is based on several factors. ASHRAE defines infiltration as the unintended flow of outdoor air into a building through cracks, openings, and exterior doors [61]. Airtightness is a related concept, referring to the amount of air infiltrating a building at a pressure difference of 50 Pa [62]. Infiltration and airtightness are distinct but related phenomena, with empirical evidence suggesting that infiltration is typically around 1/20th the value of airtightness [62].

The blower door test (BDT) is commonly used to measure airtightness by measuring air change rates under a 50 Pa pressure difference [61]. The resulting value, known as ACH50, is a measure of the infiltration of outdoor air into a building and is influenced by envelope tightness. Infiltration can contribute significantly to a building's heating and cooling loads, with estimates ranging from 25% to 50% in some studies [63,64]. Research has shown a wide range of ACH50 values in residential buildings, with values as high as 39 ACH50 in some cases [65]. However, the exact value will vary depending on various factors, including the type of window frames used.

In Saudi Arabia, a study found a lack of infiltration data on the building stock and recorded ACH50 values of 6.58 and 7.04 for 2 houses in Dhahran City [61]. These values were due to exfiltration caused by the central HVAC fan system. This study and other literature show that 20 ACH50 is not considered high for an existing residential building.

To validate the proposed energy retrofitting scenarios, a value of 20 ACH50 is used for the basic case model to improve this value to 4 ACH50, as required by the Saudi Building Code for the airtightness of residential buildings in Jeddah. This approach aligns with previous research recommendations and is based on a range of empirical evidence.

7. Results and Analysis

The energy performance simulation process follows three steps. The first step explores the highest average energy consumption of a residential building using different urban positions. The second step shows the different energy consumption results when the variables have been changed, such as the infiltration rate or how the temperature in the various rooms is controlled, which will later affect the possible energy-saving results. The third step involves performing an energy simulation for each proposed scenario in order to calculate the potential energy savings. Hence, every step will provide significant information that will help analyze the simulation results using different variables.

7.1. Step One: Building Position (Locating the Highest Energy Consumption)

In the KSA context, it is possible for a residential building to be situated in six different positions when the alone (no surrounding buildings) position faces towards the south (see Figure 17). The southwest (SW) position (see Figure 18) recorded the highest AAEC compared to other positions, as shown in Figure 19. The north position is found to have almost the same AAEC as 180 kWh/m^2/year (the value is total average energy consumption of all apartments in one building), as shown in Figures 19 and 20.

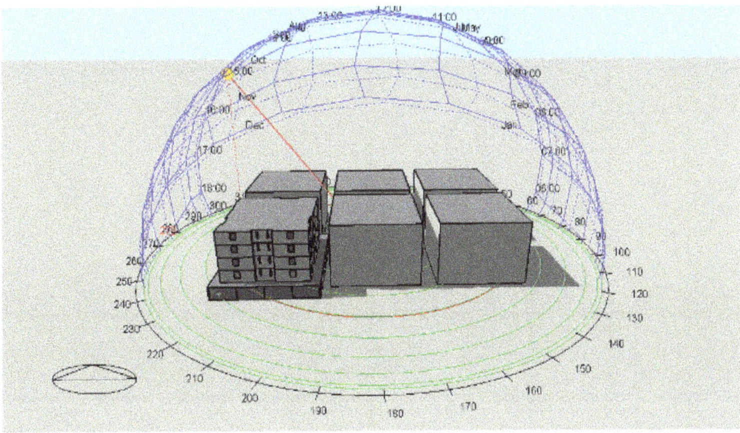

Figure 17. Rendering of a residential building (south orientation).

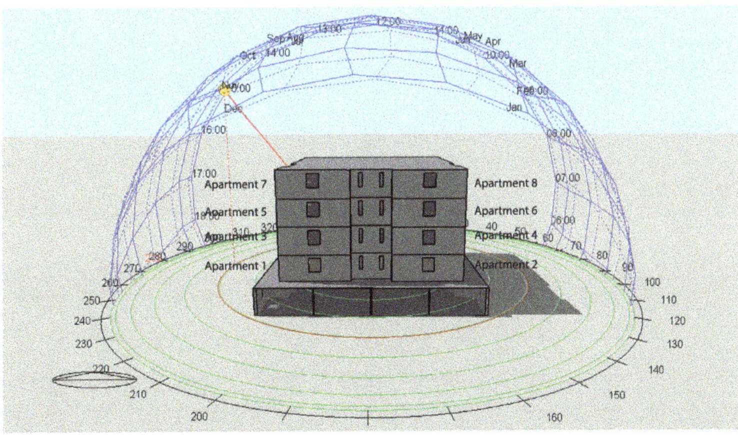

Figure 18. Rendering of a residential building (southwest orientation) within other buildings.

Figure 19. Comparison of energy consumption for different building positions that face south.

Figure 20. Comparison of energy consumption for different building positions that face north.

Note that the apartments switched sides when the building switched from north to south orientation. At the apartment level, the AAEC increased from ground-level to top-floor apartments, requiring additional energy-saving interventions in the designing stage (see Figures 19 and 20).

7.2. Step Two: Effect of Changing Infiltration Rate and Cooling Temperature on AAEC

The infiltration rate ACH50 is crucial in determining the AAEC. In this study, ACH50 values of 4, 6, 7, 8, 15, 30, and 50 were considered, with 4 ACH50 considered best practice, according to the SBC [36]. The maximum ACH50 value of 50 was determined based on previous studies that found a maximum of 39 ACH50 through monitoring methods [65]. This study includes the infiltration rate and its impact on the AAEC, with results demonstrating the significance of the ACH50 on energy consumption for each scenario and apartment. The study used 20 ACH50, calibrated with the average energy consumption bill, as reported by Aldossary for the first 2 floors of the building [32]. Hence, different infiltration rates (50 to 4 ACH50) were tested, and when applying lower infiltration rates, lower AAEC results were achieved. Figure 21 demonstrates a range of increases in AAEC when only changing the infiltration rate from 50 to 4 within the same apartment. The AAEC increase percentages range from 62% to 74%, as the top-floor apartments with higher ACH50 had the highest values compared to the lower floors.

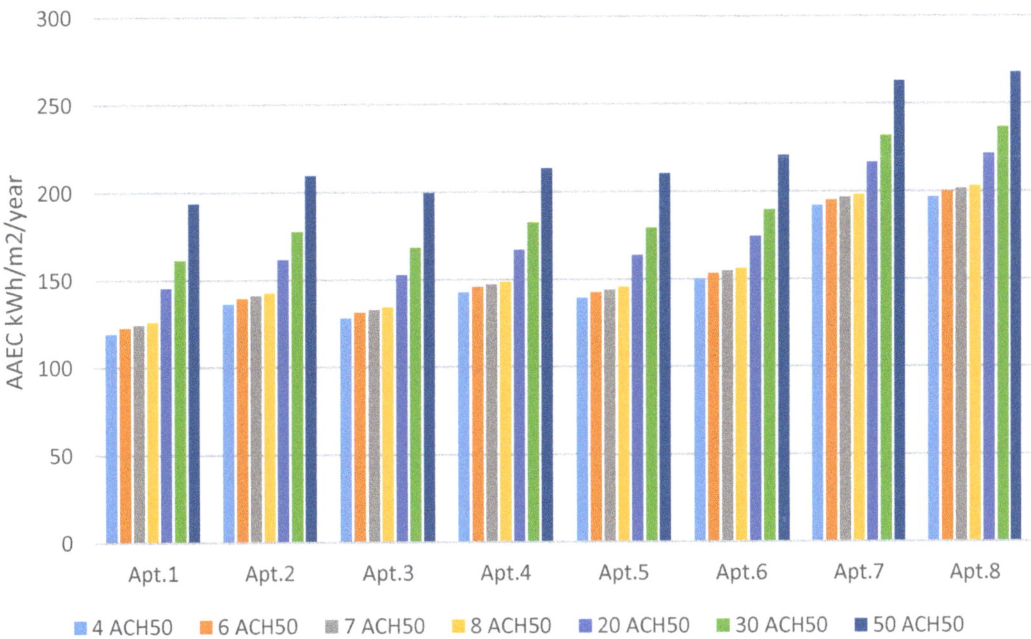

Figure 21. Changes in the impact of the ACH50 rate on AAEC for every apartment in south orientation case.

The user comfort level is another factor affecting the AAEC, as cooler temperatures increase energy consumption, requiring extra cooling. The thermal comfort preferences of occupants in Jeddah vary, with a typical cooling temperature range of 19–24 °C, according to Felimban [1]. The scenarios for changing cooling temperatures highlight the impact on the AAEC. For example, as shown in Figure 22, the AAEC for apartment 1 decreases by approximately 4 kWh/m^2/year when the cooling temperature is increased by 1 °C. However, decreasing the cooling temperature by 2 or 3 °C increases the AAEC by around 15, 33, or 51 kWh/m^2/year. However, a lower cooling set-point temperature leads to a higher AAEC in air-conditioned apartments. To conclude, both the infiltration rate and user thermal comfort levels are considered primary impact factors that contribute to the increase or decrease in the AAEC, as shown in Figures 10 and 11.

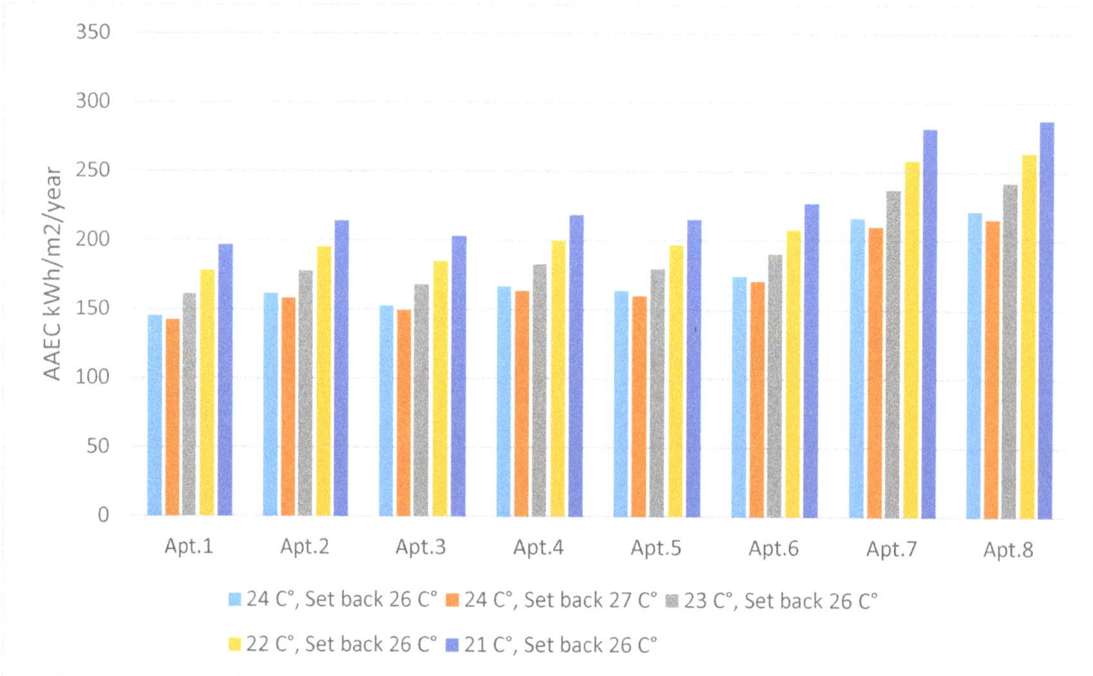

Figure 22. Impact of changing the setback point of the cooling temperature on AAEC using DesignBuilder simulation in south orientation case.

7.3. Step Three: Energy Performance Simulation and Energy Savings

The energy simulation of the basic model used 20 ACH50 infiltration rates and aimed to produce scenarios targeting a rate of 4 ACH50, as the SBC standards require. According to the simulation results, Figures 12, 13, 17 and 18 illustrate the AAEC for each apartment using infiltration rates of 20 and 4 ACH50. The following two sections illustrate the AAEC results that depend on the user scenario and the selected infiltration rate. The simulation was divided into indoor and outdoor scenarios, as explained earlier in the description of the scenarios.

A—Indoor Scenarios

As previously explained, indoor scenarios can be applied individually to any apartment. The simulation results show an extensive reduction in AAEC when using a deep energy retrofit scenario (Scenario 10); the reduction was up to 121 kWh/m^2/year. When applying a minor retrofit scenario (Scenario 2), it was possible to reduce the amount of electricity used by at least 34 kWh/m^2/year compared to the basic model. In addition, the AAEC varied from one apartment to another depending on the apartment position (floor level) and the apartment orientation in the building. All of the deep retrofit scenarios led to a more efficient AAEC for all apartments.

The most critical factor of AAEC reduction was the insulation upgrades for the walls and roofs. Adding an insulation layer to the walls and roofs resulted in a significant sharp reduction in energy use, as shown in Figures 23 and 24. Additional upgrade interventions, such as window replacement and roof upgrade, added further reduction to the energy consumption with different percentages.

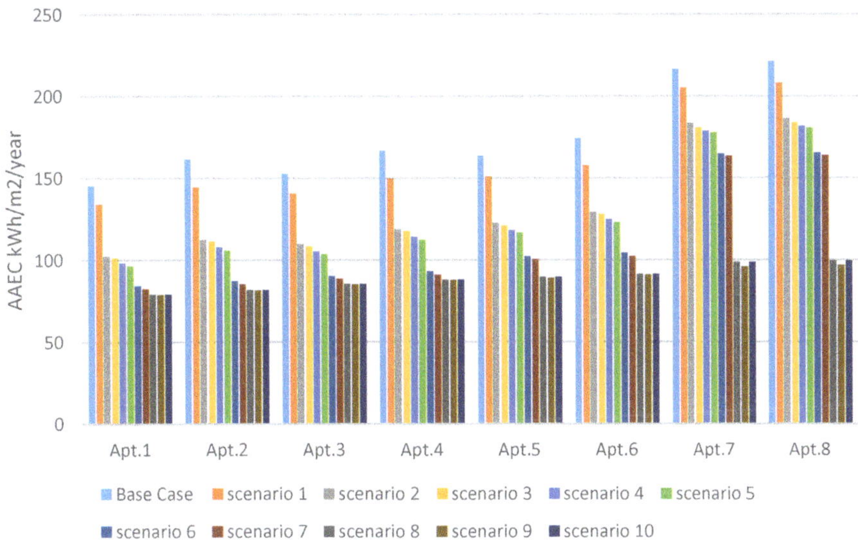

Figure 23. AAEC values for indoor energy retrofitting scenarios using 4 ACH50 for infiltration rate.

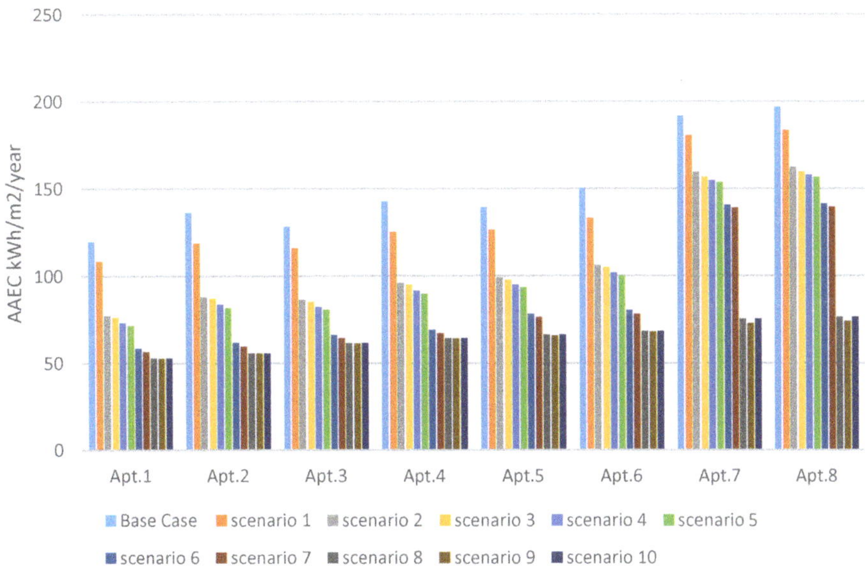

Figure 24. AAEC values for indoor energy retrofitting scenarios using 20 ACH50 for infiltration rate.

It was also shown that window and AC upgrades greatly influence the ACH50 rate, where outdoor heat is prevented from penetrating through the air gaps to the indoor space.

Energy savings gradually increased from Scenario 1 (5–10%) to Scenario 10 (45–56%), where the infiltration rate was 20 ACH50, while for the 4 ACH50 infiltration rate, Scenario 1 (6–12%) to Scenario 10 (55–65%) are illustrated in Figures 25 and 26. There were remarkable differences in energy savings between apartments when applying the different Scenarios (1, 2–5, 6–7, and 8–10).

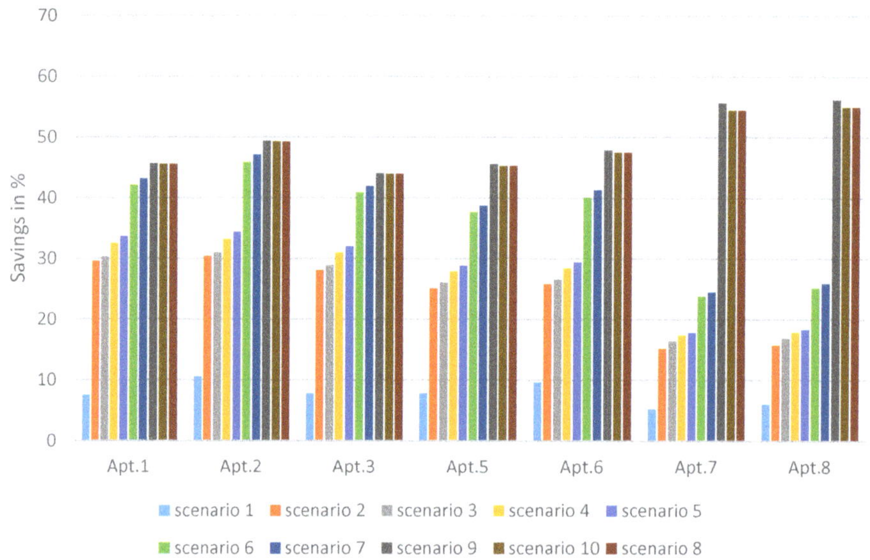

Figure 25. Possible energy savings from testing different scenarios (indoor) where the infiltration rate is 20 ACH50.

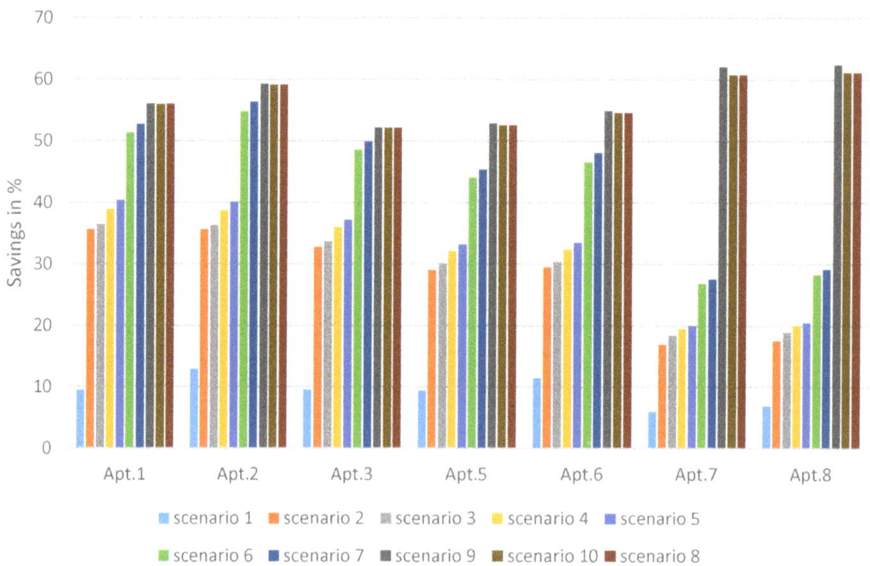

Figure 26. Possible energy savings from testing different scenarios (indoor) where the infiltration rate is 4 ACH50.

Apartments 7 and 8 recorded around 60% savings when using Scenarios 8, 9, and 10, where additional insulation was added to the roofs. However, apartments 1–6 only had a slight savings increase when applying Scenarios 8, 9, and 10 compared to Scenarios 6 and 7. Apartments 7 and 8 had less energy savings than apartments 1-6 when using Scenarios 1–7. Therefore, it is suggested that every apartment has specific properties that require different energy retrofitting scenarios, and an individual cost analysis per apartment is required.

Furthermore, more energy savings were achieved when the basic model used 20 ACH50 and the applied scenarios used 4 ACH50. The energy savings increased for Scenario 1 from 5–10% to 17–26%, and for Scenario 10, they increased from 45–56% to 63–65%, where the change in the ACH50 rate had a significant impact on the energy savings percentage (see Figure 27). The considerable energy savings show the importance of considering infiltration rate levels in energy retrofitting applications to achieve a better AAEC for all apartments.

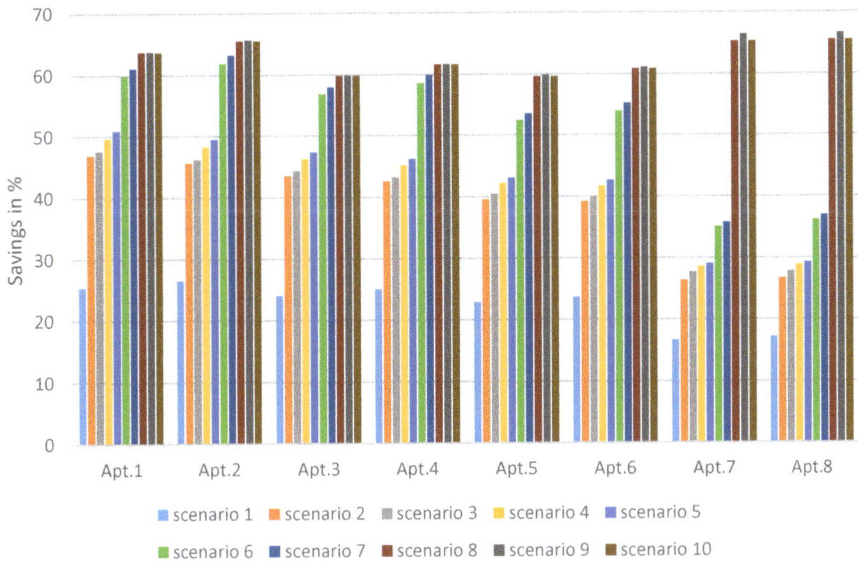

Figure 27. Possible energy savings from testing different scenarios (indoor) where the infiltration rate is 4 ACH50, with a rate of 20 ACH50 for the basic model.

The indoor scenarios are very valuable for individual decision making for energy retrofit upgrades. The only concern in these indoor scenarios is the thermal heat transfers through the concrete skeleton structure (thermal bridges), especially when the structure intersects with an indoor partition. In this study, thermal bridges have not been incorporated in the calculations, as the main objective of the study was to calculate the overall energy-saving possibilities so that the factors could be easily calculated in the future in order to help retrofit the residential buildings and ensure energy efficiency.

In summary, the indoor scenarios of energy retrofitting applications have great potential to enhance the energy efficiency of residential apartments, with energy savings ranging from 20% to 65% depending on the apartment's circumstances.

B—Outdoor Scenarios

The outdoor scenarios, as observed earlier, can only be applied to the whole building and cannot be applied to individual upgrades to individual apartments. The simulation results show a sharp reduction in AAEC when using a deep energy retrofit scenario, as can be seen with Scenario 7 shown in Figures 28 and 29. However, adding 10 cm of insulation to the outdoor wall, as shown in Scenario 1, can significantly reduce at least 50 kWh/m^2/year compared to the basic model. Figures 28 and 29 illustrate significant reductions in AAEC, each using different infiltration rates of 20 ACH50 and 4 ACH50.

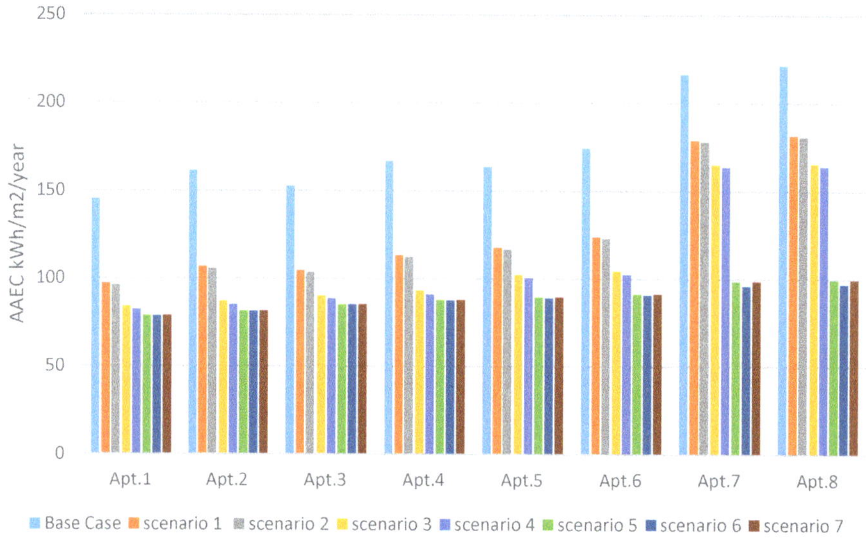

Figure 28. AAEC for outdoor energy retrofitting scenarios using 20 ACH50 for the infiltration rate.

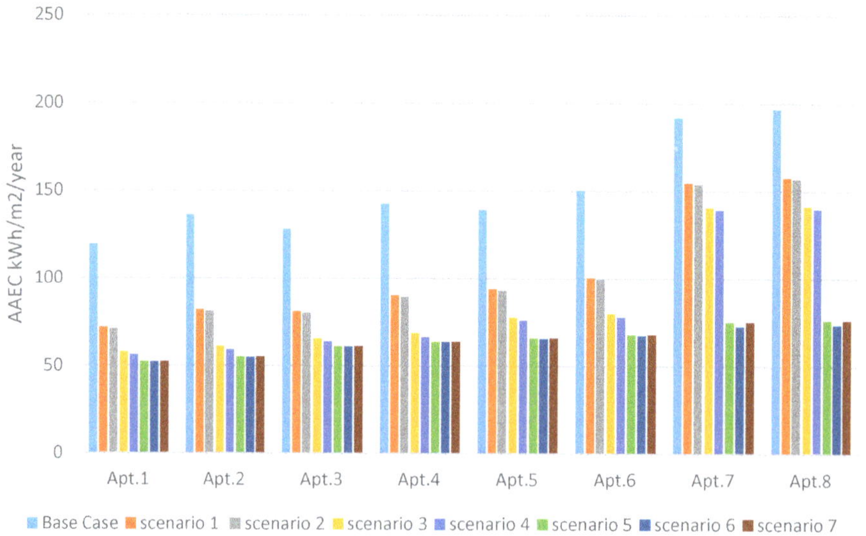

Figure 29. AAEC for outdoor energy retrofitting scenarios using 4 ACH50 for the infiltration rate.

To provide more detail, Figure 28 presents different ranges of decrease of the AAEC depending on the apartment and the applied scenario. The AAEC results for apartment 1 show a 33% reduction for Scenario 1 and a 46% reduction for Scenario 7. However, apartment 8 records an 18% reduction for Scenario 1 and a 55% reduction for Scenario 7.

Apartments 1–6 gradually increased their energy savings when applying the scenarios in order, as Figures 30 and 31 illustrate. Apartments 7 and 8 had less energy savings when using Scenarios 1–4 compared to the other apartments. However, outdoor Scenarios 5–7 significantly increased the energy savings for apartments 7 and 8. Generally, the high-resolution scenarios depend on the infiltration rate levels and the selected scenario.

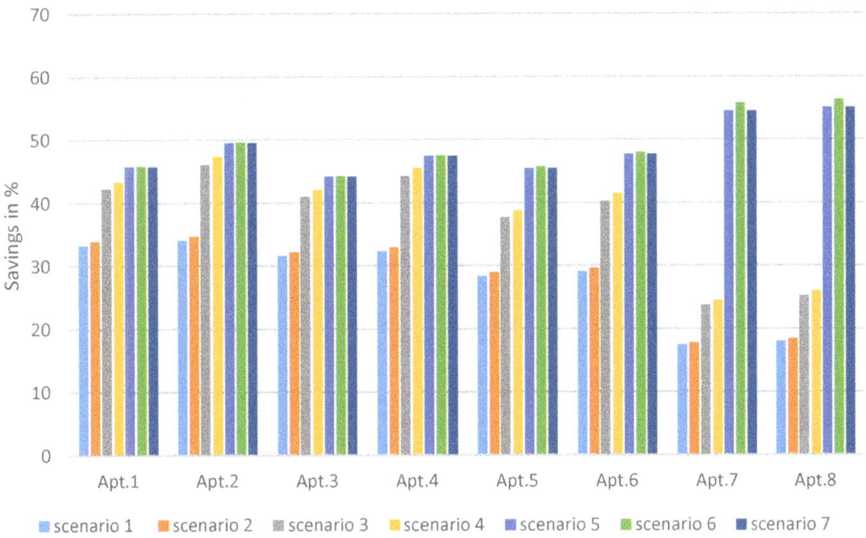

Figure 30. Possible energy savings from testing different outdoor scenarios where the infiltration rate is 20 ACH50, with a rate of 20 ACH50 for the basic model.

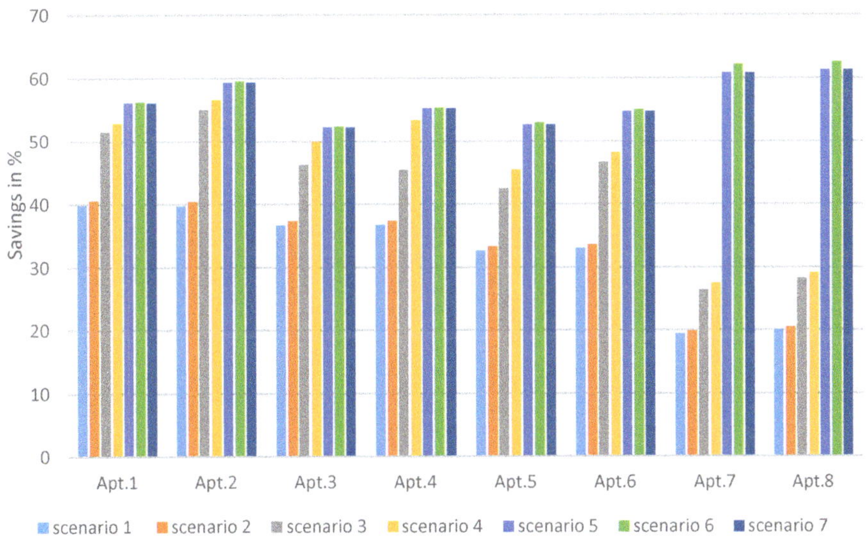

Figure 31. Possible energy savings from testing different scenarios (outdoor) where the infiltration rate is 4 ACH50, with a rate of 4 ACH50 for the basic model.

Figure 31 indicates more promising energy savings for all units when applying scenarios that include improving the infiltration rate to 4 ACH50 compared to the results in Figure 31. Figure 32 shows decreasing savings percentages from 50% to around 30% for apartments 1 and 8, respectively. However, if any of Scenarios 5–7 applied to all apartments 1, 2, 3, 4, 5, 6, 7, and 8, then AAEC could reach efficient consumption values of 52, 55, 61, 63, 66, 68, 75, and 76 kWh/m^2/year, respectively.

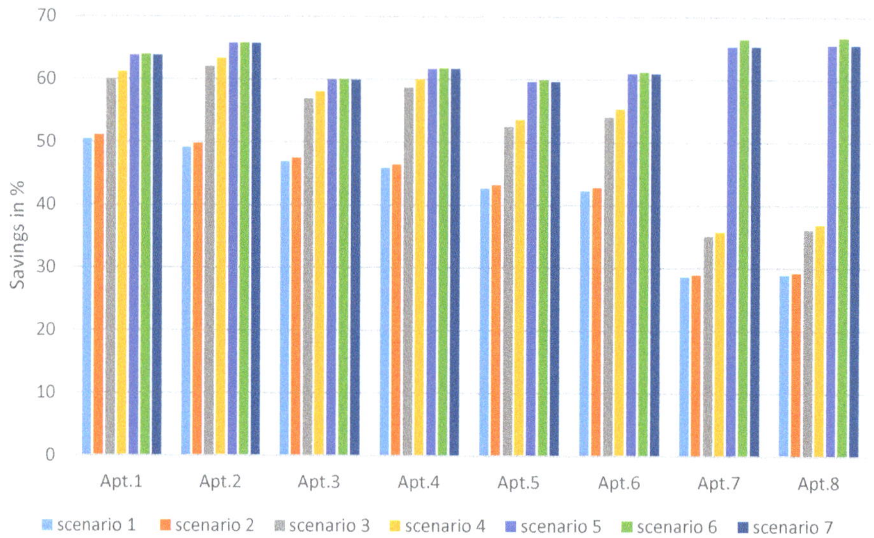

Figure 32. Possible energy savings from testing different scenarios (outdoor) where the infiltration rate is 4 ACH50, with a rate of 20 ACH50 for the basic model.

In summary, the simulation results of the energy performance for residential buildings in Jeddah validated an optimistic range of energy savings (30–60%) when applying different energy retrofit scenarios.

8. Discussion

The discussion has been divided into three main points. Initially, the AAEC is discussed in respect to the eight apartments, based on the analyzed properties; secondly, the energy-saving possibilities are discussed in respect to applying different scenarios; finally, the uncertainties and the effects on the AAEC are addressed, such as the infiltration rate (ACH50) and the user thermal comfort temperature.

8.1. Average Annual Energy Consumption

The simulation results for residential apartments range from 145 to 221 kWh/m^2/year, depending on the orientation and the floor level. Apartments situated on the upper floors consume more AAEC than apartments found lower in the building due to the heat exposure from the roof. For instance, apartments 7 and 8 recorded the highest AAEC of 216 and 221 kWh/m^2/year.

The apartments that faced the west recorded a higher AAEC than east-facing apartments when they were located on the same floor. In addition, 2 west-facing apartments, i.e., apartments 2 and 4 (161, 166 kWh/m^2/year), consumed more than the upper floor, east-facing apartments 3 and 5 (152, 163 kWh/m^2/year). The apartment location, specifically the orientation and floor level, are the main factors used to calculate the AAEC.

8.2. Energy-Saving Possibilities

In general, the simulation results demonstrate a significant impact from every scenario. The impact degree is based on the weaknesses in the envelope component design, such as the walls, the windows, or the roof. Furthermore, in respect to apartments 1–6, the weaknesses came from the walls and the windows, where different energy savings were recorded from Scenarios 1–7, ranging from 7% to 47%, whereas Scenarios 8–10 only add about 2% savings compared to Scenario 7.

The weaknesses in apartments 7 and 8 were due to all components, and the roof presented the main weakness. For instance, apartment 8 had energy savings when applying Scenarios 1–7, ranging from 6% to 26%, and 55% to 56% for Scenarios 8–10.

Every scenario has energy-saving possibilities, leading to better energy performance to achieve the main objective of extensive simulation validation.

8.3. Uncertainties

Uncertainty factors affect the AAEC, such as the actual infiltration rate and the user's thermal preferences (user thermal comfort). Each factor dramatically influences the AAEC as they can increase the energy-saving possibilities when they are known before designing the possible energy scenarios.

An actual infiltration rate (ACH50) is a significant factor that can be used to demonstrate actual energy savings, as Figure 33 illustrates. It is also important to note that the savings percentage increased when the infiltration rate was enhanced.

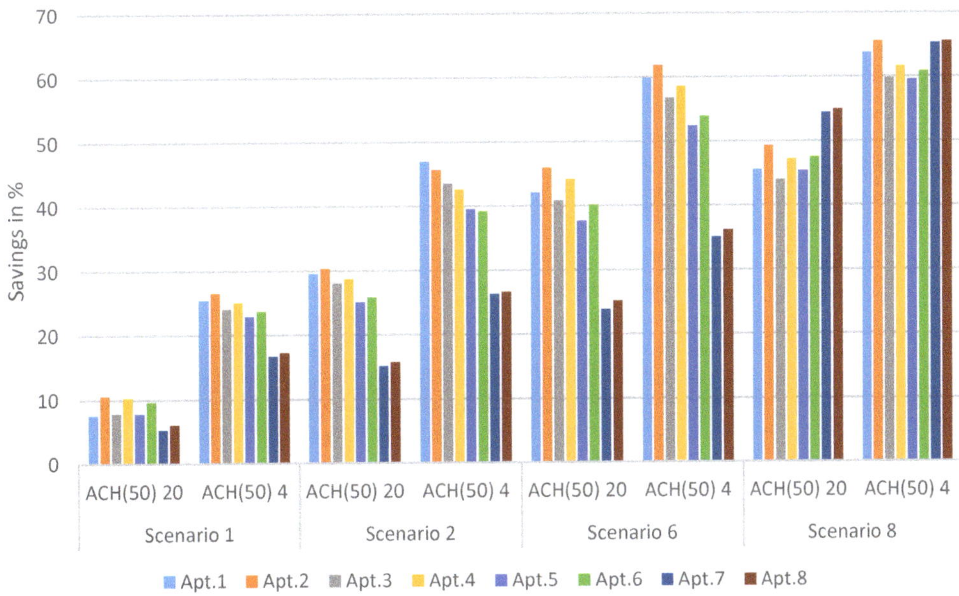

Figure 33. Possible energy savings when different infiltration rates were applied.

The existing residential buildings in Jeddah, KSA, currently require an air conditioning system every day of the year when an infiltration procedure occurs. If the infiltration rate is tested, then the air tightness of the indoor spaces could be designed better in the energy retrofitting scenarios.

The other factor is the difference in user thermal comfort. User thermal comfort varies from family to family. However, both the infiltration rate and cooling temperature affect the increasing possibility of AAEC for all apartments, as Figure 34 illustrates. Understanding the user's thermal comfort would help designers and occupants to lower their energy usage; increasing designers' awareness so that thermal comfort is considered in the design process is very important. In short, higher cooling temperatures and lower infiltration rates lead to extensive energy savings.

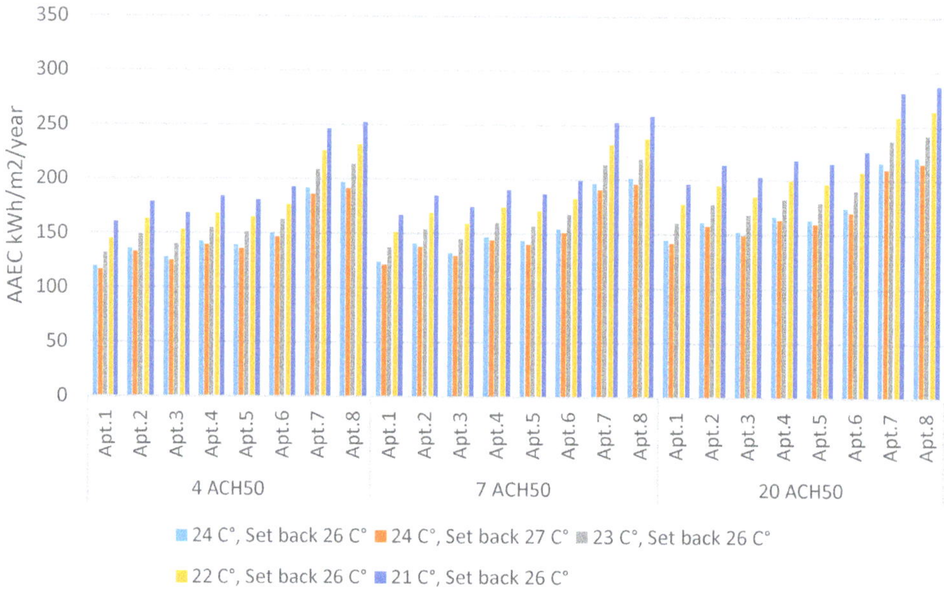

Figure 34. Impacts of cooling temperature and infiltration rate change on AAEC for all apartments.

9. Conclusions

The energy retrofit scenarios presented in this study were validated through digital simulation using DesignBuilder software to demonstrate the potential for energy savings. The baseline case model yielded AAEC values for apartments 1, 2, 3, 4, 5, 6, 7, and 8, respectively, of 145, 161, 152, 166, 163, 174, 216, and 221 kWh/m^2/year. However, it is important to note that the building's location within the urban environment influences the AAEC for all units. Additionally, the position of each apartment (orientation and floor level) results in different AAEC values.

This paper has presented a comprehensive case study that considers crucial elements, such as building location, apartment positioning, user profiles, and ownership types. Two energy-upgrade scenarios, focusing on indoor and outdoor improvements, were introduced for the eight apartments. These scenarios primarily involve upgrading building components (walls, windows, and roofs) to meet the energy benchmark level defined by the upgraded SBC energy standards. The outcomes of the analysis provide insights into key variables that can significantly impact energy savings. It is worth noting that achieving the highest energy savings depends on various factors, including interventions for improving the building envelope, enhancing the infiltration rate, and determining the desired level of thermal comfort. While the simulation encompassed different design variables, two main variables (infiltration and user thermal comfort level) can yield more accurate AAEC values if known during the scenario design phase.

However, it is important to consider not only energy savings, but also the cost aspect, when selecting the optimal scenario. Evaluating the cost associated with each scenario is crucial to determine its suitability for individual cases. This aspect will be further explored in a subsequent paper, providing a more comprehensive understanding.

In conclusion, based on the analysis of energy retrofit scenarios in Jeddah, a series of simulations was conducted to confirm the potential for energy savings, ranging from 25% to 66%. The findings emphasize the significance of implementing energy-saving measures and highlight the opportunities for improving energy efficiency in the residential building sector.

Author Contributions: Conceptualization, A.F., U.K. and T.K.; Methodology, A.F., U.K. and T.K.; Validation, A.F.; Formal analysis, A.F.; Writing—original draft, A.F.; Writing—review & editing, A.F.; Visualization, A.F.; Supervision, U.K. and T.K. All authors have read and agreed to the published version of the manuscript.

Funding: This research received no financial support for the research, authorship, and/or publication of this article.

Conflicts of Interest: The authors declare no conflict of interest with respect to the research results, authorship, and/or publication of this article.

Nomenclature

List of abbreviations	
AAEC	Average Annual Energy Consumption
ACH	Air Change per Hour
ACH50	Air Changes per Hour at 50 pascals pressure differential
CDD	Cooling Degree Days
COP	Coefficient of Performance
CSS	Concrete Skeleton Structure
EIFS	External Insulation Finishing System
EPS	Expanded Polystyrene Insulation
ERA	Energy Retrofitting Application
GCC	Gulf Cooperation Council
KPI	Key Performance Indicator
KSA	Kingdom of Saudi Arabia
SBC	Saudi Building Code
SEEC	Saudi Energy Efficiency Centre
WWR	Window-to-Wall Ratio
XPS	Extruded Polystyrene Insulation

References

1. Felimban, A.; Prieto, A.; Knaack, U.; Klein, T.; Qaffas, Y. Assessment of Current Energy Consumption in Residential Buildings in Jeddah, Saudi Arabia. *Buildings* **2019**, *9*, 163. [CrossRef]
2. Saudi Energy Efficiency Center. BUILDINGS | Saudi Energy Efficiency Center. 2018. Available online: https://www.seec.gov.sa/en/energy-sectors/buildings-sector/ (accessed on 21 October 2018).
3. Krarti, M.; Aldubyan, M.; Williams, E. Residential building stock model for evaluating energy retrofit programs in Saudi Arabia. *Energy* **2020**, *195*, 116980. [CrossRef]
4. Laponche, B.; López, J.; Raoust, M.; Devernois, N. *Energy Efficiency Retrofitting of Buildings: Challenges and Methods*; Research Program in Hubei Province, China; Agence Française de Développement: Paris, France, 2012; p. 182. Available online: http://www.afd.fr/webdav/site/afd/shared/PUBLICATIONS/RECHERCHE/Scientifiques/Focales/08-VA-Focales.pdf (accessed on 13 March 2023).
5. Aldossary, N.A.; Rezgui, Y.; Kwan, A. Domestic energy consumption patterns in a hot and arid climate: A multiple-case study analysis. *Renew. Energy* **2014**, *62*, 369–378. [CrossRef]
6. AlFaris, F.; Juaidi, A.; Manzano-Agugliaro, F. Energy retrofit strategies for housing sector in the arid climate. *Energy Build.* **2016**, *131*, 158–171. [CrossRef]
7. Aldossary, N.A.; Rezgui, Y.; Kwan, A. Establishing domestic low energy consumption reference levels for Saudi Arabia and the Wider Middle Eastern Region. *Sustain. Cities Soc.* **2017**, *28*, 265–276. [CrossRef]
8. Konstantinou, T. Supporting the Design of Residential Energy Upgrades. Ph.D. Thesis, Delft University of Technology, Delft, The Netherlands, 2014.
9. Al-Ragom, F. Retrofitting residential buildings in hot and arid climates. *Energy Convers. Manag.* **2003**, *44*, 2309–2319. [CrossRef]
10. Krarti, M. Evaluation of large scale building energy efficiency retrofit program in Kuwait. *Renew. Sustain. Energy Rev.* **2015**, *50*, 1069–1080. [CrossRef]
11. Ameer, B.; Krarti, M. Impact of subsidization on high energy performance designs for Kuwaiti residential buildings. *Energy Build.* **2016**, *116*, 249–262. [CrossRef]
12. Taleb, H.M. Using passive cooling strategies to improve thermal performance and reduce energy consumption of residential buildings in U.A.E. buildings. *Front. Arch. Res.* **2014**, *3*, 154–165. [CrossRef]
13. Rakhshan, K.; Friess, W.A. Effectiveness and viability of residential building energy retrofits in Dubai. *J. Build. Eng.* **2017**, *13*, 116–126. [CrossRef]

14. Giusti, L.; Almoosawi, M. Impact of building characteristics and occupants' behaviour on the electricity consumption of households in Abu Dhabi (UAE). *Energy Build.* **2017**, *151*, 534–547. [CrossRef]
15. Friess, W.A.; Rakhshan, K. A review of passive envelope measures for improved building energy efficiency in the UAE. *Renew. Sustain. Energy Rev.* **2017**, *72*, 485–496. [CrossRef]
16. Al-Saadi, S.N.; Al-Hajri, J.; Sayari, M.A. Energy-Efficient Retrofitting Strategies for Residential Buildings in hot climate of Oman. *Energy Procedia* **2017**, *142*, 2009–2014. [CrossRef]
17. Alalouch, C.; Al-Saadi, S.; AlWaer, H.; Al-Khaled, K. Energy saving potential for residential buildings in hot climates: The case of Oman. *Sustain. Cities Soc.* **2019**, *46*, 101442. [CrossRef]
18. Alaidroos, A.; Krarti, M. Optimal design of residential building envelope systems in the Kingdom of Saudi Arabia. *Energy Build.* **2015**, *86*, 104–117. [CrossRef]
19. Alaboud, M.; Gadi, M. Indoor environmental monitoring of residential buildings in Saudi Arabia, Makkah: A case study. *IOP Conf. Ser. Mater. Sci. Eng.* **2019**, *609*, 042044. [CrossRef]
20. Krarti, M.; Dubey, K.; Howarth, N. Energy productivity analysis framework for buildings: A case study of GCC region. *Energy* **2019**, *167*, 1251–1265. [CrossRef]
21. Webb, A.L. Energy retrofits in historic and traditional buildings: A review of problems and methods. *Renew. Sustain. Energy Rev.* **2017**, *77*, 748–759. [CrossRef]
22. Ma, Z.; Cooper, P.; Daly, D.; Ledo, L. Existing building retrofits: Methodology and state-of-the-art. *Energy Build.* **2012**, *55*, 889–902. [CrossRef]
23. Natural Resources Canada. Retrofitting. Available online: https://www.nrcan.gc.ca/retrofitting/20707 (accessed on 15 June 2020).
24. Marina Economidou. Energy Renovation Programme. European Energy Efficiency Platform (E3P). 2016. Available online: https://e3p.jrc.ec.europa.eu/articles/energy-renovation (accessed on 1 March 2020).
25. Building and Construction Authority. Existing Building Retrofit. 2010. Available online: www.bca.gov.sg (accessed on 20 March 2022).
26. What Is a Building Retrofit? Available online: https://www.melbourne.vic.gov.au/business/sustainable-business/1200-buildings/building-retrofit/Pages/building-retrofit.aspx (accessed on 1 March 2020).
27. Energy Efficient Retrofitting of Buildings. Available online: https://www.bable-smartcities.eu/explore/solutions/solution/energy-efficient-retrofitting-of-buildings.html (accessed on 1 March 2020).
28. Zhivov, A.; Lohse, R. *Deep Energy Retrofit—A Guide for Decision Makers*; Springer International Publishing: Cham, Switzerland, 2021.
29. Ministry of Human Resources and Social Development. Social Development Bank. 2019. Available online: https://www.sdb.gov.sa/en-us/our-products/personal/social-loans/sub-pages/application-rules (accessed on 16 January 2020).
30. Austah, M. Private House Renovation in Yanbu City, KSA. 2020. Available online: https://twitter.com/MoaadAustah/status/1245081929789079556 (accessed on 19 September 2020).
31. Saudi Electricity Company. Fixed Bill Service. Available online: https://www.se.com.sa/en-us/Customers/Pages/Fixed_Bill/About.aspx (accessed on 22 March 2023).
32. Aldossary, N.A.; Rezgui, Y.; Kwan, A. Domestic energy consumption patterns in a hot and humid climate: A multiple-case study analysis. *Appl. Energy* **2014**, *114*, 353–365. [CrossRef]
33. Solar Decathlon Middle East. Available online: https://www.solardecathlonme.com/ (accessed on 19 June 2020).
34. Solar Decathlon Middle East Virginia Tech Team. Available online: https://www.solardecathlonme.com/teams/team-virginia-tech (accessed on 19 June 2020).
35. Alarabiya.net. Amendments to the Requirements of the Saudi Building Code for Residential Buildings. 2021. Available online: www.alarabiya.net/aswaq/realestate/2021/10/06/%D8%AA%D8%B9%D8%AF%D9%8A%D9%84%D8%A7%D8%AA-%D9%81%D9%8A-%D9%85%D8%AA%D8%B7%D9%84%D8%A8%D8%A7%D8%AA-%D9%83%D9%88%D8%AF-%D8%A7%D9%84%D8%A8%D9%86%D8%A7%D8%A1-%D8%A7%D9%84%D8%B3%D8%B9%D9%88%D8%AF%D9%8A-%D9%84%D9%84%D9%85%D8%A8%D8%A7%D9%86%D9%8A-%D8%A7%D9%84%D8%B3%D9%83%D9%86%D9%8A%D8%A9 (accessed on 10 March 2023).
36. The Saudi Building Code National Committee (Regulations). SBC 602: Energy Conservation Requirements (Arabic). 2018. Available online: https://www.sbc.gov.sa/resources/PdfPreview/web/viewer.html?avx=PUlqTQ==&type=0 (accessed on 20 February 2020).
37. Krarti, M.; Dubey, K.; Howarth, N. Evaluation of building energy efficiency investment options for the Kingdom of Saudi Arabia. *Energy* **2017**, *134*, 595–610. [CrossRef]
38. Alrashed, F.; Asif, M. Climatic Classifications of Saudi Arabia for Building Energy Modelling. *Energy Procedia* **2015**, *75*, 1425–1430. [CrossRef]
39. Al-Hadhrami, L. Comprehensive review of cooling and heating degree days characteristics over Kingdom of Saudi Arabia. *Renew. Sustain. Energy Rev.* **2013**, *27*, 305–314. [CrossRef]
40. Ministry of Municipalities Rural Affairs & Housing (MOMRAH). Sakani. Available online: https://sakani.sa/ (accessed on 19 March 2020).

41. Alayed, E.; O'Hegarty, R.; Kinnane, O. Thermal envelope analysis for new code compliance of Saudi Arabian dwellings. *Energy Build.* **2021**, *243*, 110997. [CrossRef]
42. Thermal Insulation Material List. 2023. Available online: https://www.ewa.bh/en/contact/Documents/March2023TIPlist.pdf (accessed on 13 March 2023).
43. Kumar, D.; Alam, M.; Zou, P.X.W.; Sanjayan, J.G.; Memon, R.A. Comparative analysis of building insulation material properties and performance. *Renew. Sustain. Energy Rev.* **2020**, *131*, 110038. [CrossRef]
44. Saudi Energy Efficiency Center. The Guide for Thermal Insulation Systems and Materials According to the Requirements of the Saudi Building Code. Volume 1. 2018. Available online: www.seec.gov.sa/en/seec/ (accessed on 20 March 2020).
45. General Authority for Statistics. Annual Report 2017. 2018. Available online: https://www.stats.gov.sa/sites/default/files/annualreport2017.pdf (accessed on 17 October 2018).
46. Yearly Report 2021. 2021. Available online: https://www.seec.gov.sa/media/44alu10p/seec-ar-2021.pdf (accessed on 15 October 2022).
47. Almasri, R.A.; Alardhi, A.A.; Dilshad, S. Investigating the Impact of Integration the Saudi Code of Energy Conservation with the Solar PV Systems in Residential Buildings. *Sustainability* **2021**, *13*, 3384. [CrossRef]
48. Saudi Building Committee. Issuance of Temporary Flexibility for Thermal Insulation Coefficient. 2022. Available online: https://sbc.gov.sa/ar/MediaCenter/Pages/NewsDetails.aspx?NewstId=90 (accessed on 14 March 2023).
49. Krarti, M.; Howarth, N. Transitioning to high efficiency air conditioning in Saudi Arabia: A benefit cost analysis for residential buildings. *J. Build. Eng.* **2020**, *31*, 101457. [CrossRef]
50. Hijazi, J.; Howieson, S. Displacing air conditioning in Kingdom of Saudi Arabia: An evaluation of 'fabric first' design integrated with hybrid night radiant and ground pipe cooling systems. *Build. Serv. Eng. Res. Technol.* **2018**, *39*, 377–390. [CrossRef]
51. Theodosiou, T.; Tsikaloudaki, K.; Kontoleon, K.; Giarma, C. Assessing the accuracy of predictive thermal bridge heat flow methodologies. *Renew. Sustain. Energy Rev.* **2021**, *136*, 110437. [CrossRef]
52. Taleb, H.M.; Sharples, S. Developing sustainable residential buildings in Saudi Arabia: A case study. *Appl. Energy* **2011**, *88*, 383–391. [CrossRef]
53. Ministry of Municipalities Rural Affairs & Housing (MOMRAH). Sakani-Real State Market. Available online: https://sakani.sa/app/marketplace?land_types=moh_lands&region_id=2&unit_type=land&project_types=lands_moh_land&marketplace_purpose=buy&view_mode=list&sort=-publish_date (accessed on 14 March 2023).
54. Architecture Team. 2020. Available online: http://www.archteam.co/ (accessed on 14 March 2020).
55. Al-Khraif, R.; Salam, A.A.; Rashid, M.F.A. Family Demographic Transition in Saudi Arabia: Emerging Issues and Concerns. *SAGE Open* **2020**, *10*, 2158244020914556. [CrossRef]
56. Ministry of Municipalities Rural Affairs & Housing (MOMRAH). Mullak. Available online: https://mullak.housing.gov.sa/public-page/246 (accessed on 2 February 2021).
57. Design Builder Software. 2016. Available online: https://designbuilder.co.uk/ (accessed on 6 February 2020).
58. Magni, M.; Ochs, F.; de Vries, S.; Maccarini, A.; Sigg, F. Detailed cross comparison of building energy simulation tools results using a reference office building as a case study. *Energy Build.* **2021**, *250*, 111260. [CrossRef]
59. Pawar, B.; Kanade, P. Energy Optimization of Building Using Design Builder Software. *Int. J. New Technol. Res.* **2018**, *4*, 263152.
60. Makawi, M.A. The Impact of Air Infiltration on Energy Performance of Single-Family Detached Houses in Hot Climate. Master's Thesis, King Fahd University of Petroleum and Minerals, Dhahran, Saudi Arabia, 2020.
61. Makawi, M.A.; Budaiwi, I.M.; Abdou, A.A. Characterization of Envelope Air Leakage Behavior for Centrally Air-Conditioned Single-Family Detached Houses. *Buildings* **2023**, *13*, 660. [CrossRef]
62. Prignon, M.; Van Moeseke, G. Factors influencing airtightness and airtightness predictive models: A literature review. *Energy Build.* **2017**, *146*, 87–97. [CrossRef]
63. Sorgato, M.; Melo, A.; Lamberts, R. The effect of window opening ventilation control on residential building energy consumption. *Energy Build.* **2016**, *133*, 1–13. [CrossRef]
64. Sherman, M.H.; Wanyu, R.C. *Building Airtightness: Research and Practice*; Routledge: Oxfordshire, UK, 2004.
65. Ambrose, M.; Syme, M. Air Tightness of New Australian Residential Buildings. *Procedia Eng.* **2017**, *180*, 33–40. [CrossRef]

buildings

MDPI

Article

A Systematic Approach to Optimizing Energy-Efficient Automated Systems with Learning Models for Thermal Comfort Control in Indoor Spaces

Serdar Erişen

Department of Architecture, Atılım University, Ankara 06830, Turkey; serdar.erisen@atilim.edu.tr

Abstract: Energy-efficient automated systems for thermal comfort control in buildings is an emerging research area that has the potential to be considered through a combination of smart solutions. This research aims to explore and optimize energy-efficient automated systems with regard to thermal comfort parameters, energy use, workloads, and their operation for thermal comfort control in indoor spaces. In this research, a systematic approach is deployed, and building information modeling (BIM) software and energy optimization algorithms are applied at first to thermal comfort parameters, such as natural ventilation, to derive the contextual information and compute the building performance of an indoor environment with Internet of Things (IoT) technologies installed. The open-source dataset from the experiment environment is also applied in training and testing unique black box models, which are examined through the users' voting data acquired via the personal comfort systems (PCS), thus revealing the significance of Fanger's approach and the relationship between people and their surroundings in developing the learning models. The contextual information obtained via BIM simulations, the IoT-based data, and the building performance evaluations indicated the critical levels of energy use and the capacities of the thermal comfort control systems. Machine learning models were found to be significant in optimizing the operation of the automated systems, and deep learning models were momentous in understanding and predicting user activities and thermal comfort levels for well-being; this can optimize energy use in smart buildings.

Keywords: indoor air; thermal comfort; user occupation; artificial intelligence; machine learning; natural ventilation; building performance; building information modeling

Citation: Erişen, S. A Systematic Approach to Optimizing Energy-Efficient Automated Systems with Learning Models for Thermal Comfort Control in Indoor Spaces. *Buildings* **2023**, *13*, 1824. https://doi.org/10.3390/buildings13071824

Academic Editors: Yingdong He and Nianping Li

Received: 26 June 2023
Revised: 14 July 2023
Accepted: 17 July 2023
Published: 19 July 2023

1. Introduction

Designing buildings with automated systems is emerging research; yet, further systematic approaches to the optimization of energy use and the deployment of the smart systems and learning models for thermal comfort control and well-being in indoor spaces are needed [1,2]. Smart systems and Internet of Things (IoT) technologies, which are used for energy-efficient buildings and environments, also have a rising impact on personal comfort systems (PCS) [3]. The seminal studies have investigated adjustable air-conditioning systems to meet the desired levels of thermal comfort and well-being with regard to user preferences [4–6]. Thermal comfort levels such as mean radiant temperature [4] have become significant in the exploration of the design and deployment of automated smart systems; this is especially the case with Fanger's approach, which applies the predicted mean votes of users [1]. Considering user preferences for thermal comfort and well-being also motivates this research to explore the systematic approaches and methods used in developing and improving the energy-efficient automated smart systems that are to be applied in indoor environments. Related articles also analyze the correlational changes in thermal comfort levels by controlling the automated heating, ventilation, and air conditioning (HVAC) systems through the use and levels of thermostats and fans [5], which are tested for different seasons [6].

In the control and regulation of the thermal comfort and well-being of users, a systematic approach is followed in this research in defining and optimizing the energy usage, power, workloads, and operation of the automated systems to be applied for ventilation, cooling, and heating in indoor environments. An open-source IoT-based dataset, including the voting of users acquired from the special care context in a residential building, is analyzed and processed in the experiments concerning the user occupation and predicted activities for thermal comfort and well-being. Therefore, this study aims to derive thermal comfort parameters, generate algorithms, and develop learning models in the optimization of smart systems and energy-efficient infrastructure that will be based on IoT data for thermal comfort and well-being in the contexts of designed indoor environments.

In configuring smart spaces and systems, digital simulations are crucial in the search for the technical possibilities and the limits of building performance evaluation and in the generation of contextual data and parameters for state-of-the-art smart systems for well-being and thermal comfort. In investigating the parameters for setting up energy-efficient IoT-based automation systems in smart buildings and environments, it also crucial for the streaming data to be integrated with the collaborative and common workspace platforms such as that of building information modeling (BIM) software. In this research, the critical parameters for thermal comfort and well-being are explored and investigated through BIM simulations, algorithms, and learning models using real-time IoT data from the conditioned experiment space.

Thus, the dataset from an experimental work was analyzed, where the IoT technologies were embedded within the specific physical configuration of the indoor space in an apartment flat in Ankara, Turkey, to achieve greater infection control during the pandemic for individuals who might suffer from COVID-19 or other ailments [3,7]. Additionally, real-time learning models were applied in that experimental work to produce big data about the categorized activities and the comfort levels of users after surveying the potential applications of sensors, IoT technologies, and learning models for healthcare and well-being [3]. The study was also significant in its integration of innovative research on real-time learning systems with indoor environments for monitoring and predicting thermal comfort levels and well-being. For instance, the concentration levels of CO_2 and the air quality are critical parameters, and their changes were observed through natural ventilation experiments [3]. In this research, these correlated factors are also analyzed, along with other thermal comfort parameters, user occupation patterns, and the voting of users.

This research proposes to explore and optimize the energy use, workloads, and operation of automated systems for thermal comfort control in indoor spaces. In that regard, the building performance of the indoor space is evaluated first, based on the existing dataset derived from the thermal comfort levels and optimal energy use for ventilation and air quality. BIM software is applied to simulate the experiment environment for heating and cooling to analyze the workloads in achieving thermal comfort levels. Similarly, energy optimization algorithms and models are explored, developed, and tested with regard to the inputs from natural ventilation experiments in optimizing the energy use for zero-energy building research to deploy minimum energy for air purification and ventilation as well as for the desired thermal comfort levels and well-being of users. The survey on the related literature and the outcomes of this research further showed that the learning models could improve the efficiency of smart systems in energy optimization. Thus, automated systems are proposed to achieve thermal comfort levels and to better meet the desired well-being of users by deploying black box learning models in indoor environments with smart technologies. Accordingly, the voting of users is examined by both including and excluding them from the dataset to observe the challenges in developing artificial intelligence (AI) models such as convolutional neural networks (CNN)s for automation systems.

In brief, this research is unique in comparison to that of the related literature and applies a systematic approach that deploys learning models to explore the optimal energy use and building performance of automated smart systems that are to be developed for thermal comfort and well-being in the context of indoor spaces. The research examines an open-source dataset that allows the evaluation of the facts about the building performance, natural ventilation, air quality, and user activities and the user voting acquired via PCS. The outcomes of this systematic approach also indicate that in deploying simulations and smart systems with learning models, it is crucial to derive and process the context-based data when optimizing the thermal comfort control for each specific condition. The simulation of the experiment area by BIM software allows the generation and evaluation of context-based information on building performance, power loads, and energy optimization parameters. The data from the environment were applied to develop algorithms and unique deep learning models for the operation of smart systems, and Fanger, as well as non-Fanger, approaches were examined in the optimization of the novel, lightweight, and efficient deep learning models for state-of-the-art automation systems. Briefly, the objectives and major contributions of the research can be highlighted as follows:

- BIM simulations and energy optimization algorithms are explored to provide context-based information about the energy use and capacities of the automated systems to be installed into the experiment environment.
- Machine learning models are applied in discovering the optimization of the operation of the automated systems for thermal comfort control in the context of indoor environments.
- Lightweight and efficient deep learning models are developed for understanding the user activity and thermal comfort levels in the context of the experiment environment, in which IoT-based smart systems for thermal comfort and well-being are installed.

This article proceeds as follows: in Section 2, the related works on thermal comfort control and well-being are surveyed with regard to the research on the building performance evaluation as well as natural and artificial ventilation, smart systems and buildings, and learning models. In Section 3, the article introduces the materials and methods of the systematic approach followed in this research. Section 4 presents the experiments and results that are followed in the order introduced in Section 3. Section 5 discusses the results and evaluates the facts with regard to the possible steps to be applied in developing and operating automated smart systems in indoor environments. Section 6 briefly concludes the research.

2. Related Works

The seminal articles have reviewed the literature and systems for thermal comfort control in buildings; they have also considered the occupation and well-being of users [1,2]. One of these articles proposes a combination of models, systems, and procedures for smart applications [1]. The research emphasizes that the predicted mean vote (PMV) is the main thermal comfort modeling approach in finding the correlation between the environmental parameters and the personal factors via the vote of occupants, in reference to Fanger's pioneer work in the 1970s [1].

Many articles also investigate zero-energy buildings, the building design, and its components with regard to thermal comfort parameters such as heating, cooling, natural and artificial ventilation, and air quality. For instance, Wei et al. used *CiteSpace* (5.8.R3 SE 64-bit) to discover the relevant literature, which was also categorized with respect to the different approaches to zero-energy building research [8]. Gassar et al. explore the performance optimization parameters and models during the design of buildings for energy-efficient heating, cooling, and lighting by surveying related articles and works in the literature [9]. In another seminal article, the façade performances of the buildings in Singapore were studied with regard to climate change by reporting the major factors, such as temperature, humidity, wind speed, corrosion, degradation, material use, and vegetation, which also influencing the building performance [10].

The seminal works also consider passive building design [11–13]. For instance, Ekihouse is a solar house prototype for the Solar Decathlon Europe 2012 competition developed by examining the building design strategies for different seasons and with regard to the energy design optimization of a house in Madrid deploying photovoltaic systems [11].

Similarly, Lopez-Escamilla et al. conducted a study on a social housing prototype, which was first presented in the Solar Decathlon competition in 2015, by investigating the design of bioclimatic double skin in a tropical climate [12]. Santy et al. explored the standards for the design of a passive house, without using HVAC systems, by depending on the bioclimatic analyses of some regions in Indonesia [13].

Furthermore, the researchers have examined natural and artificial ventilation as significant thermal comfort parameters. For instance, the authors investigated the parameters that influence the optimization of energy use with regard to the specific atrium typology in buildings and divided the energy simulation periods in which the HVAC systems were applied [14]. The researchers have also explored using HVAC systems to meet thermal comfort levels in different seasons [5,6]. On the other hand, Abdullah et al. studied the performance of building components with regard to the design of windows for natural ventilation; they considered the analyzed parameters and evaluated their performance in relation to the environmental facts [15]. Similarly, window opening behavior (WOB) was inspected by Kim et al. in relation to environmental parameters of indoor and outdoor spaces; they also considered the occupant behavior and thermal comfort in "structural equation modeling" [16]. The paper argues that new human–technology relations should be followed in adopting advanced technologies and systems for successful energy-efficient design strategies [16]. In the experiments of that research, the collected data from the exterior and indoor spaces were processed with regard to temperature, CO_2 concentration, and solar radiation [16]. Monitoring the occupant behavior is also considered in the development of structural equation modeling with regard to WOB [16]. Window opening events were discretely evaluated, and the research concluded that occupant access to the systems can increase tenant satisfaction and reduce operational costs [16].

The air quality and air index values of indoor environments were also seen as equally significant in one of the recent articles, which considers the data on the air quality of indoor environments to develop learning models for health and well-being purposes [3]. In another research work, air conditioning design was studied by considering a building performance simulation of the air quality and thermal comfort of an indoor stepped hall [17]. The impact of the research on thermal comfort levels was explored by measuring the air velocity, temperature, and air change rate with the help of building modeling that simulated the airflow, temperature, and air velocity distribution in the different locations in that indoor space [17]. The user occupation patterns with the air index and gas sensor value (GSV) also provide substantial results for energy-dependent modeling in the performance evaluation of buildings and the development of IoT-based state-of-the-art HVAC systems [18,19].

There are also reviews on the related literature that evaluate the capabilities of BIM for managing the operation and maintenance of green buildings [20]. Many articles also review the smart systems and buildings with regard to digital twins and BIM in Industry 4.0 [21] and the use of neural networks to decrease errors in predicting the actual energy consumption rates in the building energy performance evaluation [22].

In another seminal research work, Lee et al. reviewed the major concepts of responsive architecture that can be evaluated within the scope of energy-efficient buildings and environments with smart systems [23]. Comprehensive surveys on smart building technologies, systems, and sensing technologies also aim to discover new approaches in human–building interactions with regard to energy use and occupation, as well as the features of buildings and components [2,14]. A deep survey was also made on the categorizing of the applied smart technologies in [2]; the work was similar to that in [24]. Correspondingly, Al-Obaidi et al. review the use of IoT technologies applied for energy-efficient buildings and cities with a comprehensive survey on the concepts of infrastructural development, models, technological potential, applications, and the challenges related to their use [25].

Sensor-based environments and IoT technologies in buildings have also increased considerably [24,26]. These technologies meet the requirements for special use and self-care [27,28], in addition to providing improvements in energy efficiency and crowdsourcing environmental data [24], which have led to an increase in the quality of smart homes [29].

They also have significant potential for smart decision making, which can be integrated into BIM platforms [30] in the monitoring of workflow and construction processes [31–33], as well as in building energy models, analyzing sensorial information [34], and energy management in smart house systems [35,36]. IoT and sensor-based technologies have also been employed for well-being and for the remote monitoring purposes of smart healthcare in buildings [24,26–29,37].

Occupant behavior and health in buildings are vital aspects that should be considered with the energy usage, well-being, and building performance evaluation [18,38–42]. There is also research on identifying the evaluation criteria of the smartness of buildings, which can be categorized into different groups with the use of scoring systems [43]. A seminal article also applied the voting system and the learning models together with data from the indoor working spaces of the Helios building with IoT-based sensors [44]. In assessing the votes, the mean and standard deviation were applied, in a similar manner to Fanger's approach [44]. Thus, predicting the complex user behavior and updating the existing dataset ubiquitously with regard to user activity remain developing research areas with higher potential. Architects continue to survey the demand for further design requirements and parameters that can be updated in real-time when considering user activities and thermal comfort levels [45,46]. Accordingly, Almusaed et al. inspected the smart building design concepts with regard to the rising impact of AI and digital twins [47]. In most recent applications, IoT technologies and AI have been used for real-time learning and monitoring [3].

The potential of smart systems was also discovered in related studies with overarching experiments using machine learning (ML) and real-time deep learning models [3]. In a comprehensive review article, the models were classified as white box, grey box, and black box models, such as machine learning models, artificial neural networks, and recurrent neural networks, which are deployed for predicting building energy usage [48]. In that research, the building envelope parameters, HVAC systems, and weather factors were surveyed accordingly [48]. Another survey was conducted on applying ML and AI models, as well those that do not apply black box models, in the estimation and prediction of thermal comfort in the classrooms of primary schools in Japan [49]. The research also conducted comprehensive analyses of the factors that define thermal comfort by considering the voting data from the occupants [49].

In another research work with a comprehensive review, the applications of machine learning and statistical models were analyzed in the evaluation of building energy performance and the prediction of energy-efficient retrofitting considerations [50]. In another study, machine learning models were applied in the generation of a design method to explore the optimal energy-saving residential form [51]. According to the research, machine learning models are used to compare and combine the results of energy optimization methods with regard to various performance indicators. These methods also aid in determining the best locations for energy-saving buildings [51]. In another related research work, machine learning models, such as Gaussian process regression (GPR), were applied in global sensitivity analyses for energy prediction and optimization by exploring the architectural typology of a courtyard house [52]. In another article, machine learning models were applied to predict building energy usage under different environmental conditions, with a greater concern for climate change and the difficulties in measuring its effects in different geographic regions and contexts [53].

Different studies also applied genetic algorithms, incorporated with neural networks, for green architecture by reviewing the studies in various research areas and different parts of buildings regarding energy usage and optimization in several countries [54]. Similarly, Fallah et al. applied artificial neural networks optimized with electrostatic discharge algorithms to calculate the parameters for evaluating the energy efficiency and thermal loads of residential buildings [55]. In another research work, artificial neural networks were applied in the prediction and analyses of the mean radiant temperature for the location of selected residential building in Sendai, Japan [56].

Accordingly, the systematic approaches are explored and deployed in our research through the surveyed thermal comfort parameters, BIM, algorithms, and learning models used in the improvement of the building performance evaluation, thermal comfort control, and well-being by the applied models and methods. The dataset from a unique experiment environment, in which IoT and smart sensor systems are applied, is processed by deploying the BIM software, energy-optimization algorithms, and step-by-step learning models that are proposed for use with state-of-the-art automation systems.

3. Materials and Methods

This article aims to derive a systematic approach to developing energy-efficient solutions for thermal comfort control and well-being in indoor spaces and to employ them in operating state-of-the-art systems. Thus, the research explores the thermal comfort parameters for the energy-efficient execution of the automated solutions by first considering the building performance and context-based data of indoor spaces. The thermal comfort parameters surveyed from the related literature that are to be explored in this research are briefly noted as follows:

- Temperature;
- Humidity;
- Air quality and CO_2 concentration;
- Airflow and ventilation;
- User occupation;
- User voting;
- Performance of the buildings and building components;
- The facts from indoor and exterior environments.

The study analyzes a dataset related to these parameters from the selected indoor experiment space and deploys the energy optimization algorithms as well as building performance evaluation models. The research also explores the performance and efficiency of "grey-box" and "black-box" learning models [48] and attempts to develop unique deep CNNs to increase the efficiency of smart systems deployment to predict the thermal comfort levels and well-being of users.

The open-source dataset used in acquiring the critical inputs about these parameters is based on the IoT cloud of the experiment environment, which has smart systems installed to acquire the user-rated well-being values with the help of personalized devices [3,7,57]. The experimental data of the observations were also processed and tested during the pandemic by the real-time learning models deployed to predict and monitor occupant behavior and well-being in the experiment environment [3].

In the parametrization of the state-of-the-art smart and automated systems for thermal comfort and well-being, the BIM simulations of this experiment environment were generated in this research using *Autodesk Revit Architecture 2023* and *Autodesk Insight* (for Revit 2023) software to evaluate the building performance and to calculate the energy use and workloads (Figures 1 and 2a). The components of the building were modeled, and their construction details and heat transfer coefficients were reported with the help of the BIM software *Autodesk Revit Architecture*, using Intel(R) Core(TM) i7-4700HQ CPU @ 2.40 GHz as the hardware resource with four cores and eight logical processors for computations, and NVIDIA GeForce 750 M as the Graphic Processing Unit (GPU) used in the simulations and renderings (Figures 1 and 2a, Table 1).

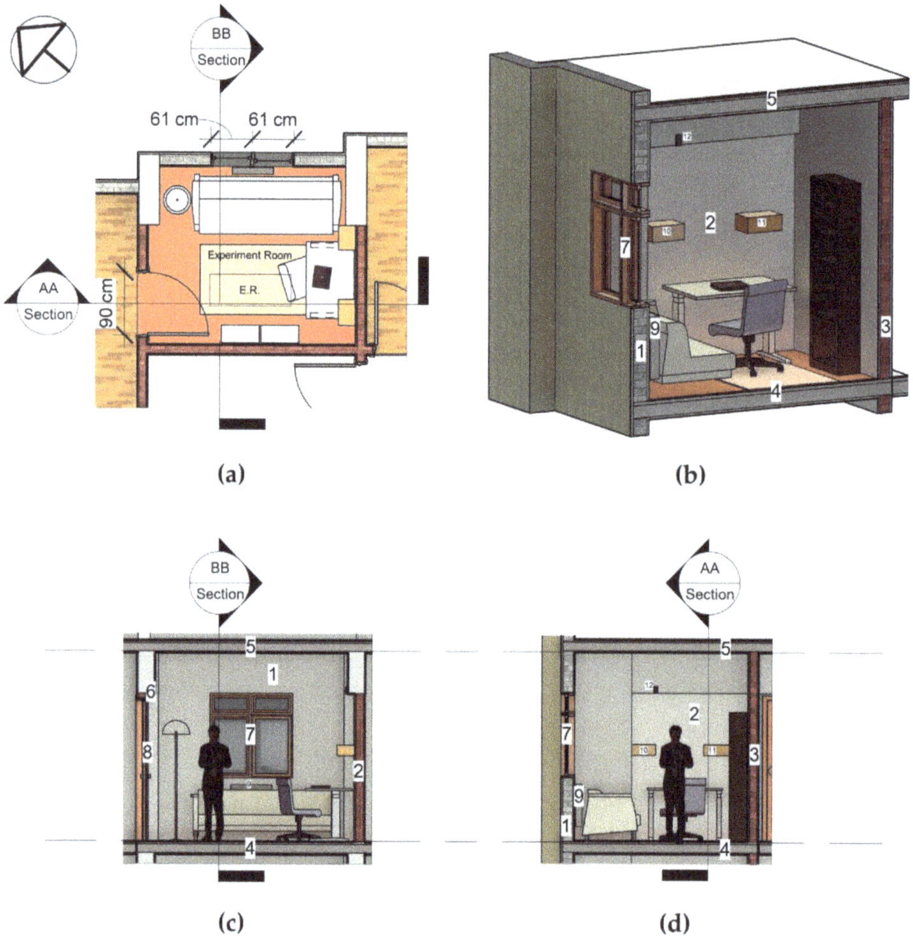

Figure 1. BIM simulation and the drawings of the experiment environment: (**a**) plan; (**b**) 3-dimensional drawing; (**c**) AA section; (**d**) BB section.

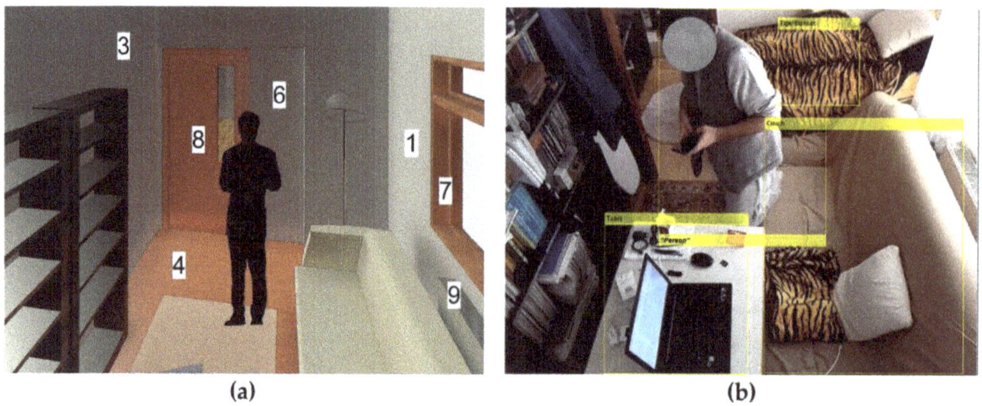

Figure 2. (**a**) Rendering from BIM simulation scene of the IoT-based camera; (**b**) faster R-CNN prediction applied in the experiment area [3].

Table 1. Building components and their features in the experiment area.

Component ID	Building Components	Dimensions		Heat Transfer Coefficient (W/(m^2·K))
		Area (sq·m)	Thickness (cm)	
1	Exterior wall	6.851	20	2.66
2	Interior wall	7.27	15	3.65
3	Interior wall	9.036	15	3.82
4	Floor	7.64	18	4.65
5	Ceiling	7.64	18	4.71
6	Interior wall	5	15	3.65
7	Exterior windows	1.45	1	3.69
8	Interior door	1.8	3.5	1.87
9	Radiator	0.366	10	-
10	Smart system	0.0875	15	-
11	Smart system	0.0875	15	-
12	IoT-based camera	-	-	-

Air index parameters that were highly related to thermal comfort levels were studied with regard to energy use and optimization through the derived models and energy optimization algorithms for smart systems that depended on the dataset from the experiment environment. The test dataset from the natural ventilation experiments was further processed to compare the results with the earlier observations and performances of the learning models with regard to the chosen parameters for thermal comfort.

Additionally, various machine learning and deep learning models were trained and tested, and their performances were compared to the training and test datasets to find the optimal learning models for the operation of smart systems based on the ventilation experiments. The dataset acquired from this experiment space also included the classified labels for user activities and thermal comfort levels. In that regard, novel and efficient deep CNNs were further developed, trained, and tested with these datasets in this research for state-of-the-art technologies which correlated with the thermal comfort and air quality levels, user activities, and user-defined voting values.

3.1. The Parameters for Models and Methods Applied in the Experiments

The challenges of the COVID-19 pandemic have inspired studies on the air quality and activity patterns of users who may require intensive care in indoor spaces [3,18,38]. The temperature, humidity, airflow, and air quality of indoor environments are absolutely vital for analyzing user occupation patterns and well-being [18,37,38]. Their parameterization for energy usage and behavior prediction through gaining knowledge about the user activity is also extremely important [18,37,38] and useful in developing the building performance evaluation models by applying BIM, energy optimization models, and algorithms and in the evaluation of advanced learning models using machine learning algorithms and even state-of-the-art deep learning models.

Since the 1970s, Fanger's approach has encouraged the processing of data about the well-being of users in spaces [1]. Thus, PMV has become the relevant method in modeling and assessing thermal comfort levels [1]. Similarly, related projects regarding thermal comfort and well-being have also considered the specific design of special care contexts for healthcare and infection control in buildings using innovative IoT technologies [3,7]. The developed projects provided big data about the activities and preferences of users, which can be processed for deciding on thermal comfort levels. Thus, the crowdsourced IoT data from indoor spaces were considered in the scope of user occupation patterns and thermal comfort parameters in this research, including the natural ventilation experiments for energy optimization methods and greater well-being at the residential scale.

3.2. The Experiment Environment and Its Simulation

During the COVID-19 pandemic, a room in a residential building in Ankara, Turkey, was designed as the experiment environment using IoT-based sensors and real-time learning models and aiming for the special care and well-being of the users [3,7]. The room that was used in the experiments has a 7.64 sq m floor area and a 2.7 m clear height from the floor to ceiling; thus, it is around 20.63 m^3. Accordingly, the BIM simulation of the experiment environment was modeled via *Autodesk Revit Architecture* to assess the building performance of the components and the thermal comfort conditions of this area (Figures 1 and 2a, Table 1).

3.2.1. Smart Systems Applied in the Experiment Environment

An IoT-based camera and smart systems (Figure 2b, Table 1) were installed in the experiment room and included ultrasonic sensors together with humidity and temperature sensors [3,7]. The smart systems also included an MQ-2 gas sensor to provide critical data regarding the air quality and the levels of particles, including carbon dioxide (CO_2) as well as butane (C_4H_{10}), liquefied petroleum gas (LPG), methane (CH_4), and smoke [3,7]. Additionally, a rating system was designed to be handled by the remote controller, enabling the users to vote on their well-being and comfort levels (Figure 3) [3,7]. Similarly, a web server based on the local area network (LAN) of this space was developed to collect the relative user data [3,7].

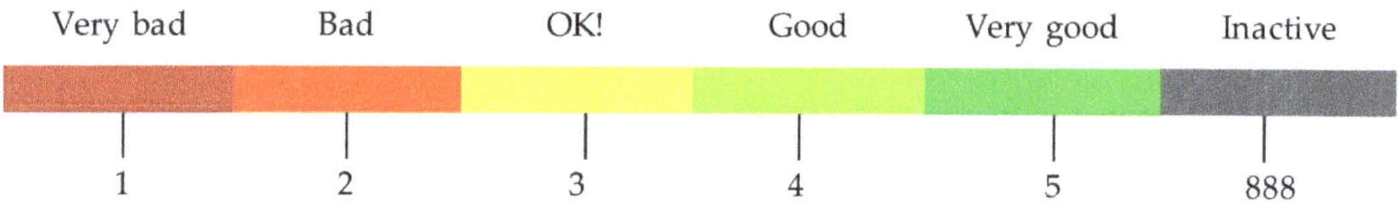

Figure 3. User-rated voting system for monitoring the well-being of the occupant.

Thus, the related parameters about the air quality and air index values, together with the recognized activities of the users and the user-rated data, were acquired as six different observations from different sensors simultaneously, generating the dataset, and fed to the IoT Cloud channel on *ThingSpeak*, the *MathWorks* IoT platform [57], via the system [3,7].

3.2.2. The Real-Time Learning System

The generated dataset from the sensors was acquired as the IoT data and used to develop a real-time learning system, including a CNN, which was trained, tested, and optimized for the monitoring and well-being of the users [3,7]. The developed real-time deep learning system predicted the labeled user occupation patterns and environmental data, including thermal comfort levels and critical behavior. It provided feedback about the user activity and critical circumstances for well-being in real-time [3]. The system also served to generate big data about the labeled predictions of user activities, which were applied as separate behavior labels; the system proved its efficiency with a 99.97% success rate in test accuracy and in real-time prediction and recognition of activity during the experiments [3]. Furthermore, faster R-CNN models were applied to recognize objects and people in the room using IoT-based imaging devices (Figure 2b) [3].

3.3. Dataset

The dataset, "IoT Channel for Real-Time Learning & Monitoring" [57], used in this research had 113.327 inputs, including sensor data and the predictions of the real-time learning system based on each sensor data item [3]. The last 100 inputs also served as open-source datasets on the public IoT channel [57]. The initially collected dataset from the IoT cloud included 2170 channel feeds; 1567 of these included raw data from the indoor environment. Six hundred and three inputs included six observation inputs about motion tracking, air quality, temperature, humidity, the well-being of users, and the correlated data about the user activities, as they were also used in training the real-time learning system in the related research [3]. The training dataset with six observations included ten categorized activities about the user occupation (1, 5, 6, 7, 9, 10) and thermal comfort levels in the indoor environment (2, 3, 4, 8), as illustrated in Table 2; these were recorded and predicted for thermal comfort and well-being.

Table 2. Examples from the dataset with ten classes.

Categorized Activities	Temperature (Celsius)	Humidity (Percent)	Gas Sensor Value (GSV)	User-Rated Well-Being	D1 (cm)	D2 (cm)
1. Visitor (or user) sits	24.6	58	502	888	41.1	57.2
	25.9	50	566	888	54.6	72.3
	25.9	50	566	888	66.27	84
	26.1	49	539	888	87.3	95
	26.1	49	536	888	87.3	98
2. Ventilation	25.4	36	481	888	76.59	78
	25.4	35	472	888	77.5	79
	25.4	36	468	888	77	80
	25.3	33	467	888	76.26	81
3. Cold, dry indoor air	25.2	33	472	888	77.02	81
	25.3	33	467	888	76.26	81
	20.6	32	469	888	76.46	81
	20.9	32	488	888	76.26	81
4. Hot, humid indoor air	26.1	49	533	888	78.31	88
	26.1	49	536	888	78.63	87
	29.8	51	502	888	69.6	86
	29.8	52	502	888	71.5	101.1
5. Going Out	26	50	542	5	88.8	124
	26	50	542	5	85.82	124
	26	50	542	5	86.25	131
	26	50	546	5	86.25	131
	24.6	48	542	5	102	149.5
	24.7	53	502	5	129	149
6. Entering In	26	48	541	888	98.43	134
	26	48	540	888	107.19	123
	27.1	50	502	888	127	143.3
	27.1	50	502	888	114.4	129.1
7. User moves into the bed	25.7	49	501	5	75.71	67
	25.8	49	424	4	78.26	69
	24.8	48	517	5	48.5	59.2
	25.3	49	502	4	23.1	33.2
8. Air quality and well-being correlation	25.8	50	556	2	75.61	102
	26	50	508	5	77.61	92
	24.5	48	555	5	75.3	94.3
	24.9	48	502	5	85.5	107.1
9. Two people move within the room	25.9	49	585	888	66.4	101
	25.9	49	578	888	66.4	103
10. User moves from the bed	25.9	49	425	4	77.83	68
	25.9	49	541	5	77.11	53
	24.5	48	552	5	34.1	42.2
	24.6	48	549	5	27.5	34.9

The dataset also included specific experiments on natural ventilation and observations on the changes in air index values and the air quality of the indoor environment. For example, changes in the room temperature, humidity, and GSV were observed by providing natural ventilation to the well-heated experiment room at noon on 18 November 2020 for 14 min 41 s, or 881 s, as shown in Figure 4, from moments A to B [3]. The average room temperature in the experiments was 25.9 °C, and the outside temperature in Ankara, Turkey, was 13.9 °C (Figure 4) [3]. There was 25–30 s period between the sensor observations, which were sent to the cloud and acquired in real-time, and 12–13 s for predicting each observation. Accordingly, the outcomes of the experiment provided a substantial basis for the evaluation of the energy-based modeling and building performance of the indoor environment. Thus, the observations on temperature, humidity, and GSV were also used as the training dataset in this research into developing machine learning and deep learning models based on these ventilation experiments (Figure 4).

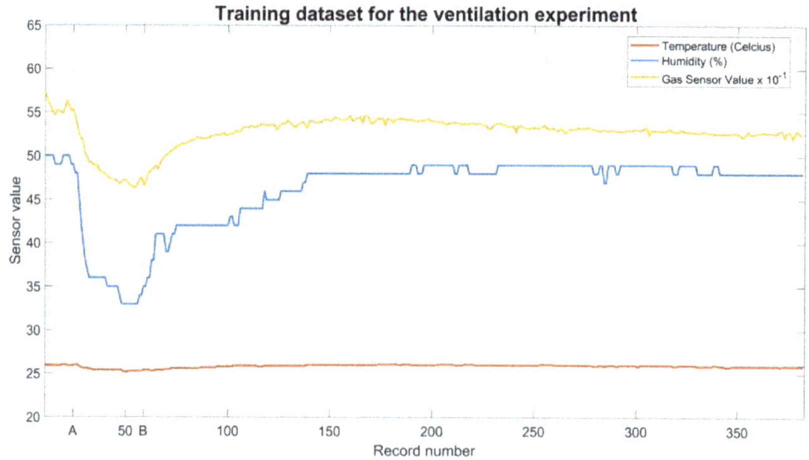

Figure 4. The training dataset for the ventilation experiment on November 18th [3,57].

3.4. Analyses of the Heat Losses and Power Loads to HVAC Systems

In the evaluation of the building performance and in finding the optimal energy use of state-of-the-art HVAC technologies, such as heaters, radiators, coolers, air conditioners, and air purifiers, building performance simulations were conducted first with the help of the BIM data of the experiment room. In the calculation of the heat losses and workloads that were observed in the simulations as well as in the real experiments, the heat coefficients of the building components in Table 1 were used, depending on the information from the BIM simulation. Thus, the power loads for the radiator or heater, to compensate for the heat losses from the exterior wall and windows in the experiment room at a given time, can be found by Equation (1).

$$H_{load} = h \times A \times (T_{out} - T_{in}) = h \times A \times \Delta T(\mathrm{t}), \quad (1)$$

where h is the heat transfer coefficient, A is the area of the building component (Table 1), and $\Delta T(\mathrm{t})$ is the temperature difference between the exterior (T_{out}) and indoor zones (T_{in}).

3.5. Algorithms for Optimizing the Energy Use of the Automated Thermal Comfort Systems for Ventilation

Given that the research investigated the optimization of energy use in buildings with automated systems, thermal comfort parameters such as natural ventilation were explored together with the contextual information from the experiment environment in deciding the sizing and capacities of the systems to be applied in the contexts of indoor spaces. This section aims to derive algorithms for deciding on the balanced energy levels and energy loss in the experiment environment during the natural ventilation experiments. Thus, the heating loads for the automated systems were aimed to be computed by comparing the used energy with regard to the heating and cooling in the experiment environment. The outcomes of the algorithms were assumed to derive critical energy levels and sensor values for automated systems. Thus, it was also significant to consider the values from the BIM simulations and the modeling in developing the algorithms and computing the building performance by learning the context-based data from the indoor environment to decide on the critical energy levels for the automated state-of-the-art smart systems.

The concern regarding the automated systems for thermal comfort control offers further potential for energy-efficient smart infrastructural developments. To this end, this study explored the design constraints and parameters for efficient automated systems based on energy optimization algorithms and optimized learning models developed with the datasets from the natural ventilation experiments together with the data provided by the BIM software *Autodesk Revit Architecture*.

Even though the flow of air by natural ventilation removes the number of particles with negative effects on the air quality and health, there is a compromise in the heat energy through the decrease in the temperature and humidity of the experiment room. Thus, novel algorithms were explored and applied to the training and test datasets to optimize the changes in GSV with regard to the energy change without directly changing and computing the room's temperature or the building components' heat constants. The energy loss or the amount of work to change GSV was correlated in this research to derive the optimal energy loss function that can be used for artificial systems to improve the air quality without decreasing temperature and heat energy.

In the development of the algorithms, thermal loads and temperature values in the experiment environment were considered balanced or in equilibrium at certain moments. Thus, Equation (2) was deployed for this analysis in the optimization of energy usage and energy loss for the conditioned areas in the buildings by natural ventilation and airflow, with decreasing GSV in the time interval from moments A to B.

$$\frac{\partial_{energy\ (per\ GSV)}}{\partial t} = \frac{H_{loss}\mid_B^A}{\frac{\partial_{GSV}\mid_B^A}{\partial t}} \tag{2}$$

The outcomes of Equation (2) can be defined with the units of *'Joule per GSV'*. Thus, the energy loss to decrease the gas level by artificial means can be optimized by Equation (2), which defines a gradient function for the optimal energy loss that is based on the change in GSV in the dataset. Using Equation (1), (A) denotes the infinitesimally thin and resisting surface area for the heat transfer, like a window opening for ventilation, and (t) is the duration of the natural ventilation experiment.

The heat loss from the wall is compensated for by heating the room with another energy source, the radiator (Figures 1 and 2, Table 1), as discussed, to keep the room temperature steady throughout the day. Thus, the heat loss during the ventilation of the room can be calculated as the direct heat transfer between the room and the exterior environment. Therefore, the heat loss during the ventilation experiments, from moments A to B (Figure 4), can be expressed in Equation (3).

$$H_{loss}\mid_B^A = h_{Air} \times A_{opening} \times \Big|_B^A \begin{matrix} ((T_{outA} - T_{inA}) - \\ (T_{outB} - T_{inB})) \end{matrix} \tag{3}$$

Thus, the changes in GSV can be calculated by Equation (4).

$$\frac{\partial_{GSV}\mid_B^A}{\partial t} = \nabla_{GSV} = \frac{|GSV_A - GSV_B|}{\Delta t\mid_B^A} \tag{4}$$

In this regard, it is significant to consider the outcomes of Equations (2)–(4) in defining and deciding the critical energy levels and optimizing the size, capacities, and energy that can be used by the smart systems in the regulation of the air quality and thermal comfort levels of indoor spaces.

3.6. Applying Machine Learning and Deep Learning Models on the Dataset

Machine learning and deep learning models have also been advanced for the recognition of human activity and thermal comfort levels; they are used in smart spaces, industrial applications, and energy-efficient smart infrastructure [7,24]. In this regard, machine learning and deep learning models were explored further in the ventilation experiments datasets to find the optimal learning models for the operation of energy-efficient and smart automated systems, such as air purifiers, by recognizing and predicting thermal comfort levels and well-being.

In the experiments of the related studies, the well-being values and GSV were also correlated with the air index values by exploring the real-time IoT data, which are considered critical for the user occupation [3]. This correlation is significant for building performance evaluation models trained through machine learning algorithms and artificial neural networks. Thus, the correlating changes in room temperature, humidity, and GSV were separately studied through ML and AI to develop the efficient and optimized learning models in this research.

The earlier related studies on the training dataset showed that the most efficient methods were applied through GPR algorithms [3]. Thus, GPR algorithms were also surveyed and applied in this research to the training as well as the new test datasets for the ventilation experiments. Accordingly, the kernel function for the exponential GPR model [58], which was applied in the experiments of this research, can be iterated in Equation (5).

$$k_e(x_n,\ x_m|\theta) = \sigma_f^2 \times exp\left(-\frac{\sqrt{(x_n - x_m)^T \times (x_n - x_m)}}{\sigma_l}\right) \tag{5}$$

In Equation (5), *theta* (θ) is the parametrization vector; x_n and x_m represent two different inputs in the training data as the observations of the measured sensor values to predict the following one with regard to the given parametrization vector. σ_l stands for the standard deviation between inputs, σ_f denotes the length scale, and T stands for the transpose operator.

The squared exponential kernel function is similarly expressed in Equation (6) [58].

$$k_{se}(x_n,\ x_m|\theta) = \sigma_f^2 \times exp\left(-\frac{1}{2}\frac{(x_n - x_m)^T \times (x_n - x_m)}{\sigma_l^2}\right) \tag{6}$$

Thus, the Matérn 5/2 GPR function can be defined as in Equation (7) [58].

$$k_{mtrn}(x_n,\ x_m|\theta) = \sigma_f^2 \times \left(1 + \frac{\sqrt{3\sqrt{(x_n - x_m)^T \times (x_n - x_m)}}}{\sigma_l}\right) \times \exp\left(-\frac{\sqrt{3\sqrt{(x_n - x_m)^T \times (x_n - x_m)}}}{\sigma_l}\right), \tag{7}$$

and the rational quadratic kernel function is expressed in Equation (8), where alpha (α) is the non-negative parameter of the covariance [58].

$$k_{rq}(x_n,\ x_m|\theta) = \sigma_f^2 \times \left(1 + \frac{\left|(x_n - x_m)^T \times (x_n - x_m)\right|}{2\,\alpha \times \sigma_l^2}\right),\ \alpha \geq 0 \tag{8}$$

In the experiments, the kernel functions and the standard deviation signals were calculated and assigned by MATLAB Regression Learner (in Statistics and Machine Learning Toolbox, Version 12.5, MATLAB R2023a) software. Thus, machine learning and deep learning models were employed using MATLAB Regression Learner (Table 3) and using the abovementioned equations on the training and test datasets.

Table 3. Learning models applied to the natural ventilation datasets.

Model Type	Number of Connected Layers	First Layer Size	Second Layer Size	Third Layer Size
GPR Squared Exponential	-	-	-	-
GPR Matérn 5/2	-	-	-	-
GPR Exponential	-	-	-	-
GPR Rational Quadratic	-	-	-	-
Neural Network—Narrow	1	10	-	-
Neural Network—Medium	1	25	-	-
Neural Network—Wide	1	100	-	-
Neural Network—Bilayered	2	10	10	-
Neural Network—Trilayered	3	10	10	10
Neural Network—Optimizable 1	1	10	-	-
Neural Network—Optimizable 2	3	10	10	10
Neural Network—Optimizable 3	3	10	10	10

In the experiments, GSV and the user activity were studied by further tracing the votes for well-being to find the optimal energy levels for the automated systems to be used for activating the ventilation and in deciding the grounding algorithms and learning models with regard to the standard deviation and mean values, as in Fanger's approach [48].

3.7. Development of Deeper and Efficient CNNs

Regarding the rising challenges in developing extremely small and efficient real-time learning models [59], the research developed novel artificial neural networks with increased depth and various kernel sizes of the CNN, which used the same dataset as that in related articles [3,7,57] (Table 4). Given that the dataset included tiny and compelling inputs about the user activities and thermal comfort levels, the kernel sizes of the neural networks needed to be adjusted to a very constrained dimension, and the models were supposed to be trained and to make fast predictions. Therefore, new deep learning models were developed to decrease the size and parameters of the CNN compared to the CNN applied in the real-time learning system in the related research [3] (Table 4). Increasing the depth and efficiency of the learning models was the research aim and the challenge of this research; the limits were overcome by adding convolution layers, increasing the channels, changing the kernel sizes of the convolution layers, and adding further global pooling layers, as illustrated in Table 4 and Figure 5. The developed neural networks were applied to the dataset with ten classes while only including the four related categories for the thermal comfort levels (2, 3, 4, and 8) in Table 1; this increased the difficulty of training the learning models and testing their performance compared to the earlier research [3]. The models are also compared to the non-Fanger and Fanger's approaches by excluding and including the user-rated well-being values in the dataset to understand the challenges and the potential of the user-defined values in developing the learning models for smart automation systems.

Table 4. Convolutional neural networks (CNNs), developed and applied to the dataset with ten classes.

Model	Number of Convolution Layers	Number of Layers	Number of Connections	Depth of Architectures	Kernel Sizes of Convolutions	Parameters	Model Size (kB)
CNN [3]	1	6	5	1	3×3	13,900	34.9
CNN-D3 (ours)	8	20	19	3	6×6 and 1×1	2500	15.9
CNN-D3_v2 (ours)	11	24	23	3	1×1 and 3×3	12,800	54.9

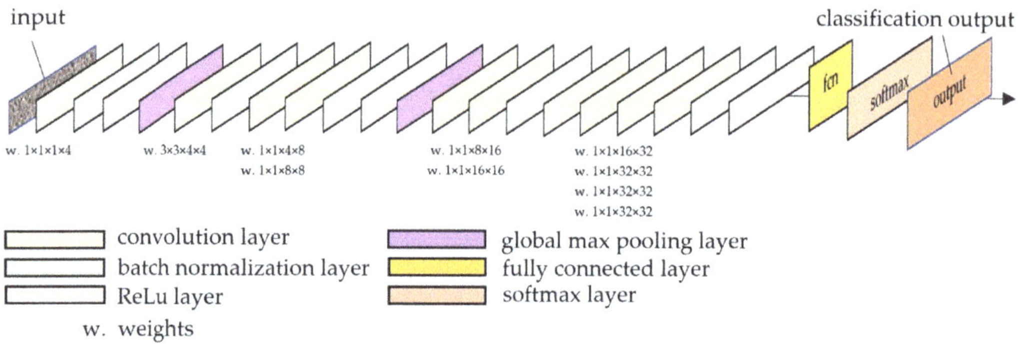

Figure 5. The network architecture of CNN-D3_v2.

4. Experiments and Results

Based on the chosen parameters, the experiments were conducted through BIM simulations, the developed algorithms, and learning models on the datasets in evaluating the building performance and thermal comfort levels with regard to the introduced systematic approach. The methods were applied in order to decide the optimization of the automated smart systems for thermal comfort control and well-being in indoor spaces and special care contexts.

4.1. Experiments and Results on the Heat Losses and Thermal Conductivity

In the experiments, BIM simulation and the insights for the heat losses and power loads were deployed first in order to have proper modeling parameters for the sizing of the HVAC systems. Depending on the digital model of the experiment room, developed via *Autodesk Revit Architecture, Autodesk Insight* software was applied to simulate the hottest and coolest days of one year and to calculate the peak loads for the cooling and heating of the location of the experiment area (Table 5).

Table 5. The calculated power loads and building performance analyses for thermal conductivity.

Experiment Conditions	Instant Sensible Load (W)	Delayed Sensible Load (W)	Power Load (W) (Heat Loss)
21 January at 00:15, outside: −8.0 °C, inside: 21.1 °C (Heating—BIM simulation)			
Glass—Conduction	−89	-	-
Wall	-	−102	-
21 July at 14:45, outside: 37.8 °C, inside: 23.9 °C (Cooling—BIM simulation)			
Glass—Conduction	12	-	-
Wall	-	132	-
18 November at 12:30, outside: 13.9 °C, inside: 25.9 °C (Ventilation—Real experiment)			
Glass—Conduction	-	-	−64.06
Wall	-	-	−218.68
14 March at 14:30, outside: 15.0 °C, inside: 26 °C (Ventilation—Real experiment)			
Glass—Conduction	-	-	−58.72
Wall	-	-	−200.46

A similar methodology was followed in the ventilation experiments to evaluate the building performance in real-world conditions (Table 5). Accordingly, the ventilation experiment, which was similar to the one on 18 November (Figure 4), was repeated on 14 March 2021. The experiments were conducted in the same room with an average room temperature of 26 °C and the outside temperature of 15 °C at noon, at 14:30 (Figure 6).

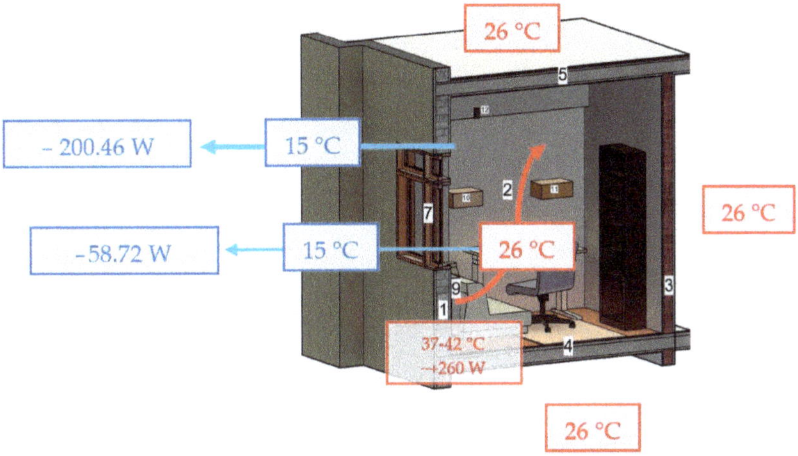

Figure 6. BIM visualization of the experiment room for simulating the building performance of thermal conductivity on 14 March.

In the calculations of the heat loads using Equation (1), it was decided that all the interior zones and other neighboring rooms would have the same temperature as the experiment room (Figure 6). Given that the heat coefficients for the building components were in Kelvin (K), the temperature values were converted into K during the calculations. As a result, the heating load for the radiator was found to be 259.18 Watts (W), at least, to compensate for the heat loss from the exterior wall, 200.46 W, and from the windows, 58.72 W, for this real experiment on 14 March (Table 5). Similarly, 282.74 W was needed for the radiator to compensate for the heat loss from the wall, 218.68 W, and the windows, 64.06 W, on 18 November.

The observations from the natural ventilation experiments on 14 March were also arranged as the test dataset, as illustrated in Figure 7, to examine the performance of the developed learning models. Thus, the observation values were also planned to derive the values of the minimal energy loss for a certain amount of betterment in the air quality.

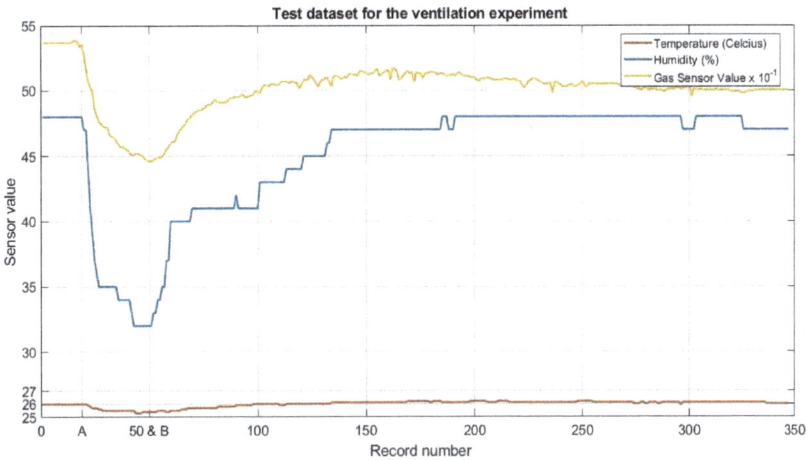

Figure 7. Test dataset for the ventilation experiments on 14 March.

4.2. Energy Optimization of the Automated HVAC Systems with Regard to the Energy Lost by the Natural Ventilation

In the calculation of the minimal energy loss for a certain amount of betterment in the air quality, the abovementioned Equations (2)–(4) were employed together on the given temperature and GSV acquired on 18 November and 14 March (Table 6). Accordingly, the temperature values given in Celsius were converted into K during the calculations (Table 6).

Table 6. Values from the natural ventilation experiments to calculate the energy loss.

Experiment Date	h_{air} (W/(m^2·K)) [60]	$A_{opening}$ (sq·m)	T_{outA} (K)	T_{outB} (K)	T_{inA} (K)	T_{inB} (K)	GSV_A	GSV_B	$\Delta t\,|_B^A$·(s)
18 November [3]	25.5	0.48	287.05	287.05	299.05	298.45	557	463	881
14 March	25.5	0.48	288.15	288.15	299.15	298.55	536	446	828

Accordingly, the heating loads were both equal to −7.344 W (*Joule/s*) during the natural ventilation experiments. The energy used to improve the air quality in the experiments on 18 November was calculated as 6470.064 *Joule* and 68.831 *Joule per GSV*. Similarly, 6080.832 *Joule* of energy was used for the experiments on 14 March, resulting in 67.565 *Joule per GSV*, indicating the energy needed to decrease the GSV per unit.

4.3. Experiments on Machine Learning and Deep Learning Models

Given that the exact conditions were not purely linear and solvable by simpler equations, machine learning and deep learning models were further explored on the training and test datasets of the ventilation experiments (Figures 4 and 7). Thus, the aim was to optimize the operation of the smart systems, such as the air purifiers and conditioners, by deploying the learning models in controlling the thermal comfort and air quality and taking action before the needed energy and desired thermal comfort levels were exceeded. Accordingly, MATLAB Regression Learner software was run to deploy the learning models on the training and test datasets of the natural ventilation experiments that were executed when determining the best computation method for a precise calculation in simulating the thermal conductivity and change in thermal comfort levels. In all the experiments with the learning models, Intel(R) Core(TM) i7-4700HQ CPU @ 2.40 GHz was used as the hardware resource with four cores and eight logical processors. The use of a parallel pool was activated with four local parallel pool workers during the experiments. For the regression experiments, GSV or temperature was selected as the response value, and the remaining two observations, including humidity, were arranged as inputs. To prepare the training data, MATLAB Regression Learner was set through ten cross-validation folds; principal component analysis (PCA) was disabled, and the optimizer was not applicable.

Table 7 illustrates the results of the experiments on the selected machine learning and deep learning models applied to the partial training dataset, which only included the observation values from the sensors from moment A, when the natural ventilation was allowed, to moment B, when the natural ventilation was ended (Figure 4). In these experiments, the inputs were the temperature and humidity values to predict GSV as the response. Accordingly, Figure 8 illustrates the prediction results of the wide neural network on the partial training dataset.

The trained learning models were then tested through the test dataset (Figure 7, Table 7). Figure 9 illustrates the prediction results of the medium neural network, which was trained on the partial training dataset and tested on the partial test dataset and only included the observations from moments A to B (Figure 7).

Table 7. Training and test results of learning models on the partial training dataset for the ventilation experiment. Inputs: temperature and humidity. Response: gas sensor value (GSV).

Model Type	RMSE (Validation)	RMSE (Test)	Number of Iterations	Model Size (kB)	Training Time (s)	Prediction Speed (obs/s)
GPR Squared Exponential	0.477	1.648	-	9	19.113	410
GPR Matérn 5/2	0.446	1.802	-	9	18.398	660
GPR Exponential	0.445	2.004	-	9	17.857	850
GPR Rational Quadratic	0.459	1.846	-	9	17.237	720
Neural Network—Narrow	0.584	1.358	-	4	15.973	470
Neural Network—Medium	0.477	1.314	-	5	20.639	600
Neural Network—Wide	**0.441**	1.843	-	7	20.332	300
Neural Network—Bilayered	0.471	2.246	-	6	20.152	510
Neural Network—Trilayered	0.474	2.476	-	8	24.284	1200
Neural Network—Optimizable 1	2.503	1.793	30	4	33.325	880
Neural Network—Optimizable 2	0.602	**1.288**	30	76	250.21	640
Neural Network—Optimizable 3	0.612	1.405	100	4	406.3	720

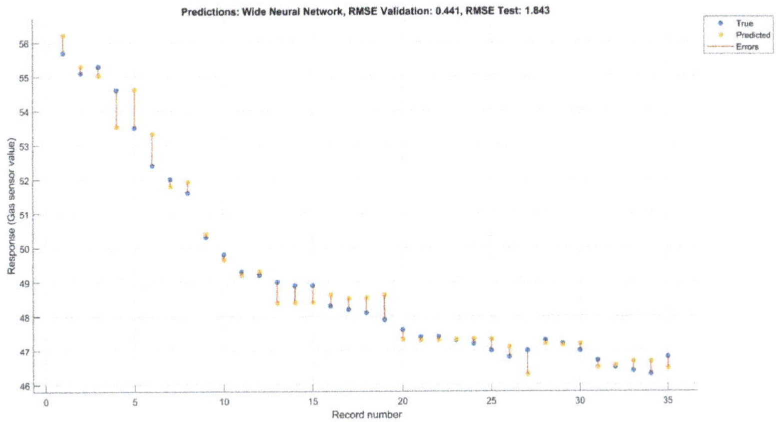

Figure 8. Prediction results of wide neural network on the partial training dataset for the ventilation experiment. Inputs: temperature and humidity. Response: GSV.

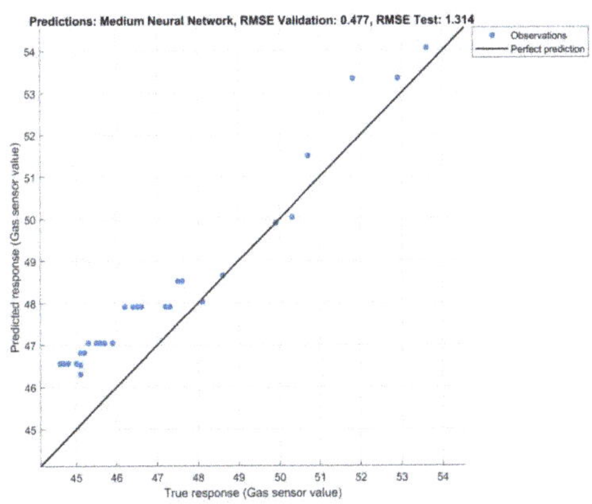

Figure 9. Prediction results of medium neural network on the partial test dataset for the ventilation experiment. Inputs: temperature and humidity. Response: GSV.

The developed learning models that were trained and tested on the selected datasets, and they were also utilized using randomly selected inputs to compare the prediction results with the ground truth (Table 8).

Table 8. Predicted GSV by the trained wide neural network.

Inputs		Output	Ground Truth
Temperature (°C)	**Humidity (Percent)**	**Predicted (GSV) ***	**Real (GSV)**
26	47	53.3	51.8
25.9	44	51.5	50.7
25.8	41	50.9	50.3
25.5	35	48.5	47.3
25.5	35	48.5	47.2

* error < 0.05.

Based on the same partial training dataset, all the selected learning models were also trained by setting the inputs as humidity and GSV in order to predict the temperature as the response (Table 9). The trained models were then tested on the partial dataset using the selected inputs. Figure 10 illustrates the prediction results of the trilayered neural network, tested on the partial test dataset for the ventilation experiment.

Table 9. Training and test results of learning models on the partial training dataset for the ventilation experiment. Inputs: humidity and GSV. Response: temperature.

Model Type	RMSE (Validation)	RMSE (Test)	Number of Iterations	Model Size (kB)	Training Time (s)	Prediction Speed (obs/s)
GPR Squared Exponential	0.056	0.119	-	9	3.9408	680
GPR Matérn 5/2	0.051	0.121	-	9	6.4251	670
GPR Exponential	0.060	0.137	-	9	5.7391	560
GPR Rational Quadratic	**0.052**	0.123	-	9	4.5206	760
Neural Network—Narrow	0.067	0.225	-	4	8.6182	760
Neural Network—Medium	0.067	0.236	-	5	9.5076	820
Neural Network—Wide	0.055	0.170	-	7	11.817	810
Neural Network—Bilayered	0.063	0.159	-	6	11.588	920
Neural Network—Trilayered	0.078	**0.112**	-	8	14.179	960
Neural Network—Optimizable 1	0.110	0.247	30	4	341.47	740
Neural Network—Optimizable 2	0.070	0.181	30	4	47.279	1500
Neural Network—Optimizable 3	0.591	0.652	100	112	185.44	1600

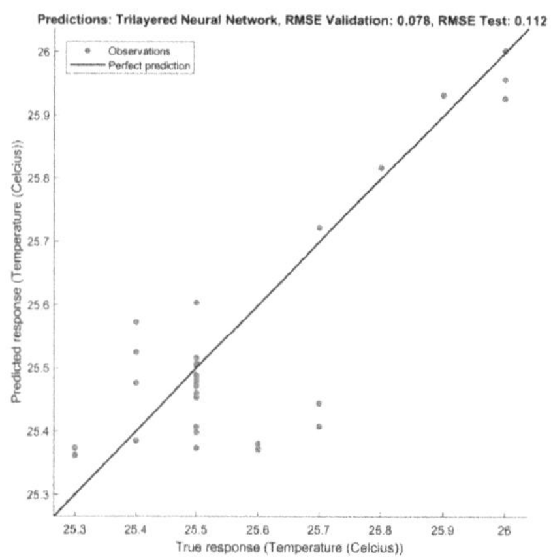

Figure 10. Prediction results of Trilayered Neural Network on the partial test dataset for the ventilation experiment. Inputs: humidity and GSV. Response: temperature.

All the selected learning models were also trained and tested on the whole training and test datasets (Figures 4 and 7) to observe the performances of the learning models in predicting the values for the ventilation experiments. Table 10 illustrates the results of the learning models on the training dataset by setting the inputs as temperature and humidity values and the response as GSV. Accordingly, Figure 11 illustrates the prediction results of the exponential GPR model on the selected training dataset.

Table 10. Training and test results of learning models on the training dataset for the ventilation experiment. Inputs: temperature and humidity. Response: GSV.

Model Type	RMSE (Validation)	RMSE (Test)	Number of Iterations	Model Size (kB)	Training Time (s)	Prediction Speed (obs/s)
GPR Squared Exponential	0.358	3.121	-	17	6.0042	4500
GPR Matérn 5/2	0.355	3.195	-	17	5.3396	5700
GPR Exponential	**0.346**	3.149	-	17	8.7419	6500
GPR Rational Quadratic	0.349	3.157	-	17	14.765	6400
Neural Network—Narrow	0.701	3.108	-	4	6.6549	7300
Neural Network—Medium	0.602	3.434	-	5	11.471	7500
Neural Network—Wide	0.390	3.401	-	7	17.75	9500
Neural Network—Bilayered	0.537	3.135	-	6	15.43	8200
Neural Network—Trilayered	0.431	3.312	-	8	19.729	13,000
Neural Network—Optimizable 1	0.973	**2.508**	30	4	45.322	14,000
Neural Network—Optimizable 2	0.734	2.863	30	4	236.92	14,000
Neural Network—Optimizable 3	0.703	2.858	100	11	599.37	17,000

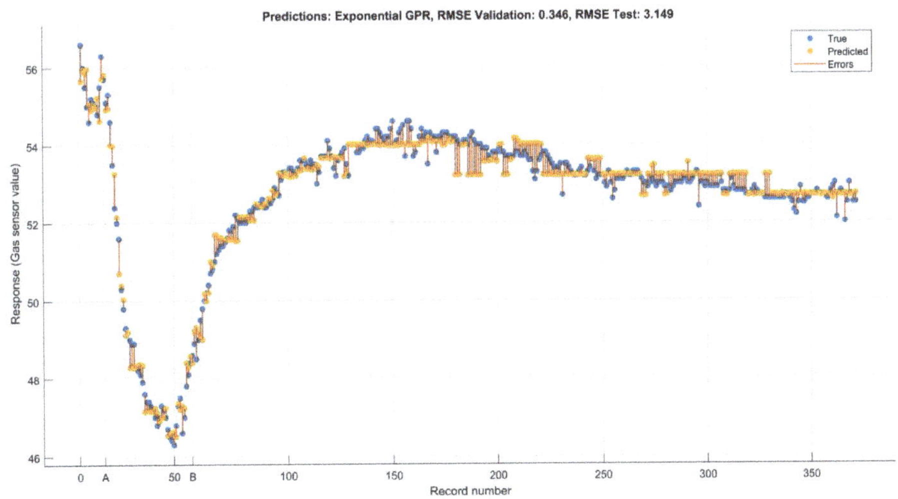

Figure 11. Prediction results of exponential GPR on the training dataset for the ventilation experiment. Inputs: Temperature and humidity. Response: GSV.

The whole training dataset was also prepared by setting the inputs as humidity and GSV in order to predict temperature as the response. The trained learning models were also tested on the test dataset with selected inputs and responses. Accordingly, Table 11 illustrates the training and test results of the learning models. The prediction results of the rational quadratic GPR on the training dataset are illustrated in Figure 12, and the prediction results of optimizable neural network 3 on the test dataset are illustrated in Figure 13.

Table 11. Training and test results of learning models on the training dataset for the ventilation experiment. Inputs: humidity and GSV. Response: temperature.

Model Type	RMSE (Validation)	RMSE (Test)	Number of Iterations	Model Size (kB)	Training Time (s)	Prediction Speed (obs/s)
GPR Squared Exponential	0.039	0.293	-	17	6.5397	8600
GPR Matérn 5/2	0.039	0.308	-	17	8.877	7500
GPR Exponential	0.039	0.335	-	17	9.6738	7900
GPR Rational Quadratic	**0.038**	0.324	-	17	14.116	7300
Neural Network—Narrow	0.061	0.251	-	4	12.466	8900
Neural Network—Medium	0.051	0.344	-	5	19.061	6200
Neural Network—Wide	0.041	0.244	-	7	23.841	11,000
Neural Network—Bilayered	0.052	0.437	-	6	21.347	10,000
Neural Network—Trilayered	0.073	0.389	-	8	26.61	11,000
Neural Network—Optimizable 1	0.301	0.393	30	4	43.231	15,000
Neural Network—Optimizable 2	0.070	0.246	30	13	88.246	19,000
Neural Network—Optimizable 3	0.099	**0.157**	100	4	360.81	17,000

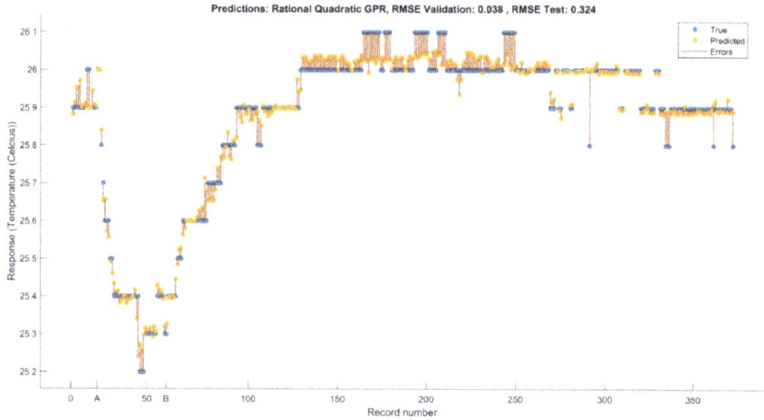

Figure 12. Prediction results of rational quadratic GPR on the training dataset for the ventilation experiment. Inputs: humidity and GSV. Response: temperature.

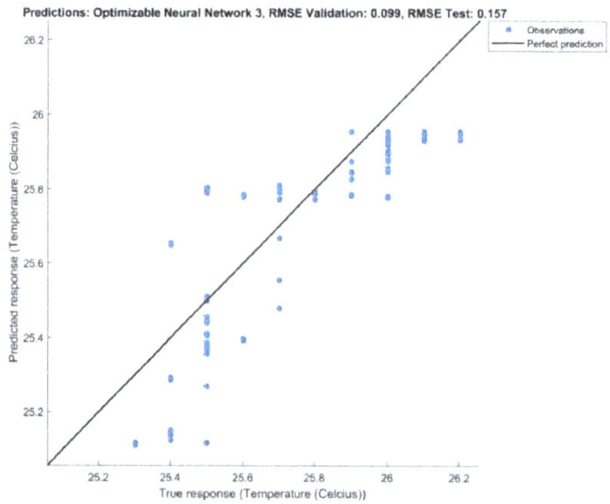

Figure 13. Prediction results of optimizable neural network on the test dataset for the ventilation experiment. Inputs: humidity and GSV. Response: temperature.

4.4. Experiments and Results of Deeper and Efficient CNNs on the Selected Datasets

The results of the natural ventilation experiments revealed that training precise learning models was another challenging task. Thus, the deep learning models were further developed by considering the complex datasets and inputs for smart systems and automated technologies using less energy to predict user activity and thermal comfort levels.

The developed deep learning models (Table 4), considering the challenges and needs related to efficient smart systems and the users' thermal comfort levels and well-being, were tested through the selected datasets. Table 12 illustrates the results of the developed neural networks applied to the dataset with four classes, excluding the user-rated well-being values. Table 13 illustrates the results of the same CNN models, which were also trained and tested to grasp the challenging role of Fanger's approach by including the user-rated well-being values in the training dataset with four classes, categorized by considering the user occupation and thermal comfort levels through six sensor observations for each input.

Table 12. Results of CNNs on the dataset with four selected classes without user-rated well-being values.

Models *	Training Time (s)	Final Training Accuracy (Percent)	Validation Accuracy (Percent)	Test Accuracy (Percent)	Training Loss	Validation Loss
CNN [3]	38	100	100	100	0.001300	0.001300
CNN-D3 (ours)	71	100	100	100	0.001500	0.001500
CNN-D3_v2 (ours)	88	100	100	100	**0.000485**	**0.000485**

* All models were trained for 5000 iterations with a 0.001 initial learning rate.

Table 13. Results of CNNs on the dataset with four selected classes, including the user-rated well-being values.

Models *	Training Time (s)	Final Training Accuracy (Percent)	Validation Accuracy (Percent)	Test Accuracy (Percent)	Training Loss	Validation Loss
CNN [3]	32	77.78	77.78	80	0.3967	0.3967
CNN-D3 (ours)	73	**100**	**100**	**100**	**0.0019**	**0.0019**
CNN-D3_v2 (ours)	88	**100**	**100**	**100**	**0.0301**	**0.0301**

* All models were trained for 5000 iterations with a 0.001 initial learning rate.

The results reveal that the accuracy of CNNs, used for the real-time learning system in the related studies [3,7], sharply decreased when including the user-rated voting values in the dataset (Tables 12 and 13). On the other hand, the CNNs that were developed specifically with regard to the size of the datasets in this research performed much better when compared to the earlier and shallower models, even if they were lightweight and smaller than 16 and 55 kB (Tables 4, 12 and 13).

5. Discussion

Automated systems in buildings are expected to be designed to be energy-efficient. The consideration of environmental facts in the development of energy optimization algorithms and computing technologies offers greater potential for automated systems. The energy-efficient automated systems with deep learning and real-time learning models, developed through the IoT data from the environments, are also gaining increasing attention. Regarding the related seminal works in the literature, this study aimed to deploy a systematic approach to discover the parameters and constraints in optimizing the energy usage of automated solutions for thermal comfort control. In this regard, building performance simulations, energy optimization algorithms, and grey box and black box learning models, such as ML and deep learning models, were explored for the control of thermal comfort and well-being. The chosen methods were conducted through BIM simulations in the evaluation of the building performance (Table 1, Figures 1 and 2), as well as for energy optimization in the heat loss and ventilation experiments via the equations and learning models developed through the IoT-based datasets, including the user-rated voting in accordance with Fanger's seminal approach (Figures 3, 4, 6 and 7, Table 2). Thus, the design of automated systems can be considered in relation to the parameters and the challenges of optimizing the cost of operations and energy usage, as well as the performance of the learning models in thermal comfort control (Tables 3 and 4).

Smart automation systems should consider the parameters for thermal comfort levels, such as temperature, humidity, and air quality. In this research, BIM software was used to derive context-based information about the building components and to simulate the thermal comfort levels with regard to the essential performance parameters. The building performance simulations (Figures 1 and 2, Table 1) for the evaluation of the energy loss and usage in the ventilation experiments (Figures 4 and 7) also made it possible to decide on the functional parameters and electricity features of energy-efficient air purifier and conditioner systems. Thus, based on the facts from the developed BIM simulations, using *Autodesk Revit Architecture* and *Autodesk Insight* software, calculations for the thermal conductivity of the building components and the peak loads of heating, cooling, and ventilation were made (Figures 1 and 2, Tables 1, 5 and 6). Thus, to keep the room temperature steady, the peak load to the HVAC system for cooling was calculated as 144 Won 21 July and 191 W for heating on 21 January, based on the BIM simulations and modeling (Table 5).

It can also be concluded from the experiments that 58.72 W of heat was calculated as being lost from the windows, and 200.46 W was calculated as the heat loss from the wall that the heater should compensate for in order to keep the room temperature constant during and after the ventilation experiments on 14 March (Table 5, Figure 6). Similarly, the experiments on 18 November revealed that the peak load to the radiator was calculated as 282.74 W; to compensate for the heat losses from the wall, it was as 218.68 W; and for the windows, it was as 64.06 W (Table 5). The attempt was made to further optimize the energy usage and the electricity features of the systems via Equations (2)–(4), without tackling the temperature and humidity changes in the experiment area. For instance, on 14 March, a state-of-the-art air purifier needed to have a lowest fan setting of at least 7.344 W (*Joule/s*) to take action instead of natural ventilation; thus, further energy loss from the openings and windows can be minimized by the system without decreasing the temperature of the indoor spaces (Table 6). If the purifier was allowed to apply 40 W for a maximum fan setting, then the system needed to run for at least 161.75 s for the conditions on 18 November and 152 s for the conditions on 14 March for the purification of the indoor air using the same energy as that lost during the experiments (Table 6).

The energy optimization algorithms were only expressed for the simpler solutions in deciding on the needed capacities of the smart systems, whereas the conditions were not purely linear in this research. Thus, the related real-time IoT data from the sensors on the thermal comfort levels were also processed through machine learning and deep learning models to decide on the optimization of the operation of the automated systems, which can be activated or stopped before the needed energy and desired values are exceeded. In other words, apart from the fixed algorithmic solutions, regression models such as machine learning and deep learning models were further explored in finding the optimal model to operate the automated systems (Table 3). A series of experiments was conducted by deploying MATLAB Regression Learner to develop machine learning and deep learning models and to predict the desired sensor values and states that could stimulate the systems to take action or to decrease and terminate the energy usage. Comparing the results of the root mean square error (RMSE) values of learning models revealed that the complex algorithms and deep learning models (Table 4) could be further evaluated for more sophisticated conditions and robust calculations in energy optimization (Tables 7–11).

For instance, in predicting the ventilation experiments, the wide neural network returned the best results for the partial training dataset (Figure 8), and the optimizable (2) neural network had the best performance for the partial test dataset once GSV was set as the response (Table 7). Thus, the wide neural network was also developed and tested to predict GSV by randomly selecting the given inputs (Table 8).

In the following experiments, however, the rational quadratic GPR returned the best RMSE result for the partial training dataset, and the trilayered neural network performed the best for the partial dataset once the temperature of the room was set as the response parameter (Figure 10, Table 9). In predicting the responses for a longer period, exponential GPR returned the best RMSE result for the training dataset (Figure 11), and the optimizable (1) neural network returned the best RMSE for the test dataset once GSV was set as the response (Table 10). By setting temperature as the response (Table 11), rational quadratic GPR returned the best RMSE for the training dataset (Figure 12), even though most of the neural networks also returned similar results. On the other hand, the test dataset allowed for a better examination of the efficiency of the trained networks, and the optimizable (3) neural network returned the best RMSE result by far for the test dataset (Figure 13, Table 11).

In brief, various machine learning and deep learning models yielded the best results for different datasets and response parameters. It can be concluded from these results that different learning models can be trained and optimized for each different action and parameter. On the other hand, it should be noted that most of the optimizable neural networks were more lightweight and returned more consistent results for the training datasets and better results in predicting the test datasets compared to the machine learning algorithms (Tables 7 and 9–11).

Better performance results for the unknown test datasets were much more significant for the real-time learning and real-world applications. Moreover, the prediction accuracies of the neural networks improved by increasing the depth and training iteration; they perform much better on complex test datasets with longer periods. They also had the potential to be improved by increasing the number of fully connected layers (Tables 10 and 11). In this regard, the novel deep neural networks were developed and further trained in this research for energy-efficient automated systems predicting thermal comfort levels and well-being (Table 4).

Regarding the user-rated well-being values and Fanger's approach, the attempt was made to advance the calculation methods by adopting the deep neural networks, such as CNN, in recognizing and correlating the thermal comfort levels and user activities in the datasets (Table 2, Figures 3, 4 and 7). On the other hand, it is observed from the results that developing learning models for ideal prediction is yet another challenge and a difficult research problem, as only a few learning models predicted the ideal responses in this research (Figures 9, 10 and 13).

Thus, the experiments were also conducted to develop new classification-based neural networks by increasing the depth of the learning models (Figure 5), and they were tested on the dataset with the classified user activities (Table 2). Accordingly, the new CNNs with different depths and the CNNs of the earlier related research [3] were used in the experiment on the training dataset by only including four classes of thermal comfort levels (2, 4, 5, 8 in Table 2). The networks were also trained with regard to both non-Fanger and Fanger's approaches by excluding and including the user-rated well-being votes in the dataset.

The results show that the earlier version of the CNN, used in the real-time learning system, needed more depth and accuracy in predicting the dataset with four classes and the user-rated well-being values (Tables 12 and 13). The experiments also revealed that involving the user-rated well-being values also increased the difficulty and challenges for the learning models. In this regard, it can be concluded that the consideration of Fanger's approach in developing deep learning models also needs novel and efficient solutions to be applied to energy-efficient automated systems; the attempt to investigate this was made in this research (Tables 4 and 13).

Limitations and Proposals of the Research

The limitations and future proposals of this work can be briefly discussed. There are still limited data about the building performance and natural ventilation experiments for the explored context. For instance, there are no data from other spaces as there is a lack of smart systems or corresponding IoT data. Thus, only hypothetical models could be developed with regard to the adjacent areas, with the help of BIM simulation; the heat coefficients of the building components of each area were simulated. On the other hand, the developed algorithms allowed the calculation of the optimization of energy use for natural ventilation without depending on the heat coefficients. Instead, machine learning and deep learning models were explored for the more complex conditions to be adapted to the internal dynamics and correlations among the indoor facts and usage patterns and were considered to be much more significant in this research.

Be that as it may, the usage of IoT systems can be encouraged and multiplied in various environments to calculate heat losses and energy efficiency in the surrounding areas. Additionally, the usage periods in this research were not divided into seasons as a limitation of the dataset once the related studies in the literature were considered [6]. Comprehensive datasets from the experiment environments with affordable new hardware resources and scalable computing technologies can be developed and trained for energy-efficient solutions and sustainable environments in different contexts for future work [24,61].

All in all, a systematic approach for defining the parameters and energy-efficient solutions for automated systems in thermal comfort control was explored and executed through the existing datasets in a special care context. The BIM simulations provided abundant resources about the context in evaluating the building energy performance, energy use, and optimization for heating, cooling, and ventilation of the experiment area. These provided the basis for the sizing and electrical features of the state-of-the-art automated systems to be applied in this context. The contextual data from the experiment area were also processed to develop energy-optimization algorithms, ML, and deep learning models in order to predict thermal comfort levels and classify activities so that the efficiency of the automated systems, by applying learning models, could be improved by decreasing the energy usage and the runtime needed to take action. In brief, from the experiments and results, a combination of different optimization methods, algorithms, and learning models can be proposed for developing and deploying energy-efficient state-of-the-art technologies to take action for thermal comfort control with regard to the specific context of indoor environments and the occupations of the users:

- BIM simulations should be encouraged in providing context-based information from the areas in which IoT-based smart systems are to be installed.
- Energy optimization algorithms should be explored and applied with regard to the thermal comfort parameters, such as air quality and context-based data, to be evaluated in the building performance, energy usage, and thermal conductivity when deciding on the optimal electricity features of the automated systems without changing other thermal comfort levels.
- Machine learning models that fit best when optimizing the energy use and operation of the automated systems should also be discovered and applied with regard to the thermal comfort parameters and contextual information.
- Finally, more complex, yet lightweight, deep learning models that are to be trained quickly and are good at accurate prediction should be explored and applied to smart systems that take action by recognizing the classified activities needed to control thermal comfort and well-being.

6. Conclusions

Smart systems and technologies in buildings offer well-being and thermal comfort control, and they have the potential to be developed systematically with regard to thermal comfort parameters, user preferences, and energy-efficient concerns. Based on the experiments conducted, it can be inferred that BIM simulations and IoT systems are highly effective in acquiring, generating, and processing comprehensive data from indoor spaces. Additionally, black box models were found to be highly useful for processing sensor data through advanced algorithms and learning models, thereby enabling effective thermal comfort control and enhancing overall well-being. These findings have significant implications for the development of advanced building management systems and can pave the way for more efficient and sustainable indoor environments in the future.

It is clear from the results that context-based information and the outcomes of computations of the building performances of indoor spaces, acquired with the help of BIM simulation and modeling, are absolutely crucial in the evaluation of the experiments related to thermal comfort parameters, such as natural ventilation, in order to develop zero-energy buildings as well as state-of-the-art and energy-efficient automated systems for thermal comfort control. Moreover, energy-optimization algorithms that are developed with the help of context-based information from BIM simulations and smart systems allow for a decision on the critical thermal comfort parameters, energy usage, and power loads for the capacities and electricity features of the automated systems. Nevertheless, the energy-optimization algorithms are only considered when the heating loads and thermal conditions are caught in a balance during the ventilation experiments, even though the phenomena of airflow and natural ventilation need to be considered through the changing states. In this regard, the development of machine learning and deep learning models is understood to be extremely significant in the optimization of the operation of energy-efficient automated systems that can predict the critical energy levels and user activity when the temperature and heating loads in the indoor spaces are changed and not in equilibrium, or once they cannot be resolved by simpler equations. The experiments on learning models showed that machine learning and deep learning models that fit best to the conditional changes might facilitate and optimize the operation of the automation systems efficiently with regard to the varying circumstances and might help to decide how and when these intelligent systems should take action. In this regard, the use of artificial intelligence models in the development of state-of-the-art automated systems is also seen as crucial in understanding user activity and thermal comfort levels in indoor spaces. Thus, user-defined ratings and Fanger's approach were considered in developing novel learning models to decide on the thermal comfort levels during the operation of smart systems, which were found to be absolutely crucial in recognizing and predicting the user activity and thermal comfort levels that can significantly decrease the energy used for thermal comfort control.

The systematic approaches applied in this research can enable the development of energy-efficient state-of-the-art smart systems for thermal comfort control and well-being. Consequently, thermal comfort levels in buildings can be optimized by the automated systems with regard to the contextual information about the building components and the evaluation of their performances and also by considering the user activities and preferences in indoor environments in designing energy-efficient buildings with smart systems.

Funding: This research received no external funding.

Data Availability Statement: Publicly available datasets were analyzed in this study. These data can be found here: [https://thingspeak.com/channels/1229234 (accessed on 16 July 2023)].

Conflicts of Interest: The author declares no conflict of interest.

Abbreviations

AI	Artificial Intelligence
BIM	Building Information Modeling
°C	Celsius
CH_4	Methane
C_4H_{10}	Butane
cm	Centimeter
CNN	Convolutional Neural Network
CO_2	Carbon dioxide
GPR	Gaussian Process Regression
GPU	Graphic Processing Unit

GSV	Gas Sensor Value
HVAC	Heating, Ventilation, Air Conditioning
IoT	Internet of Things
K	Kelvin
kB	Kilobyte
LAN	Local Area Network
LPG	Liquefied Petroleum Gas
m	Meter
min	Minute
ML	Machine Learning
obs/s	Observations per second
PCA	Principal Component Analysis
PCS	Personal Comfort Systems
PMV	Predicted Mean Vote
R-CNN	Region-Based Convolutional Neural Network
RMSE	Root Mean Squared Error
s	Second
sq m	Square Meter
W	Watt
WOB	Window Opening Behavior

References

1. Grassi, B.; Piana, E.A.; Lezzi, A.M.; Pilotelli, M. A Review of Recent Literature on Systems and Methods for the Control of Thermal Comfort in Buildings. *Appl. Sci.* **2022**, *12*, 5473. [CrossRef]
2. Ji, W.; Yang, L.; Liu, Z.; Feng, S. A Systematic Review of Sensing Technology in Human-Building Interaction Research. *Buildings* **2023**, *13*, 691. [CrossRef]
3. Erişen, S. Real-Time Learning and Monitoring System in Fighting against SARS-CoV-2 in a Private Indoor Environment. *Sensors* **2022**, *22*, 7001. [CrossRef] [PubMed]
4. Xing, D.; Li, N. Numerical Prediction of Mean Radiant Temperature in Radiant Cooling Indoor Environments. *J. Therm. Sci.* **2022**, *31*, 359–369. [CrossRef]
5. He, Y.; Li, N.; Zhang, H.; Han, Y.; Lu, J.; Zhou, L. Air-conditioning use behaviors when elevated air movement is available. *Energy Build.* **2020**, *225*, 110370–110382. [CrossRef]
6. He, Y.; Li, N.; Lu, J.; Li, N.; Deng, Q.; Tan, C.; Yan, J. Meeting thermal needs of occupants in shared space with an adjustable thermostat and local heating in winter: An experimental study. *Energy Build.* **2021**, *236*, 110776–110790. [CrossRef]
7. Erişen, S. IoT-Based Real-Time updating multi-layered learning system applied for a special care context during COVID-19. *Cogent Eng.* **2022**, *9*, 2044588. [CrossRef]
8. Wei, J.; Li, J.; Zhao, J.; Wang, X. Hot Topics and Trends in Zero-Energy Building Research—A Bibliometrical Analysis Based on CiteSpace. *Buildings* **2023**, *13*, 479. [CrossRef]
9. Gassar, A.A.A.; Koo, C.; Kim, T.; Cha, S.H. Performance Optimization Studies on Heating, Cooling and Lighting Energy Systems of Buildings during the Design Stage: A Review. *Sustainability* **2021**, *13*, 9815. [CrossRef]
10. Chew, L.W.; Li, X.-X.; Chew, M.Y.L. Climate Change Projection and Its Impacts on Building Façades in Singapore. *Sustainability* **2023**, *15*, 3156. [CrossRef]
11. Irulegi, O.; Torres, L.; Serra, A.; Mendizabal, I.; Hernández, R. The Ekihouse: An energy self-sufficient house based on passive design strategies. *Energy Build.* **2014**, *83*, 57–69. [CrossRef]
12. Lopez-Escamilla, A.; Herrera-Limones, R.; Leon-Rodríguez, A.L. Evaluation of environmental comfort in a social housing prototype with bioclimatic double-skin in a tropical climate. *Build. Environ.* **2022**, *218*, 109119. [CrossRef]
13. Santy; Matsumoto, H.; Tsuzuki, K.; Susanti, L. Bioclimatic Analysis in Pre-Design Stage of Passive House in Indonesia. *Buildings* **2017**, *7*, 24. [CrossRef]
14. Ji, Y.; Xu, M.; Zhang, T.; He, Y. Intelligent Parametric Optimization of Building Atrium Design: A Case Study for a Sustainable and Comfortable Environment. *Sustainability* **2023**, *15*, 4362. [CrossRef]
15. Abdullah, H.K.; Alibaba, H.Z. A Performance-Based Window Design and Evaluation Model for Naturally Ventilated Offices. *Buildings* **2022**, *12*, 1141. [CrossRef]
16. Kim, A.; Wang, S.; Kim, J.-E.; Reed, D. Indoor/Outdoor Environmental Parameters and Window-Opening Behavior: A Structural Equation Modeling Analysis. *Buildings* **2019**, *9*, 94. [CrossRef]
17. Chen, R.; Liu, Y.-T.; Tsay, Y.-S. An Air Conditioning Design Strategy of the Stepped Hall Based on Building Performance Simulation. *Buildings* **2022**, *12*, 1269. [CrossRef]

18. Guerra-Santin, O.; Itard, L. Occupants' behaviour: Determinants and effects on residential heating consumption. *Build. Res. Inf.* **2010**, *38*, 318–338. [CrossRef]
19. Carpino, C.; Loukou, E.; Heiselberg, P.; Arcuri, N. Energy performance gap of a nearly zero energy building (nZEB) in Denmark: The influence of occupancy modelling. *Build. Res. Inf.* **2020**, *48*, 899–921. [CrossRef]
20. Cao, Y.; Kamaruzzaman, S.N.; Aziz, N.M. Building Information Modeling (BIM) Capabilities in the Operation and Maintenance Phase of Green Buildings: A Systematic Review. *Buildings* **2022**, *12*, 830. [CrossRef]
21. Hu, W.; Lim, K.Y.H.; Cai, Y. Digital Twin and Industry 4.0 Enablers in Building and Construction: A Survey. *Buildings* **2022**, *12*, 2004. [CrossRef]
22. Hammad, A.W. Minimising the Deviation between Predicted and Actual Building Performance via Use of Neural Networks and BIM. *Buildings* **2019**, *9*, 131. [CrossRef]
23. Lee, J.H.; Ostwald, M.J.; Kim, M.J. Characterizing Smart Environments as Interactive and Collective Platforms: A Review of the Key Behaviors of Responsive Architecture. *Sensors* **2021**, *21*, 3417. [CrossRef] [PubMed]
24. Erişen, S. Incremental transformation of spatial intelligence from smart systems to sensorial infrastructures. *Build. Res. Inf.* **2021**, *49*, 113–126. [CrossRef]
25. Al-Obaidi, K.M.; Hossain, M.; Alduais, N.A.M.; Al-Duais, H.S.; Omrany, H.; Ghaffarianhoseini, A. A Review of Using IoT for Energy Efficient Buildings and Cities: A Built Environment Perspective. *Energies* **2022**, *15*, 5991. [CrossRef]
26. Claypool, M.; Garcia, M.; Retsin, G.; Soler, V. Introduction: Architecture in the age of automation. In *Robotic Building. Architecture in the Age of Automation*; Claypool, M., Garcia, M., Retsin, G., Soler, V., Eds.; Detail Business Information GmbH: Munich, Germany, 2019; pp. 10–20.
27. Shang, C.; Chang, C.-Y.; Liu, J.; Zhao, S.; Roy, D. FIID: Feature-based implicit irregularity detection using unsupervised learning from IoT data for homecare of elderly. *IEEE Internet Things* **2020**, *7*, 10884–10896. [CrossRef]
28. Deen, M.J. Information and communications technologies for elderly ubiquitous healthcare in a smart home. *Pers. Ubiquitous Comput.* **2015**, *15*, 573–599. [CrossRef]
29. Kocaturk, T. Intelligent building paradigm and data-driven models of innovation. *Archit. Eng. Des. Manag.* **2019**, *15*, 311–312. [CrossRef]
30. Dave, B.; Buda, A.; Nurminen, A.; Främling, K. A framework for integrating BIM and IoT through open standards. *Autom. Constr.* **2018**, *95*, 35–45. [CrossRef]
31. Li, C.Z.; Xue, F.; Li, X.; Hong, J.; Shen, G.Q. An Internet of Things-enabled BIM platform for on-site assembly services in prefabricated construction. *Autom. Constr.* **2018**, *89*, 146–161. [CrossRef]
32. Louis, J.; Dunston, P.S. Integrating IoT into operational workflows for real-time and automated decision-making in repetitive construction operations. *Autom. Constr.* **2018**, *94*, 317–327. [CrossRef]
33. McGlinn, K.; Brennan, R.; Debruyne, C.; Meehan, A.; McNerney, L.; Clinton, E.; Kelly, P.; O'Sullivan, D. Publishing authoritative geospatial data to support interlinking of building information models. *Autom. Constr.* **2021**, *124*, 103534. [CrossRef]
34. Martín-Garín, A.; Millán-García, J.; Baïri, A.; Millán-Medel, J.; Sala-Lizarraga, J. Environmental monitoring system based on an Open Source Platform and the Internet of Things for a building energy retrofit. *Autom. Constr.* **2018**, *87*, 201–214. [CrossRef]
35. Pedrasa, M.A.A.; Spooner, T.D.; MacGill, I.F. Coordinated scheduling of residential distributed energy resources to optimize smart home energy services. *IEEE Trans Smart Grid* **2010**, *1*, 134–143. [CrossRef]
36. Al-Ali, A.; Zualkernan, I.A.; Rashid, M.; Gupta, R.; AliKarar, M. A smart home energy management system using IoT and Big Data analytics approach. *IEEE Trans. Consum. Electron.* **2017**, *63*, 426–434. [CrossRef]
37. Brookfield, K.; Fitzsimons, C.; Scott, I.; Mead, G.; Starr, J.; Thin, N.; Tinker, A.; Thompson, C.W. The home as enabler of more active lifestyles among older people. *Build. Res. Inf.* **2015**, *43*, 616–630. [CrossRef]
38. Roulet, C.-A.; Flourentzou, F.; Foradini, F.; Bluyssen, P.; Cox, C.; Aizlewood, C. Multicriteria analysis of health, comfort and energy efficiency in buildings. *Build. Res. Inf.* **2006**, *34*, 475–482. [CrossRef]
39. Eker, S.; Zimmermann, N.; Carnohan, S.; Davies, M. Participatory system dynamics modelling for housing, energy and wellbeing interactions. *Build. Res. Inf.* **2018**, *46*, 738–754. [CrossRef]
40. Sharpe, T. Mainstreaming building performance evaluation for the benefit of users. *Build. Res. Inf.* **2018**, *47*, 251–254. [CrossRef]
41. Tweed, C.; Zapata-Lancaster, G. Interdisciplinary perspectives on building thermal performance. *Build. Res. Inf.* **2018**, *46*, 552–565. [CrossRef]
42. Zhang, Y.; Tzortzopoulos, P.; Kagioglou, M. Healing built-environment effects on health outcomes: Environment-occupant-health framework. *Build. Res. Inf.* **2019**, *47*, 747–766. [CrossRef]
43. Gunatilaka, R.N.; Abdeen, F.N.; Sepasgozar, S.M.E. Developing a Scoring System to Evaluate the Level of Smartness in Commercial Buildings: A Case of Sri Lanka. *Buildings* **2021**, *11*, 644. [CrossRef]
44. Tardioli, G.; Filho, R.; Bernaud, P.; Ntimos, D. An Innovative Modelling Approach Based on Building Physics and Machine Learning for the Prediction of Indoor Thermal Comfort in an Office Building. *Buildings* **2022**, *12*, 475. [CrossRef]
45. Kim, T.W.; Kim, Y.; Cha, S.H.; Fischer, M. Automated updating of space design requirements connecting user activities and space types. *Autom. Constr.* **2015**, *50*, 102–110. [CrossRef]
46. Vischer, J. Towards a user-centred theory of the built environment. *Build. Res. Inf.* **2008**, *36*, 231–240. [CrossRef]
47. Almusaed, A.; Yitmen, I. Architectural Reply for Smart Building Design Concepts Based on Artificial Intelligence Simulation Models and Digital Twins. *Sustainability* **2023**, *15*, 4955. [CrossRef]

48. Yu, J.; Chang, W.-S.; Dong, Y. Building Energy Prediction Models and Related Uncertainties: A Review. *Buildings* **2022**, *12*, 1284. [CrossRef]
49. Lala, B.; Hagishima, A. A Review of Thermal Comfort in Primary Schools and Future Challenges in Machine Learning Based Prediction for Children. *Buildings* **2022**, *12*, 2007. [CrossRef]
50. Anastasiadou, M.; Santos, V.; Dias, M.S. Machine Learning Techniques Focusing on the Energy Performance of Buildings: A Dimensions and Methods Analysis. *Buildings* **2022**, *12*, 28. [CrossRef]
51. Guo, H.; Duan, D.; Yan, J.; Ding, K.; Xiang, F.; Peng, R. Machine Learning-Based Method for Detached Energy-Saving Residential Form Generation. *Buildings* **2022**, *12*, 1504. [CrossRef]
52. Guo, J.; Li, M.; Jin, Y.; Shi, C.; Wang, Z. Energy Prediction and Optimization Based on Sequential Global Sensitivity Analysis: The Case Study of Courtyard-Style Dwellings in Cold Regions of China. *Buildings* **2022**, *12*, 1132. [CrossRef]
53. Mohammadiziazi, R.; Bilec, M.M. Application of Machine Learning for Predicting Building Energy Use at Different Temporal and Spatial Resolution under Climate Change in USA. *Buildings* **2020**, *10*, 139. [CrossRef]
54. Elshafei, G.; Vilceková, S.; Zeleňáková, M.; Negm, A.M. An Extensive Study for aWide Utilization of Green Architecture Parameters in Built Environment Based on Genetic Schemes. *Buildings* **2021**, *11*, 507. [CrossRef]
55. Fallah, A.M.; Ghafourian, E.; Sichani, L.S.; Ghafourian, H.; Arandian, B.; Nehdi, M.L. Novel Neural Network Optimized by Electrostatic Discharge Algorithm for Modification of Buildings Energy Performance. *Sustainability* **2023**, *15*, 2884. [CrossRef]
56. Xie, Y.; Hu, W.; Zhou, X.; Yan, S.; Li, C. Artificial Neural Network Modeling for Predicting and Evaluating the Mean Radiant Temperature around Buildings on Hot Summer Days. *Buildings* **2022**, *12*, 513. [CrossRef]
57. IoT Channel for Real-Time Learning & Monitoring. Available online: https://thingspeak.com/channels/1229234 (accessed on 16 June 2023).
58. Zhang, N.; Xiong, J.; Zhong, J.; Leatham, K. Gaussian Process Regression Method for Classification for High-Dimensional Data with Limited Samples. In Proceedings of the 2018 Eighth International Conference on Information Science and Technology (ICIST), Cordoba, Granada, Seville, Spain, 30 June–6 July 2018.
59. Strickland, E. Andrew Ng, AI Minimalist: The Machine-Learning Pioneer Says Small is the New Big. *IEEE Spectr.* **2022**, *59*, 22–25. [CrossRef]
60. The Engineering Toolbox.com. Available online: https://www.theengineeringtoolbox.com/air-properties-viscosity-conductivity-heat-capacity-d_1509.html (accessed on 10 March 2021).
61. Emmitt, S. *Architectural Technology: Research & Practice*; Wiley-Blackwell: Chichester, UK, 2013.

buildings

Article

An Experimental Study on Human Thermal Comfort with Thermal-Conductive Bed during Sleep in Summer

Jinhua Hu [1,2], Xiaoli Hao [1,2,*], Minhua Huang [1], Yingdong He [3,*], Nianping Li [3], Yaolin Lin [4] and Shiqiang Chen [1]

1 College of Civil Engineering, Hunan University of Science and Technology, Xiangtan 411201, China; hujinhua@hnust.edu.cn (J.H.); huangm_h@163.com (M.H.); shiqiangchen@hnust.edu.cn (S.C.)
2 Hunan Engineering Research Center for Intelligently Prefabricated Passive House, Hunan University of Science and Technology, Xiangtan 411201, China
3 College of Civil Engineering, Hunan University, Changsha 410082, China; linianping@126.com
4 School of Environment and Architecture, University of Shanghai for Science and Technology, Shanghai 200093, China; yaolinlin@gmail.com
* Correspondence: haoxiaoli2002@aliyun.com (X.H.); heyingdong2022@hnu.edu.cn (Y.H.)

Abstract: Sleep is vital for human health, while sleep quality is affected by indoor thermal environments. This study aims to investigate human thermal comfort with a thermal-conductive bed during sleep in summer. A series of experiments were conducted in a climatic chamber of a university. Subjects slept on a thermal-conductive bed, with or without the supply of water cooling them by flowing through the pipes embedded in the bed. The supply water was set at 20, 25, 28, and 30 °C. The indoor temperature was set at 32 °C. The bed surface temperature (back, buttock, thigh, and calf) and the subject's skin temperature were measured. Questionnaires after waking up the next morning were conducted. The results show that when there was no water supply in the pipes of the thermal-conductive bed, the bed surface temperature was 0–1.5 °C higher than the indoor temperature, and subjects felt slightly warm. When the supply water temperature was 28 or 30 °C, subjects felt thermally comfortable during a night's sleep. This study demonstrates that the proposed thermal-conductive bed with supply water temperatures of 28–30 °C can create a comfortable sleep environment for residents who have no air-conditioning systems in summer, which can also help save building energy.

Keywords: thermal comfort; thermal-conductive bed; sleep comfort; thermal environment

Citation: Hu, J.; Hao, X.; Huang, M.; He, Y.; Li, N.; Lin, Y.; Chen, S. An Experimental Study on Human Thermal Comfort with Thermal-Conductive Bed during Sleep in Summer. *Buildings* **2023**, *13*, 1936. https://doi.org/10.3390/buildings13081936

Academic Editor: Cinzia Buratti

Received: 3 July 2023
Revised: 24 July 2023
Accepted: 27 July 2023
Published: 29 July 2023

1. Introduction

People spend approximately one-third of their lifetime sleeping. Sleep can help people recover from physical and mental fatigue caused by daytime activities, which is essential to help the body restore energy to maintain physical function [1,2]. Sleep quality and sleep thermal comfort could be strongly affected by indoor thermal environments [3–5]. A comfortable thermal environment is vital to short sleep latency and obtaining deep sleep [6]. Thus, to obtain higher work or study efficiency during the daytime, air conditioners are adopted in residential bedrooms in many buildings [7]. According to a survey in Shanghai, 90% of 800 families used air conditioners during the whole sleep time in summer [8]. In Hongkong, where the hot and humid summer lasts for over four months, 68% of 554 respondents kept their air conditioners running all night [9]. As a result, a large amount of energy was consumed. In China mainland, the electricity consumption of air conditioners accounted for over 30% of the total power consumption of residential buildings in summer [10]. In Hongkong, the proportion of electricity usage from residential air conditioners in 1971 accounted for 14.6%, while it increased to 30.4% in 1996 and 45% in 2011, respectively [11,12].

Nevertheless, many discomfort complaints were found when air conditioning systems were used, especially in summer [13,14]. A study in Hongkong showed that around 60% of the respondents experienced waking up even if they turned on air conditioners. They felt cold or warm during sleep [9]. Kim et al. [15] showed similar results that occupants in Korea were exposed to either high or low indoor temperatures during sleep. Air temperatures that deviated from the neutral temperature would raise the frequency and duration of wakefulness and reduce slow wave sleep (SWS) and rapid eye movement sleep [16]. By estimating thermal comfort and sleep quality before and after sleep, Pan et al. [17] found that the requirements for thermal comfort of sleeping people differ from that of waking people. It has been investigated that sleep quality was affected by cold exposure or moderate heat in a constant thermal environment [18].

Undoubtedly, heating, ventilation, and air conditioning (HVAC) systems play a vital role in creating comfortable indoor thermal environments in modern buildings. However, the high level of thermal comfort certainly needs much more energy [7]. For the reason of economic development and environmental protection, energy consumption in buildings has attracted people's attention [19,20]. Hoyt et al. [21] and Zhang et al. [22] pointed out that every 1 °C increase in the set-point temperature for cooling can roughly reduce 10% energy consumption of an air conditioning system. Under these circumstances, methods to both save energy and improve thermal comfort have been explored, such as a personalized ventilation system. In China, electric fans are widely accepted as an effective method to improve the thermal environment and save energy in summer [23,24]. In a study conducted by He et al. [24], the fan-use rate was 80% at the environmental temperature of 30 °C.

Furthermore, the methods to improve thermal comfort during sleep have attracted the attention of some researchers in recent years. Okamoto-Mizuno et al. [25] performed a series of experiments to confirm the effect of head cooling on human sleep stages and body temperature. The obtained results show that the whole-body sweating rate was significantly decreased by the use of a cooling pillow during sleep under humid heat conditions. Irshad K. et al. [26] conducted a field study on sleeping comfort in a test chamber with a thermoelectric air-cooling system. The results show that the subjects felt comfortable and had a better sleep quality in a thermoelectrically cooled room at a power input of 720 W. Yu et al. [23] studied the proper electric fan control strategy in summer sleep thermal environments. The results indicate that the dynamic air supply control strategy of the fan, which was stabilizing–falling–stabilizing–rising–stabilizing, can improve sleep quality. Another similar study found that when the indoor temperature was below 32 °C, providing appropriate air velocities could alleviate the thermal discomfort of occupants during sleep [27]. Furthermore, He et al. [24,28] conducted a series of experiments on subjects' thermal comfort with desk fans in a hot–humid environment. They found that desk fans made more than 80% of subjects feel acceptable at 28 and 30 °C. In addition, He et al. [29,30] also found that the radiant cooling desk reduced warm sensation and increased both comfort and acceptability of subjects in hot environments.

Moreover, Pan et al. [31] designed a bed-based task/ambient conditioning (TAC) system, which was defined as the environmental conditioning system that allowed occupants to independently control their local thermal micro-environments in buildings. The results indicate that the application of the bed-based TAC system would help to attain obvious energy savings. In order to investigate the effect of TAC on the local body of sleeping people, Lan et al. [32] designed a bedside personalized ventilation (PV) system to provide fresh and cool air directly to the head and face of sleepers. After using the bedside PV system, people felt cool, and their skin temperature dropped. As reported in Nunneley SA et al.'s studies [33,34], cooling the head reduces thermal stress more effectively than cooling any other area of the body. Furthermore, Krauchi et al. [35] indicated that the skin temperature of the foot was important for good sleep. It has to be maintained at a relatively high temperature to facilitate the rapid onset of sleep.

As indicated by the above-mentioned studies, local cooling can improve human thermal comfort during sleep in summer. However, some problems remain unresolved. For example, the application of the TAC system to sleeping environments in real buildings helps reduce energy consumption; however, the bulky air ducts and ventilation plenums of the TAC system may cause inconvenience to occupants [31]. Moreover, although the TAC system was improved by removing the air ducts [36–38], it might still cause a cold draft. Furthermore, during the application of the bedside PV system to sleepers, the ventilation time could be continuous for 7 h or more. In a study by Xia et al. [39], headache or dizziness were reported when people were exposed to air currents for a long time. Additionally, some diseases, such as pains in joints, rheumatism, or colds, might be caused by air drafts during night sleep [40]. For air conditioning systems based on ceiling radiant cooling panels (CRCP), the main way of exchanging heat with internal heat sources in a sleep environment is radiation. Although the CRCP-based air conditioning systems could avoid the cold draft, condensation was observed from time to time due to the high level of humidity [41]. Moreover, radiant heat transfer was not as efficient as convective or conductive heat transfer, and the radiant cooling systems consumed more energy. For example, He et al. [42,43] indicated that the radiant cooling desk consumed cooling energy 89.9, 104.4, and 130.7 W at 28, 30, and 32 °C, respectively.

In this paper, a thermal-conductive bed that directly cools sleeping users was proposed. The main purpose aims to study the effects of the proposed bed on human thermal comfort during sleep in summer, as well as its energy-saving potential. Firstly, the configuration of the thermal-conductive bed was described. Then, a series of experiments were performed in a climatic chamber during the hot summer of 2021. During the experiments, subjects' thermal sensation and thermal comfort votes were recorded by answering questionnaires. Furthermore, the skin temperature of different body areas of the subjects was measured. Finally, the energy-saving potential of the thermal-conductive bed was discussed. This study provides a potential solution for energy efficiently improving human thermal comfort during sleep in summer, especially for those people who have no air conditioning indoors.

2. Methodology

2.1. Description of Thermal-Conductive Bed

Figure 1 shows the configuration of a thermal-conductive bed. A water tank (No. 21 in Figure 1a) was installed to supply the required water to the thermal-conductive bed. The supply of water sent to the thermal-conductive bed was provided by a circulating water pump (No. 18 in Figure 1a). The bedstead was equipped with both bedhead pipe and bed-end pipe with a diameter of 15 mm for supply water circulation. The stainless-steel pipes, which have the characteristics of good thermal conductivity and low cost, were used for the pipes. The pipes were installed in the upper panel of the bed to circulate cold water so that the bed could cool the human body lying on it. The bottom of the bed panel was empty. The inlet-water pipe was divided into two branches after passing through the main solenoid valve (SV). One branch was connected with the inlet of the bedside pipe. The other was connected to the inlet of the bed-end pipe. The outlet pipes of the two branches merged after the SV. Thermometers were installed on the supply and return pipes to measure the water temperature. Moreover, a bed frame (No. 3), which was made of sturdy and corrosion-resistant wood, was used to support the pipes.

The space between the pipes was filled with heat insulation material (No. 14 in Figure 1c), and the bottom of the pipes was padded with a heat insulation plate (No. 15 in Figure 1c) to reduce the transfer of cold energy to the cavity of the bed. The temperature of water in the return pipe can be adjusted by adjusting the opening of valves 9 and 10.

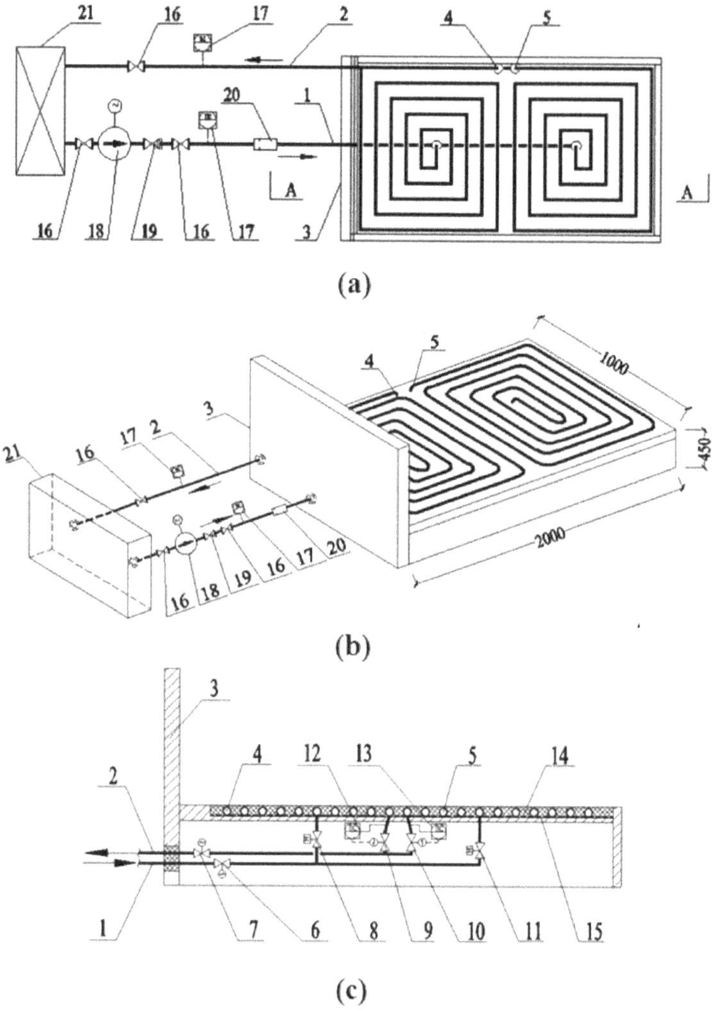

Figure 1. The configuration of the thermal-conductive bed used in this study: (a) the layout; (b) the axonometric view; and (c) the A-A sectional view. The numbers in the figure and the names are one-to-one correspondence are as follows: 1. Supply pipe; 2. Return pipe; 3. Bed frame; 4. Bedside water pipes; 5. Bed-end water pipes; 6. Solenoid valve on inlet pipe; 7. Solenoid valve on outlet pipe; 8. Solenoid valve on bedside supply pipe; 9. Solenoid valve on bedside return pipe; 10. Solenoid valve on bed-end return pipe; 11. Solenoid valve on bed-end supply pipe; 12. A thermometer in bedside return pipe; 13. A thermometer in bed-end return pipe; 14. Insulation material; 15. Heat insulation plate; 16. Shutoff valve; 17. Thermometer; 18. Water pump; 19. Non-return valve; 20. Flowmeter; 21. Water tank.

The size of the thermal-conductive bed was 2 m (length) × 1 m (width) × 0.45 m (height). The thickness of the bed plate was set at 0.05 m, and the pipe spacing was 0.085 m. The thickness of the human body was assumed to be 0.18 m. The water flows along the direction over the length of the bed, and the heat conduction of the body is vertical to the direction of water flow. The heat between the human body and the bed was mainly transferred by heat conduction, and then, the heat was carried away by water in the form of convection heat transfer, assuming that the physical body came into contact with the bed surface completely. The mathematical description of the problem can be expressed as follows:

$$c_p \times m \times \delta t_w = q \times A \tag{1}$$

$$q = K \times \left(t_s - t_w\right) \tag{2}$$

In which

$$K = \frac{1}{R_t + \frac{1}{h}} \tag{3}$$

$$R_t = \sum\nolimits_{i=1}^{n} \frac{\delta_i}{\lambda_i} \tag{4}$$

$$h = \frac{Nu \times \lambda_w}{H} \tag{5}$$

$$Nu = 0.023Re^{0.8}Pr^{0.4} \tag{6}$$

where q is heat dissipation between the human body and the bed, in W/m^2; δt_w is a temperature rise of cold water, in °C; A is the contact area between the human body and bed, in m^2; t_w is average water temperature, in °C; t_s is average skin temperature, in °C; h is the convection heat transfer coefficient between the bottom of the bedplate and water, in $W/(m^2{\cdot}K)$; c_p is the specific heat at a constant pressure of water, in J/(kg·K); m is flow mass of water, in kg/s; R_t is the total thermal resistance of bed cushion, in $(m^2{\cdot}K)/W$; λ_i and δ_i are thermal conductivity (in W/(m·K)) and thickness of mattress (in m), respectively; *Nu*, *Re*, *Pr* are Nusselt number (dimensionless), Reynolds number (dimensionless), Prandtl number (dimensionless), respectively; λ_w is a thermal conductivity of water, in W/(m·K); H is the equivalent diameter, that is, the height of the aqueduct, in m. The mattress consisted of one single material with thermal conductivity of 0.15 W/(m·K), and the thickness was 0.05 m.

2.2. Experiment Conditions

The experiment was conducted in a thermal-conductive bed placed in a climatic chamber with dimensions of 4.2 m × 3 m × 2.6 m in summer, as shown in Figure 2. The climatic chamber was located at the College of Resource and Environment and Safety Engineering at Hunan University of Science and Technology, Xiangtan, China. Xiangtan is located in the central south of China. According to the code for thermal design of civil buildings in China (GB 50176-2016) [44], Xiangtan belongs to the Hot-Summer and Cold-Winter (HSCW) Zone. The chamber was equipped with an air conditioning system to maintain constant indoor air temperature, humidity, and air velocity throughout the experiment.

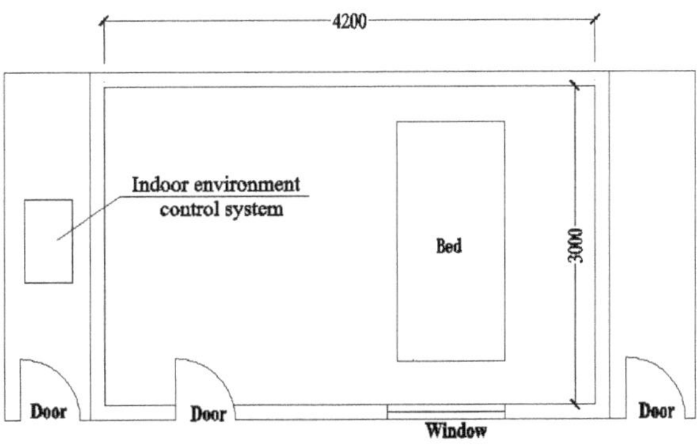

Figure 2. The layout of the climatic chamber.

A warm indoor environment (air temperature: 32 ± 0.5 °C, RH: 70 ± 5%, air velocity: 0.1 ± 0.02 m/s) was set in the chamber. The lights were turned off after the subjects fell asleep. The acoustical environment was acceptable, according to the subjects' responses. Five conditions with different supply water temperatures were defined, as listed in Table 1. It should be noted that the water temperature refers to the temperature of supplied water in pipes at the inlet of the bed head, obtained from thermometer No. 17, as shown in Figure 1.

Table 1. Experimental information.

	Indoor Air Temperature (°C)	Indoor RH (%)	Air Velocity (m/s)	Supply Water Temperature (°C)
Case 1	32 ± 0.5	70 ± 5	0.1 ± 0.02	without water
Case 2	32 ± 0.5	70 ± 5	0.1 ± 0.02	20
Case 3	32 ± 0.5	70 ± 5	0.1 ± 0.02	25
Case 4	32 ± 0.5	70 ± 5	0.1 ± 0.02	28
Case 5	32 ± 0.5	70 ± 5	0.1 ± 0.02	30

2.3. Subjects

A total of 12 subjects (age: 22.5 ± 2.5 years old; height: 170.5 ± 4.5 cm; weight: 68.7 ± 14.8 kg) participated in the experiments. They were all male students of Hunan University of Science and Technology. Since wearing only a few clothes and attaching temperature sensors to the skin (especially the chest area) during the experiment were very inconvenient for females, no female students were recruited for the experiments. Before the experiment began, each subject was inquired whether he was willing to participate in the experiment. Oral consent was obtained from each subject who was willing to participate in the experiment, and they were told that they could quit at any time during the experiment. The subjects have lived in Xiangtan, China, for more than one year. The Pittsburgh Sleep Quality Index (PSQI) questionnaire was used to evaluate the sleep quality and sleep disorders of each subject within one-month intervals [45]. The results showed that each participant was without sleep disorders. A candidate who had a PSQI global score of less than 5, which indicated no sleep disorder, was suitable for the experiment. All the subjects were in healthy condition. In the 24 h before participating in the experiments, they did not smoke and do strenuous exercise, or drink alcohol or coffee. They wore short-sleeved T-shirts and shorts without bed coverings. The clothing insulation was 0.2 clo. The thermal-conductive bed was covered with a thin blanket, and the upper layer was covered with a mat. Moreover, a simple oral explanation of the purpose and procedure of the experiment was given to each subject by the researchers before the experiments were conducted.

2.4. Measurement and Questionnaires

During the experiments, subjects' local skin temperature, bed surface temperature, thermal environment parameters, inlet and outlet water temperature (the temperature of the water entering and exiting the pipes), and inlet water flow rate were continuously measured and recorded throughout the whole night of the experiments. As for the measurement of bed surface temperature, four points on the thermal-conductive bed surface which contacted with the back, buttock, thigh, and calf of the human body were selected as the measurement points. Table 2 illustrates the detail of the instruments used in this experiment. All of the instruments had been calibrated before the measurement.

In this study, the human body was divided into five areas: shoulder; chest; abdomen; thigh; and calf. The body surface of each area included the upper part exposed to indoor air and without covering with the quilt (U), the back part in touch with the mattress (B), and the parts on both sides of the physical body (S). On the upper surface of the body, the skin temperature of the chest, abdomen, thigh, and calf were measured. On the back of the body, the skin temperatures of the back, right thigh, and right calf were measured. On the side of the body, the skin temperatures of the left shoulder, left abdomen, left thigh, and left calf were measured. The iButton sensors were attached to the skin surface by medical adhesive tape. Figure 3 shows one of the investigated subjects lying on the thermal-conductive bed in the climatic chamber after the sensors were attached. It should be noted that after attaching the iButton sensors to the human body, subjects wore short-sleeved T-shirts and shorts, as Section 2.3 described.

Table 2. The measured parameters and used sensors.

Parameter	Sensor	Type	Accuracy	Measurement Range
Skin temperature	iButton sensors	DS 1922L	±0.5 °C	−40–85 °C
Bed surface temperature; Inlet/outlet pipe water temperature	K-type thermocouples with data logger	LabQuest2	0.5 °C	0–45 °C
Air temperature	Thermo-hygrometer	TR-72Ui	±0.3 °C	0–50 °C
Relative humidity	Thermo-hygrometer	TR-72Ui	±5%	10–95%
Inlet water flow rate	Floater flowmeter	LZT-1002 M	±0.5 L/min	1–7 L/min

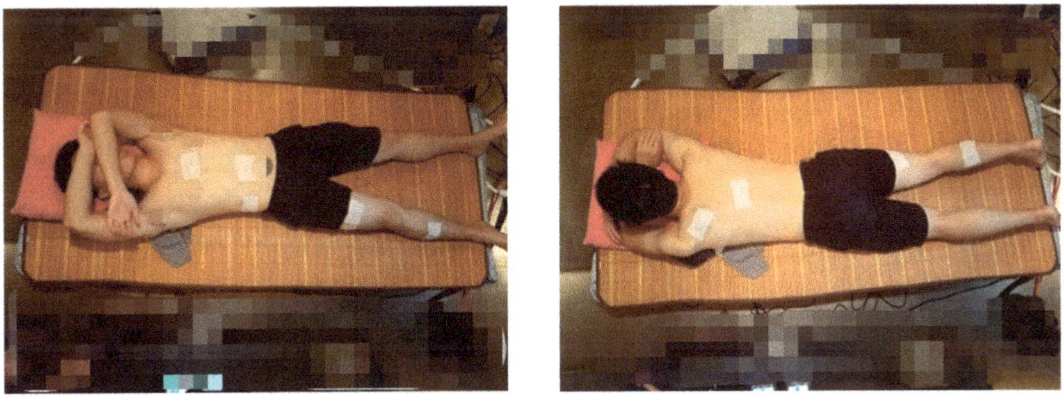

Figure 3. One of the investigated subjects lay on the thermal-conductive bed after the sensors were attached (after attaching the iButton sensors to the human body, subjects wore short-sleeved T-shirts and shorts).

The temperatures of the bed surface beneath the back, buttocks, thigh, and calf of subjects were measured with temperature sensors, and the water temperatures of the inlet and outlet were monitored. During the experiment, two sets of independent data acquisition equipment (LabQuest2) were used for measurement. Data loggers (Thermo Recorder TR-72Ui) were used to measure the indoor air temperature and RH around the bed, located 0.5 m from the bed surface. By regulating the opening of valves No. 9 and 10, as shown in Figure 1, the temperature of the supply water can be adjusted accordingly.

2.5. Questionnaires

Human thermal comfort was affected by the adaptability and expectation of the thermal environment [46,47]. Therefore, the proper indoor temperature could not be assessed simply by thermal sensation votes (TSV). In this study, the questionnaires mainly consisted of the following two parts: (1) a pre-sleep survey; and (2) a survey after waking up the next morning. In part (1), the subjects reported their physical and psychological state during the continuously measured day. It was conducted after the subjects reached the climatic chamber and adapted to the environment. In part (2), the subjects reported their thermal sensations and thermal comfort when they woke up the next morning. Table 3 presents the scales of subjective responses in the questionnaire. A seven-point scale suggested by ASHRAE Standard 55 [48] was used to rate subjects' thermal sensations. A five-point scale was adopted to assess subjects' thermal comfort.

Table 3. Scales of subjective responses in the questionnaire.

Thermal Sensation Votes (TSV)						
Cold −3	Cool −2	Slightly cool −1	Neutral 0	Slightly warm +1	Warm +2	Hot +3
Thermal comfort votes (TCV)						
Comfortable 0	Slightly uncomfortable 1	Uncomfortable 2	Very uncomfortable 3	Extremely uncomfortable 4		

2.6. Experimental Procedure

Measurement sensors were settled before the subjects arrived at the climatic chamber. The sleep schedule was consistent with the subjects' usual sleep time. The experimental procedure is described below:

(1) 22:00 Subjects reached the climatic chamber and were accommodated in the thermal environment for 30 min;
(2) 22:30 Subjects filled out the before-sleep survey questionnaire for physical and psychological status during the day;
(3) 22:35 The iButton sensors were attached to specified areas of the skin of subjects, as shown in Figure 3;
(4) 22:40 Check and adjust the equipment to make them display and record normally;
(5) 23:00 Turn off the lights after observing that the subject did not move their body frequently;
(6) 7:00 After the subjects woke up, they were asked to answer questions about their thermal sensation and thermal comfort for sleep.

2.7. Data Analysis

In this study, the mean skin temperature of the upper area (MST_U), the back area (MST_B), and the side area (MST_S) of a human body were calculated by weighing the area coefficients of each surface area. The calculation formulas are listed below:

$$MST_U = a_1 T_{chest\text{-}u} + a_2 T_{abdomen\text{-}u} + a_3 T_{thigh\text{-}u} + a_4 T_{calf\text{-}u} \tag{7}$$

$$MST_B = b_1 T_{back\text{-}b} + b_2 T_{thigh\text{-}b} + b_3 T_{calf\text{-}b} \tag{8}$$

$$MST_S = s_1 T_{shoulder\text{-}s} + s_2 T_{abdomen\text{-}s} + s_3 T_{thigh\text{-}s} + s_4 T_{calf\text{-}s} \tag{9}$$

where $T_{chest\text{-}u}$, $T_{abdomen\text{-}u}$, $T_{thigh\text{-}u}$, and $T_{calf\text{-}u}$ are the skin temperatures of the upper parts of the chest, abdomen, thigh, and calf, respectively; $T_{back\text{-}b}$, $T_{thigh\text{-}b}$, and $T_{calf\text{-}b}$ represent the skin temperatures of the back, the back parts of thigh and calf, respectively, and $T_{shoulder\text{-}s}$, $T_{abdomen\text{-}s}$, $T_{thigh\text{-}s}$, and $T_{calf\text{-}s}$ represent the skin temperatures of around the shoulder, the side parts of abdomen, thigh, and calf, respectively. Table 4 illustrates the ratios of body surface area (a, b, s) according to references [49,50].

Table 4. Surface area ratios of the calculated skin temperature.

No.	Upper (a)	Back (b)	Side (s)
1	0.27	0.54	0.03
2	0.22	0.26	0.26
3	0.36	0.2	0.36
4	0.15	/	0.33

3. Results

3.1. Bed Surface Temperature under Different Water-Supplied Conditions

The bed surface temperature of the thermal-conductive bed was measured under experimental conditions described in Section 2.2. Figure 4 shows the bed surface temperature distribution in different cases. As for the same sites of the bed surface, the bed surface temperature when there was no water flowing through the pipes was 1.8–9.3 °C higher than when the supply water temperature was set at 20, 25, 28, or 30 °C. Specifically, when there was no water flowing through the pipes, the bed surface temperature varied from 32 to 33.5 °C, which was 0–1.5 °C higher than the indoor temperature (32 °C) due to the heat from the subjects. When the supply water temperature was raised from 20 °C to 25 °C, the surface temperature of the thermal-conductive bed changed significantly. Specifically, when the supply water temperature was 20 °C, the bed surface temperature varied from 23 to 26 °C. When the supply water temperature was 25 °C, the bed surface temperature varied from 29 to 30 °C. However, the variation in bed surface temperature was slight when the water temperature increased from 25 to 30 °C. Specifically, when the supply water temperature was set at 25, 28, and 30 °C, the bed surface temperature varied from 29 to 31 °C. Furthermore, a decreasing trend of the surface temperature from the bedhead to the bed-end was observed. The results indicate that when the supply water temperature was from 25 to 30 °C, the bed surface temperature was relatively stable (29–31 °C).

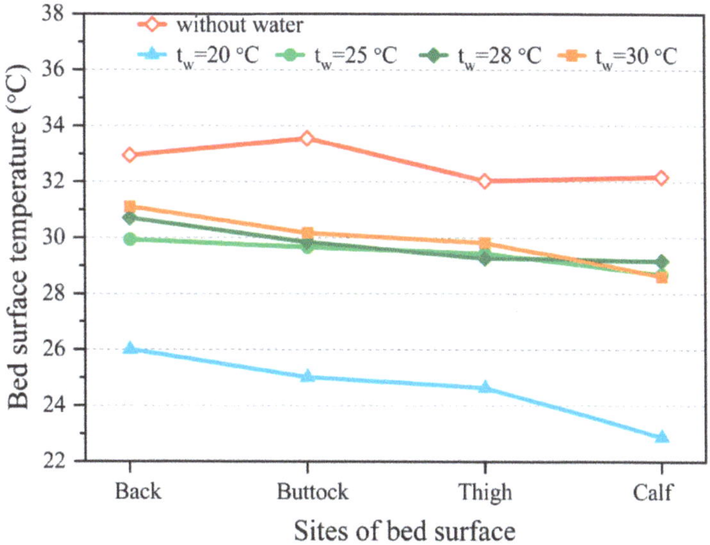

Figure 4. The surface temperature distribution of thermal-conductive bed under different cases.

3.2. Skin Temperature under Different Supply Water Temperatures

The mean skin temperature (*MST*) of different body segments (the upper, the back, and the left side of the human body) under experimental conditions calculated from

Equations (7)–(9) is shown in Figure 5. The linear fitting lines of *MST* throughout the night are also presented.

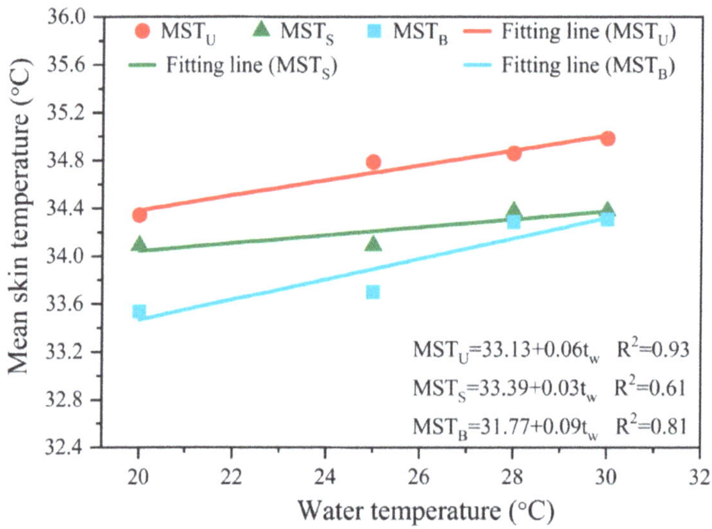

Figure 5. The correlation between the mean skin temperature and the water supply temperature.

As illustrated in Figure 5, MST_U varied from 34.3 to 35.0 °C, and MST_B varied in a range of 34.1–34.4 °C. MST_S was more fluctuating with a difference of 0.8 °C under experimental conditions. It can be seen that MST_U, MST_B, and MST_S increased stably with the rise of supply water temperature. Both MST_B and MST_S were lower than MST_U under various test cases. When t_w was 20 °C, the difference between MST_U and MST_S reached 1.08 °C. As t_w increased to 28 °C, the difference between MST_U and MST_S decreased to 0.58 °C. Moreover, when t_w increased to 28 °C, MST_S and MST_B tended to be consistent, which could be due to the side and back areas of the body being closer to the bed than the upper body areas.

Based on the linear regression functions developed for *MST* and supply water temperature (t_w), the correlation between the two parameters was found. The sensitivity between the two correlated parameters can be reflected by the slope value. It was shown that t_W was most sensitive to MST_B, followed by MST_U and MST_S.

3.3. Correlation between TSV, TCV, and Supply Water Temperature

The temperature of supply water in the pipes of the thermal-conductive bed has an important impact on the thermal sensation. Too high or too low a temperature of supply water makes people feel uncomfortable. In this study, subjects who voted "Cold", "Cool", or "Slightly cool" were regarded as having cool sensations. Then, the percentages of cool sensation (the proportion of the subjects having cool sensations to the total subjects) in the back, buttock, and thigh was calculated. Moreover, the linear fitting method was used to determine the correlation between thermal sensation vote (TSV) and supply water temperature (t_w). The results are presented in Figure 6. The regression model is listed below:

$$\O_{cool} = 2.67\text{-}0.089t_w,\ R^2 = 0.89 \tag{10}$$

where $\O_{cool}$ is the percentage of the subjects having cool sensations in the back, buttock, and thigh to the total subjects, and R^2 is the coefficient of determination.

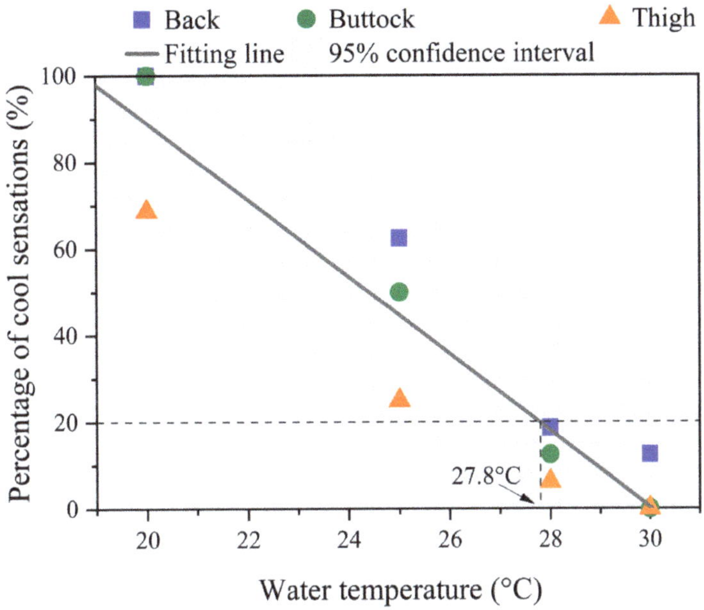

Figure 6. The correlation between the percentage of cool sensations and the supply water temperature. The percentage of cool sensations refers to the percentages of subjects voting "Cold", "Cool", or "Slightly cool".

As shown in Figure 6, the percentage of subjects with cool sensations showed a strong negative correlation with the supply water temperature (R^2 = 0.89). When the supply water temperature increased by 1 °C, the percentage of subjects with cool sensations decreased by 8.9%. Specifically, when the supply water temperature was set at 20 °C, 100% of subjects had cool sensations in the back and buttock areas, and 70% of subjects had cool sensations in the thigh. When the supply water temperature was set at 28 °C, less than 20% of subjects had cool sensations in the back, buttock, and thigh. In other words, the lowest temperature of supply water was expected to be above 27.8 °C. Furthermore, it is indicated that the back area of the human body was more sensitive to the supply water temperature. For example, when the supply water temperature was 28 °C, the percentage of cool sensations in the back was higher than the percentages in the buttock and thigh (back:19%; buttock:13%; thigh: 6%). This could be explained by a larger contact area around the back with the thermal-conductive bed than the other two body areas.

Furthermore, subjects who voted "Slightly uncomfortable", "Uncomfortable", and "Very uncomfortable" were regarded as feeling thermal discomfort. Then, the percentages of thermal discomfort (the proportion of the subjects feeling thermal discomfort to the total subjects) in the back, buttock, and thigh were calculated. Moreover, the correlation between thermal discomfort and the supply water temperature is illustrated in Figure 7. The regression model is listed below:

$$\varnothing_{dsicomfort} = 2.76\text{-}0.091t_w,\ R^2 = 0.91 \quad (11)$$

where $\varnothing_{discomfort}$ is the percentage of the subjects feeling thermal discomfort in the back, buttock, and thigh to the total subjects.

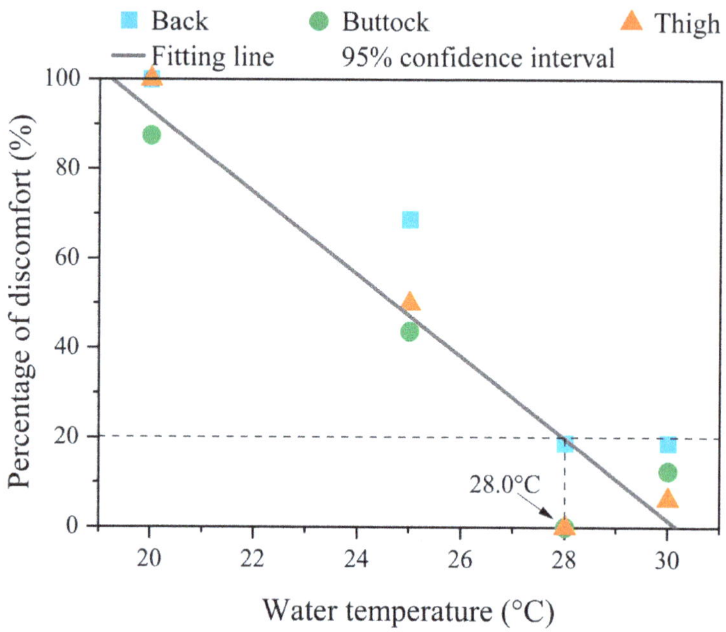

Figure 7. The correlation between the percentage of thermal discomfort and the supply water temperature. The percentage of thermal discomfort refers to the percentages of subjects voting "Slightly uncomfortable", "Uncomfortable", or "Very uncomfortable".

As shown in Figure 7, the percentage of subjects feeling thermal discomfort showed a strong negative correlation with the supply water temperature (R^2 = 0.91). When the supply water temperature increased by 1 °C, the percentage of subjects with cool sensations decreased by 9.1%. To be specific, when the supply water temperature was set at 20 °C, 100% of subjects had cool sensations in the back and buttock areas, and 88% of subjects had cool sensations in the thigh. When the supply water temperature was set at 28 °C, less than 20% of subjects had thermal discomfort in the back, buttock, and thigh. Namely, the lowest supply water temperature was expected to be greater or equal to 28 °C.

According to the above results, subjects felt thermally comfortable when their back was cool at a water supply temperature of about 28 °C. Therefore, it is suggested that t_w around 28–30 °C is suitable for a thermal-conductive bed in this study. In such conditions, MST_U, MST_B, and MST_S could be, respectively, in the range of 34.7–35 °C, 34.1–34.3 °C, and 34.2–34.4 °C.

3.4. Thermal Sensation and Comfort

Similar to the results presented in Section 3.3, the proportions of the subjects' local and overall thermal sensation and comfort votes were calculated. Figure 8 illustrates the proportions of local and overall thermal sensation votes. The thermal sensation votes varied under different water-supplied conditions. The subjects voting "Neutral" accounted for the highest proportion when the supply water temperature was 28 or 30 °C. Specifically, in terms of no-water conditions, more than 75% of the subjects felt slightly warm or warm in the areas of the calf, thigh, buttock, back, and shoulder, and 75% of the subjects felt neutral in the part of the abdomen. The overall thermal sensation votes were slightly warm (40%) and warm (60%). When the supply water temperature was 20 or 25 °C, more than 55% of the subjects felt slightly cool in the areas of the calf, thigh, buttock, back, abdomen, and shoulder. When the supply water temperature was 28 or 30 °C, more than 85% of the subjects felt neutral in the areas of the calf, thigh, buttock, back, abdomen, and shoulder.

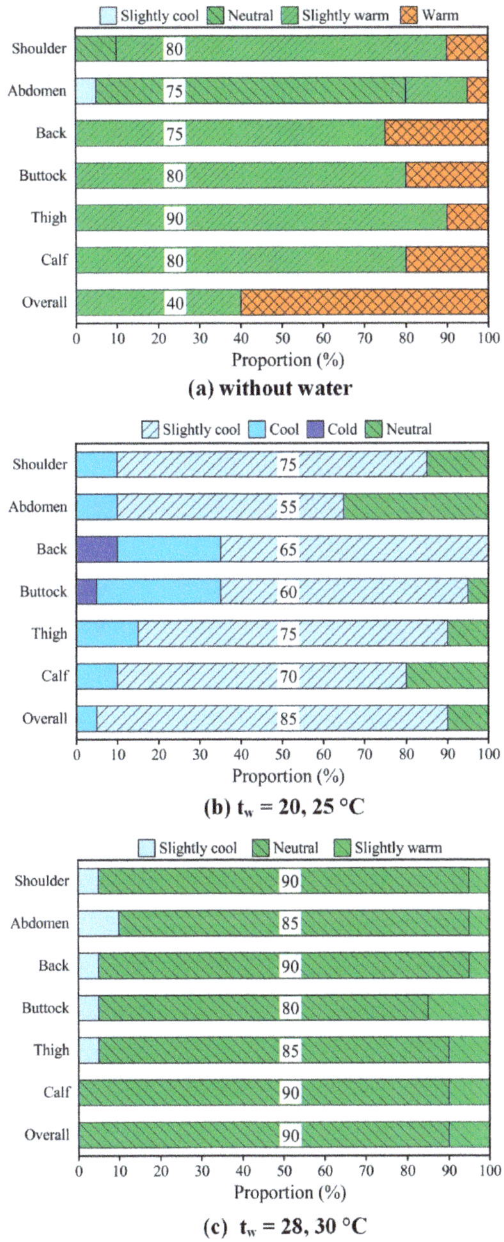

Figure 8. The proportions of thermal sensation vote with different supply water temperatures.

The proportion of thermal comfort votes under different water-supplied conditions is presented in Figure 9. The proportion of the subjects feeling comfortable with the supply water temperature at 28 or 30 °C was much higher than when without water or the supply water temperature at 20 or 25 °C. Specifically, in terms of no-water condition, only 5% of the subjects felt comfortable in the areas of the shoulder and abdomen. The overall thermal comfort votes were uncomfortable (65%). When the supply water temperature was 20 or 25 °C, less than 30% of the subjects felt comfortable in the shoulder, buttock, back, thigh, and calf. Especially 80% of the subjects felt uncomfortable in the back. However, when the supply water temperature was 28 or 30 °C, more than 85% of the subjects felt comfortable in the shoulder, abdomen, back, buttock, thigh, and calf, and the proportion of overall thermal comfort votes reached 85–90%. The results indicate that the thermal-conductive bed with the supply water temperature at 28 or 30 °C satisfied the requirements of subjects' thermal comfort during sleep.

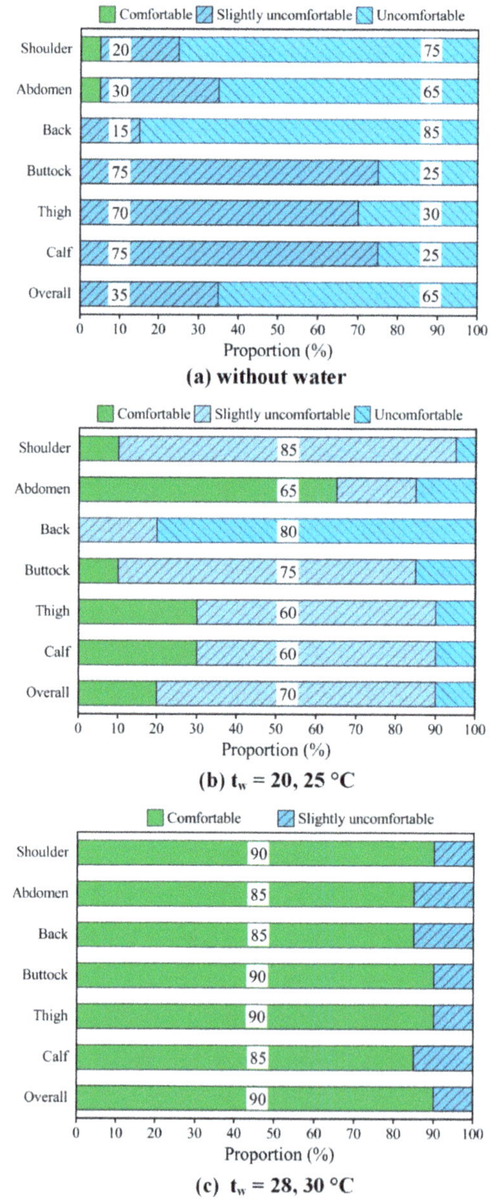

Figure 9. The proportion of thermal comfort votes with different supply water temperatures.

4. Discussion

The results in Section 3.4 indicate that the thermal-conductive bed helps improve subjects' thermal comfort during sleep in summer. Furthermore, the thermal-conductive bed also exerted great energy-saving potential. Firstly, the thermal-conductive bed maintains human thermal comfort with a background temperature of 32 °C during sleep. The upper limit of the set-point temperature of air conditioners in actual buildings at night in summer could be increased. It was reported by Zhang et al. [22] that approximately 10% energy of an air-conditioning system can be reduced when the indoor set temperature in summer increases by 1 °C. In order to obtain a comfortable zone proposed by ASHRAE Standard [48], the indoor temperature of air-conditioning buildings during summer is generally below 26 °C [41,51,52]. When occupants use this type of thermal-conductive bed for sleep, the indoor temperature could be improved to 32 °C, and the energy consumption of the air conditioning system could be reduced by over 60%. In addition, the supply water temperature of the thermal-conductive bed was high (>25 °C), which makes it possible to directly use natural cooling sources in summer and to achieve a super high COP value if any chiller is needed.

Despite the effects of the thermal-conductive bed in terms of thermal comfort and energy-saving during sleep in hot summer, some problems remain unsolved:

(1) This study only recruited male subjects aged 22.5 ± 2.5 years old for testing. Females were probably more sensitive to the thermal environment [53,54]. Moreover, subjects of other age groups may have different reactions to the thermal-conductive bed. It would be better to explore this issue in the future experimental environment;
(2) This study only investigated subjects' thermal comfort with the thermal-conductive bed when the indoor temperature was 32 °C. If the bed is effective in maintaining thermal comfort in extreme conditions, it is reasonable to believe that it is also effective in less extreme conditions, although the findings with other indoor temperatures may be different. While more experiments are needed to demonstrate the thermal comfort of people with the thermal-conductive bed in different indoor temperatures;
(3) Psychological parameters, such as electroencephalogram, electrooculogram, and electroencephalogram, were not measured to evaluate human sleep quality. The measurements of these parameters can provide basic information for examining the sleep process [55] and can be investigated in the future;
(4) It must be noted that the thermal-conductive bed was still crudely made. Future research should focus on retrofits of design, structure, and appearance to better meet the requirements of sleeping people;
(5) The optimal operating strategy of the thermal-conductive bed and the air conditioners was not discussed.

5. Conclusions

In this study, a series of experiments on human thermal comfort with a thermal-conductive bed during sleep were conducted in the summer. The main conclusions are summarized as follows:

(1) The bed surface temperature was 0–1.5 °C higher than the indoor temperature when there was no water supply in the pipes of the thermal-conductive bed. When the supply water temperature was 20 °C, the bed surface temperature was lower than 26 °C. When the supply water temperature varied between 25 to 30 °C, the bed surface temperature was relatively stable (29–31 °C);
(2) The mean skin temperature on the back of the human body was more sensitive to the supply water temperature than the upper and the side areas of the human body. When the supply water temperature varied between 28 to 30 °C, the mean skin temperature on the back and the side of the human body was consistent.
(3) Subjects perceived a slightly warm feeling when there was no water supply in the pipes of the thermal-conductive bed. When the supply water temperature was 28 or 30 °C, subjects felt thermally comfortable. Under such conditions, the mean skin

temperature of the upper, the back, and the side of the human body varied from 34 to 35 °C.

The thermal-conductive bed could be used for people who have no air conditioning systems in summer or as an alternative bed for sleep by extending the indoor set-point temperature range. It can also help save a lot of energy in buildings. However, the limitations of this study should not be ignored; for example, only male subjects aged 22.5 ± 2.5 years old participated in the experiment, while the findings of male subjects at other ages or female subjects may be different. Therefore, the experiment conditions and target subjects must be improved in future studies to verify the thermal comfort of the thermal-conductive bed for improving night sleep quality in summer. In addition, the design of the bed is required to be improved before applying it to actual buildings.

Author Contributions: Conceptualization, X.H., Y.H., N.L. and S.C.; methodology, J.H. and M.H.; formal analysis, J.H. and M.H.; investigation, J.H. and M.H.; data curation, M.H.; writing—original draft preparation, J.H. and Y.H.; writing—review and editing, X.H., Y.H. and Y.L.; funding acquisition, J.H., X.H. and Y.H. All authors have read and agreed to the published version of the manuscript.

Funding: This research was funded by the Scientific Research Fund of Hunan Provincial Education Department, China (Grant No. 21B0460), the Natural Science Foundation of Hunan Province, China (Grant No. 2022JJ50127 and 2022JJ30251), and the Fundamental Research Funds for the Central Universities of China (Grant No. 531118010826).

Data Availability Statement: Data sharing is not applicable to this article.

Conflicts of Interest: The authors declare no conflict of interest.

References

1. Opp, M.R. Sleeping to fuel the immune system: Mammalian sleep and resistance to parasites. *BMC Evol. Biol.* **2009**, *9*, 8. [CrossRef] [PubMed]
2. Walker, M.P. Sleep to Remember. *Am. Sci.* **2006**, *12*, 410–424. [CrossRef]
3. Zhang, N.; Cao, B.; Zhu, Y. An effective method to determine bedding system insulation based on measured data. *Build. Simul.* **2023**, *16*, 121–132. [CrossRef]
4. Tsuzuki, K.; Okamoto-Mizuno, K.; Mizuno, K. The effects of low air temperatures on thermoregulation and sleep of young men while sleeping using bedding. *Buildings* **2018**, *8*, 76. [CrossRef]
5. Tsuzuki, K.; Sakoi, T.; Sakata, Y. Effect of seasonal ambient temperature on sleep and thermal comfort in older people living in public elderly facilities. *Buildings* **2021**, *11*, 574. [CrossRef]
6. Cao, T.; Lian, Z.; Zhu, J.; Xu, X.; Du, H.; Zhao, Q. Parametric study on the sleep thermal environment. *Build. Simul.* **2022**, *15*, 885–898. [CrossRef]
7. Hu, J.; He, Y.; Wang, Q.; Wang, B.; Hao, X.; Li, N.; Yin, W.; Liu, L. Resilient or resistant to non-neutral environments? A comparative study on occupant thermal needs in buildings under natural ventilation, fee-free heating, and fee-charged heating modes. *J. Build. Eng.* **2023**, *72*, 106651. [CrossRef]
8. Zhang, X.; Dai, C.; Lan, L.; Lian, Z. Correlation Analysis of Residents' Health and Dwelling Environment in Shanghai. *J. Shanghai Jiaotong Univ.* **2014**, *48*, 798–803.
9. Lin, Z.; Deng, S. A questionnaire survey on sleeping thermal environment and bedroom air conditioning in high-rise residences in Hong Kong. *Energy Build.* **2006**, *38*, 1302–1307. [CrossRef]
10. National development and reform commission. The new national air conditioner energy efficiency standard comes into effect on June 1. *Resour. Dev.* **2010**, *1*, 45.
11. Braun, J.; Montgomery, K.; Chaturvedi, N. Evaluating the Performance of Building Thermal Mass Control Strategies. *HVAC R Res.* **2001**, *7*, 403–428. [CrossRef]
12. Lam, J.C. Residential sector air conditioning loads and electricity use in Hong Kong. *Energy Convers. Manag.* **2000**, *41*, 1757–1768. [CrossRef]
13. Ameen, A.; Bahrami, A.; Elosua Ansa, I. Assessment of thermal comfort and indoor air quality in library group study rooms. *Building* **2023**, *13*, 1145. [CrossRef]
14. Imanari, T.; Omori, T.; Bogaki, K. Thermal comfort and energy consumption of the radiant ceiling panel system: Comparison with the conventional all-air system. *Energy Build.* **1999**, *30*, 167–175. [CrossRef]
15. Kim, M.; Chun, C.; Han, J. A study on bedroom environment and sleep quality in Korea. *Indoor Built Environ.* **2010**, *19*, 123–128. [CrossRef]
16. Muzet, A.; Libert, J.P.; Candas, V. Ambient temperature and human sleep. *Experientia* **1984**, *40*, 425–429. [CrossRef]

17. Pan, L.; Lian, Z.; Li, L. Investigation of sleep quality under different temperatures based on subjective and physiological measurements. *HVACR Res.* **2012**, *18*, 1030–1043.
18. Lan, L.; Pan, L.; Lian, Z.; Huang, H.; Lin, Y. Experimental study on thermal comfort of sleeping people at different air temperatures. *Build. Environ.* **2014**, *73*, 24–31. [CrossRef]
19. Mobaraki, B.; Castilla Pascual, F.J.; García, A.M.; Mellado Mascaraque, M.Á.; Vázquez, B.F.; Alonso, C. Studying the impacts of test condition and nonoptimal positioning of the sensors on the accuracy of the in-situ U-value measurement. *Heliyon* **2023**, *9*, e17282. [CrossRef]
20. Mobaraki, B.; Castilla Pascual, F.J.; Lozano-Galant, F.; Lozano-Galant, J.A.; Porras Soriano, R. In situ U-value measurement of building envelopes through continuous low-cost monitoring. *Case Stud. Therm. Eng.* **2023**, *43*, 102778. [CrossRef]
21. Hoyt, T.; Arens, E.; Zhang, H. Extending air temperature setpoints: Simulated energy savings and design considerations for new and retrofit buildings. *Build. Environ.* **2015**, *88*, 89–96. [CrossRef]
22. Zhang, H.; Arens, E.; Zhai, Y. A review of the corrective power of personal comfort systems in non-neutral ambient environments. *Build. Environ.* **2015**, *91*, 15–41. [CrossRef]
23. Yu, W.; Zhang, Y.; Du, C.; Li, B.; Liu, H.; Zhang, Y.; Wei, S. An innovative fan control strategy aimed at responding to human physiological characteristics for comfort sleeping. *Therm. Sci. Eng. Prog.* **2022**, *35*, 101470. [CrossRef]
24. He, Y.; Chen, W.; Wang, Z.; Zhang, H. Review of fan-use rates in field studies and their effects on thermal comfort, energy conservation, and human productivity. *Energy Build.* **2019**, *194*, 140–162. [CrossRef]
25. Okamoto-Mizuno, K.; Tsuzuki, K.; Mizuno, K. Effects of head cooling on human sleep stages and body temperature. *Int. J. Biometeorol.* **2003**, *48*, 98–102. [CrossRef] [PubMed]
26. Irshad, K.; Algarni, S.; Jamil, B.; Ahmad, M.T.; Khan, M.A. Effect of gender difference on sleeping comfort and building energy utilization: Field study on test chamber with thermoelectric air-cooling system. *Build. Environ.* **2019**, *152*, 214–227. [CrossRef]
27. Du, C.; Lin, X.; Yan, K.; Liu, H.; Yu, W.; Zhang, Y.; Li, B. A model developed for predicting thermal comfort during sleep in response to appropriate air velocity in warm environments. *Build. Environ.* **2022**, *223*, 109478. [CrossRef]
28. He, Y.; Parkinson, T.; Arens, E.; Zhang, H.; Li, N.; Peng, J.; Elson, J.; Maranville, C. Creating alliesthesia in cool environments using personal comfort systems. *Build. Environ.* **2022**, *209*, 108642. [CrossRef]
29. He, Y.; Li, N.; Li, N.; Li, J.; Yan, J.; Tan, C. Control behaviors and thermal comfort in a shared room with desk fans and adjustable thermostat. *Build. Environ.* **2018**, *136*, 213–226. [CrossRef]
30. He, Y.; Wang, X.; Li, N.; He, M.; He, D.; Wang, K. Cooling ceiling assisted by desk fans for comfort in hot-humid environment. *Build. Environ.* **2017**, *122*, 23–34. [CrossRef]
31. Pan, D.; Chan, M.; Xia, L.; Xu, X.; Deng, S. Performance evaluation of a novel bed-based task/ambient conditioning (TAC) system. *Energy Build.* **2012**, *44*, 54–62. [CrossRef]
32. Lan, L.; Lian, Z.; Xin, Z.; Sun, C.; Huang, H.; Lin, Y.; Zhao, J. Pilot study on the application of bedside personalized ventilation to sleeping people. *Build. Environ.* **2013**, *67*, 160–166. [CrossRef]
33. Nunneley, S.A.; Maldonado, R.J. Head and/or torso cooling during simulated cockpit heat stress. *Aviat. Space Environ. Med.* **1983**, *54*, 496–499.
34. Nunneley, S.A.; Reader, D.C.; Maldonado, R.J. Head-temperature effects on physiology, comfort, and performance during hyperthermia. *Aviat. Space Environ. Med.* **1982**, *53*, 623–628. [PubMed]
35. Kräuchi, K.; Cajochen, C.; Werth, E.; Wirz-Justice, A. Warm feet promote the rapid onset of sleep. *Nature* **1999**, *401*, 36–37. [CrossRef] [PubMed]
36. Mao, N.; Pan, D.; Deng, S.; Chan, M. Thermal, ventilation and energy saving performance evaluations of a ductless bed-based task/ambient air conditioning (TAC) system. *Energy Build.* **2013**, *66*, 297–305. [CrossRef]
37. Ning, M.; Pan, D.; Chan, M.; Deng, S. Performance evaluation of an air conditioning system with different heights of supply outlet applied to a sleeping environment. *Energy Build.* **2014**, *77*, 281–291.
38. Ning, M.; Pan, D.; Chan, M.; Deng, S. Experimental and numerical studies on the performance evaluation of a bed-based task/ambient air conditioning (TAC) system. *Appl. Energy* **2014**, *136*, 956–967. [CrossRef]
39. Xia, Y.Z.; Niu, J.L.; Zhao, R.Y.; Burnett, J. Effects of turbulent air on human thermal sensations in a warm isothermal environment. *Indoor Air* **2010**, *10*, 289–296. [CrossRef]
40. Griefahn, B.; Mehnert, P.; Brödp, E.; Forsthoff, A. Working in moderate cold: A possible risk to health. *J. Occup. Health* **1997**, *39*, 36–44. [CrossRef]
41. He, Y.; Li, N.; Huang, Q. A field study on thermal environment and occupant local thermal sensation in offices with cooling ceiling in Zhuhai, China. *Energy Build.* **2015**, *102*, 277–283. [CrossRef]
42. He, Y.; Li, N.; Wang, X.; He, M.; He, D. Comfort, Energy Efficiency and Adoption of Personal Cooling Systems in Warm Environments: A Field Experimental Study. *Int. J. Environ. Res. Public Health* **2017**, *14*, 1408. [CrossRef] [PubMed]
43. He, Y.; Li, N.; Zhou, L.; Wang, K.; Zhang, W. Thermal comfort and energy consumption in cold environment with retrofitted Huotong (warm-barrel). *Build. Environ.* **2017**, *112*, 285–295. [CrossRef]
44. *GB 50176-2016*; Code for Thermal Design of Civil Building. China Architecture and Building Press: Beijing, China, 2016; p. 29.
45. Buysse, D.J.; Reynolds, C.F.; Monk, T.H.; Berman, S.R.; Kupfer, D.J. The Pittsburgh sleep quality index: A new instrument for psychiatric practice and research. *Psychiatry Res.* **1989**, *28*, 193–213. [CrossRef] [PubMed]

46. Cheong, K.W.D.; Yu, W.J.; Sekhar, S.C.; Tham, K.W.; Kosonen, R. Local thermal sensation and comfort study in a field environment chamber served by displacement ventilation system in the tropics. *Build. Environ.* **2007**, *42*, 525–533. [CrossRef]
47. Zhang, H.; Huizenga, C.; Arens, E.; Wang, D. Thermal sensation and comfort in transient non-uniform thermal environments. *Eur. J. Appl. Physiol.* **2004**, *92*, 728–733. [CrossRef] [PubMed]
48. *ASHRAE 55-2017*; Thermal Environmental Conditions for Human Occupancy. American Society of Heating. Refrigeration and Air-Conditioning Engineers: Atlanta, GA, USA, 2017; p. 3.
49. Houdas, Y.; Ring, E.J. *Human Body Temperature*; Plenum Press: New York, NY, USA, 1982; p. 15.
50. Song, C.; Liu, Y.; Zhou, X.; Wang, X.; Li, J.; Liu, J. Temperature field of bed climate and thermal comfort assessment based on local thermal sensations. *Build. Environ.* **2016**, *95*, 381–390. [CrossRef]
51. Mui, K.W.H.; Chan, W.T.D. Adaptive comfort temperature model of air-conditioned building in Hong Kong. *Build. Environ.* **2003**, *38*, 837–852. [CrossRef]
52. Wong, L.T.; Mui, K.W.; Fong, N.K.; Hui, P.S. Bayesian adaptive comfort temperature (BACT) of air-conditioning system in subtropical climate. *Build. Environ.* **2007**, *42*, 1983–1988. [CrossRef]
53. Lan, L.; Lian, Z.; Liu, W.; Liu, Y. Investigation of gender difference in thermal comfort for Chinese people. *Eur. J. Appl. Physiol.* **2008**, *102*, 471–480. [CrossRef]
54. Hu, J.; He, Y.; Hao, X.; Li, N.; Su, Y.; Qu, H. Optimal temperature ranges considering gender differences in thermal comfort, work performance, and sick building syndrome: A winter field study in university classrooms. *Energy Build.* **2022**, *254*, 111554. [CrossRef]
55. Lan, L.; Lian, Z. Ten questions concerning thermal environment and sleep quality. *Build. Environ.* **2016**, *99*, 252–259. [CrossRef]

buildings

MDPI

Article

Climatic Variability in Altitude: Architecture, Thermal Comfort, and Safety along the Facade of a Residential Tower in the Mediterranean Climate

Soultana (Tanya) Saroglou [1,*], Hofit Itzhak-Ben-Shalom [2,3] and Isaac A. Meir [1]

1 Department of Civil and Environmental Engineering, Faculty of Engineering Sciences, Ben-Gurion University of the Negev, Beer-Sheva 8410501, Israel; sakis@bgu.ac.il
2 Department of Architecture, Sami Shamoon College of Engineering (SCE), Beer-Sheva 84100, Israel; hofitit@sce.ac.il
3 Department of Environmental Studies, The Porter School of the Environment and Earth Sciences, Tel Aviv University, Tel Aviv 52621, Israel
* Correspondence: saroglou@post.bgu.ac.il

Abstract: This research is part of a wider study on the sustainability of high-rise buildings within the urban environment. The focus here is on wind and temperature stratifications per different height through in situ monitoring on a high-rise residential tower in the Mediterranean climate of Tel Aviv, and their impact on thermal comfort and user safety. The appropriateness of design is discussed in relation to the direct exposure to higher-up wind velocities, the thermal perception of the residents through questionnaires, and the safety and usability of the outdoors space according to height. The potential for advancing the energy efficiency of the structure is also discussed. The study covers a hot and a cold season, focusing on the specificities of the wind regime in the specific climate, through seasonal variations. Results from the monitoring of data confirmed increased wind and gust velocities per building height all year round, reaching the level of danger for the occupants during winter, cancelling, thus a successful operation of the outdoor balcony space. The occupants' perception and use of the outdoor balcony space per building height were in direct relationship to the increased wind velocities per height. Discussion and conclusions critically evaluate the residential high-rise building typology in the Mediterranean climate through the design of the outdoor balcony space along the height of the envelope. The results set an initial understanding and delineation for future studies, while underlining the complications of designing and occupying tall buildings and the level of detailing required.

Keywords: residential high-rise design; wind climate per building height; thermal perception per building height; Mediterranean climate; seasonal variations

Citation: Saroglou, S.; Itzhak-Ben-Shalom, H.; Meir, I.A. Climatic Variability in Altitude: Architecture, Thermal Comfort, and Safety along the Facade of a Residential Tower in the Mediterranean Climate. *Buildings* **2023**, *13*, 1979. https://doi.org/10.3390/buildings13081979

Academic Editors: Yingdong He and Nianping Li

Received: 26 June 2023
Revised: 20 July 2023
Accepted: 25 July 2023
Published: 2 August 2023

1. Introduction

The numbers of high-rise buildings around the world have been growing rapidly over the last decades, with the 21st century portraying increasingly vertical urban environments. However, while the high-rise typology is embraced in many of the world's cities, the performance of such buildings, in relation to sustainable strategies, is lacking the scientific background that it requires. One such consideration relates to the high-energy consumption in high-rise buildings in comparison to all other construction [1–3] and the need to comply with strict energy regulations around the world [4–6]. Building energy efficiency has become especially challenging in today's energy intensive era, while most of the high-rise building stock is mechanically supported, consuming large amounts of energy, relative to the buildings' increased volume and scale. As far as 'green credentials' are concerned, the percentage of buildings that pursue this procedure and aim towards improving the tall-building performance is growing; however, these numbers are still small on a global scale, and mainly refer to office environments [7–13].

However, 'building sustainability' is a wide term, and it does not necessarily mean that projects that employ a number of sustainable strategies or solutions or present a number of sustainability credentials are on the whole successful [2,14], with the actual energy needs of the structure after completion being an important parameter. In addition, advanced green technologies have not yet yielded the success they were meant to have, while their realization can potentially skyrocket a project's budget, e.g., the installation of wind turbines for electricity production. At the same time, regulations like the 2010/31/EU Energy Performance of Buildings Directive (EPBD) state that all new buildings must be nearly Zero Energy Buildings (nZEBs) by 2020 [4]. The implementation of passive strategies towards the achievement of this goal is a must, i.e., natural ventilative cooling in reducing energy loads. The success of natural ventilation is relative to building orientation, seasonal variations [15,16], and building height, while its implementation also maintains a thermally comfortable, healthy indoor environment [13,17].

Nevertheless, designing a high-performance high-rise building that incorporates renewable energy sources (RES) and non-renewable ones is a complex matter, that requires a much higher flexibility on budget and time [8,18]. The Council of Tall Buildings and Urban Habitat (CTBUH) defines a building of 14 or more floors, or more than 50 m, being within the threshold of the 'tall building' definition [19]. But, while the focus on sustainable tall-buildings mainly points towards iconic architecture of technological and structural innovation in the height range of 300–400 m [20], an increasing range of 50–100 m high buildings is designed and built without green credentials, certification, or any form of prior studies through simulation and modelling. So, despite an increase in building height, these higher-than-average buildings most likely do not satisfy the requirements of sustainable construction and lack studies relating to their successful operation, while design guidelines for tall buildings are generally lacking.

Whereas research on appropriate strategies to bridge the gap between the high-rise buildings and the new era of energy codes is vital [21,22], further issues that arise with the increase in building height within the urban fabric are equally pressing and concern social, environmental, economic factors, and more. In this process, site-specific studies in neighborhoods and urban areas become important in order to start mapping the interconnections between environmental variables and urban morphology [23–27]. The topic of urban winds and high-rise buildings, for example, has gained popularity in recent years being of growing relevance to the increasing complexity of this building typology within the urban environment [28–35]. In addition, urban winds and high-rise buildings have been studied from an urban ventilation perspective and the reduction of urban heat islands [36–39].

This study explores the impact of urban winds on a 100 m tall residential building envelope and the effect of the changing microclimate per building height. A critical evaluation is conducted on the design of balconies on the building envelope. The appropriateness of building design according to height is discussed in relation to the direct exposure of balconies to higher-up wind flow speeds, the thermal perception of the residents, and the potential for energy efficiency through a more climate responsive design. Conclusions highlight the need for a holistic understanding of the design of high-rise buildings that goes beyond the pressing issues of energy efficiency. Results form part of a wider research program on the sustainability of the high-rise typology aiming at the successful formation of sustainable design strategies. The objective is a holistic understanding on tall building design by combining issues of energy efficiency and thermal comfort. The studies are based on on-site monitoring of case-studies located in the Mediterranean climate of Tel Aviv.

2. Background Studies

The performance of high-rise buildings is influenced both by internal and external conditions. The microclimate of high-rise buildings changes with increasing altitude above ground, more specifically, wind speed increases, while dry bulb temperature drops. Urban studies have also shown that the opposite is often the case, as the Venturi effect and downward drafts may cause higher speeds at the base of tall buildings [40,41], whereas overshading the pedestrian level may cause significantly lower temperatures there than at the sunny top of a tower [42–44]. The above relationships have a direct effect on heating and cooling loads in relation to height [45,46]. Previous research on the high-rise typology highlighted the issue of increased height in relation to all other construction, and how it affects the structure's energy performance [47–50], the wind regime created at ground level and the surrounding urban fabric; affecting pedestrian thermal perception and movement [51].

Height and arrangement of neighboring buildings play an important role in high-rise construction, where urban wind velocities and solar radiation, or sun shading from surrounding structures may influence their microclimate, thermal behavior, and indoor/outdoor thermal comfort levels [52–54]. While simulations are vital in drawing conclusions on the building's behavior, these are mainly conducted on stand-alone structures, and possibly fail to capture important interconnections and mutual influences with the surrounding urban fabric. Urban studies using on-site monitoring also become important for a more accurate representation of the urban environment, and an understanding of the complexity that governs the design of tall buildings from a variety of angles, i.e., environmental, social, etc.

This paper studies the microclimate that is created along the envelope of high-rise buildings in Tel Aviv, with a special emphasis on wind and gust velocities at different height balconies. The methodology is based on in situ monitoring. Tel Aviv's increasing high-rise building activity over the last few years [55] makes the city a good case for studying the microclimate around such buildings, and assessing the potential for upgrading tall-building design. The city's Planning and Construction Committee issued the 2025 City Master Plan, setting new guidelines for further skyrise development [55], while an increasing number of buildings considerably higher than the existing urban canopy, i.e., 150 m tall and higher, is already massively changing the city's skyline.

Tel Aviv is located in the coastal zone defined as Climatic Zone A in Israel Standard SI 1045 [56]. The climate is characterized by rainless summers and mild, wet winters. Average annual temperatures range around 20 °C, with a minimum average of 13 °C in January, and a maximum average of 27 °C in August. Prevailing wind direction is W and NW, with wind speed averages fluctuating between 3–5 m/s. Global solar radiation can reach 3.43 MJ/m^2 in the summer, and 1.53 MJ/m^2 in winter. Relative humidity is high throughout the year, with annual averages of 60–67%, so even when summer temperatures are not extreme, there is still heat stress, especially during the months of July and August. The city's long hot season spans from about mid-March to the end of October [57], making cooling and ventilation an integral part of annual indoor acclimatization in buildings in Israel.

Preliminary monitoring studies were conducted in two prominent high-rise office buildings. One is Electra Tower, 165 m tall, and the other one is Midtown Tower, 200 m tall, located in close proximity to the main case-study, the residential tower. Wind and gust velocities were compared between the on-site measurements and the recorded data at the weather station located at the Tel Aviv beach. Gust wind speed is wind speed measured over a smaller period of time, i.e., 2–3 s wind speed, referring to smaller character fluctuations, and represents the maximum gust speed values that people feel the most, while these values are usually double those of wind speed [33,58,59]. Data collection covered two monitoring periods. The first one took place during early-March 2019, this period being considered a transitional winter–spring period with temperatures ranging between 13–18 °C. The second monitoring period was conducted during August (2019) which is considered the heart of the summer. The on-site monitoring stations were located on the rooftops of the high-rise buildings.

Figure 1 depicts the results of the first monitoring period, 6–12 March 2019. Results are given as wind and gust speed averages between the depicted timeframes. Data confirm that gust speeds are usually 2–2.5 times and higher compared to the wind speed. Generally, gust speeds resemble more the wind speed recordings at the Tel Aviv weather station at the beach, but with higher values at the tower tops. Increases in the wind velocities occur from midday to approximately 21:00 h, while during the night they subside. Gust speed maxima are in the range of 6–7 m/s. On the last day of the monitoring period gust speed averages reached 8.5 m/s, while individual recordings were about 10 m/s. The second monitoring period took place between 7–16 August 2019. Here, again, gust speed averages resemble more closely the recordings at the Tel Aviv beach weather station. In this set of data, one more on-site weather station was located on the balcony of Midtown Tower approximately halfway of its total height, at an approximate height of 100 m, located on the side of the tower, and at an approximate distance of 30 m from its neighboring residential tower (Figure 2, right). Wind and gust velocity averages are in the range of 4–5 m/s, while individual gust recordings at the Midtown balcony reached 8 m/s during 14 August 2019.

The summer recordings presented lower wind velocities (Figure 3). The lowest average recordings are at the balcony level, confirming that wind speed increases with building height. However, gust velocities at the balcony level accelerate to as high as those at the tower tops, and at times reach even higher values. Data indicate that the Venturi effect is at play here, with the narrower cross section of the urban canyon between the towers presenting higher gust velocities than the ones on the roof. A possible explanation for this is that gust velocities are influenced both by the distributions of the average wind speed and the turbulence intensity. As a result, an area with high wind velocities may also have high gust velocities, while an area with low average wind velocities may present high gust velocities if the turbulence is high [60].

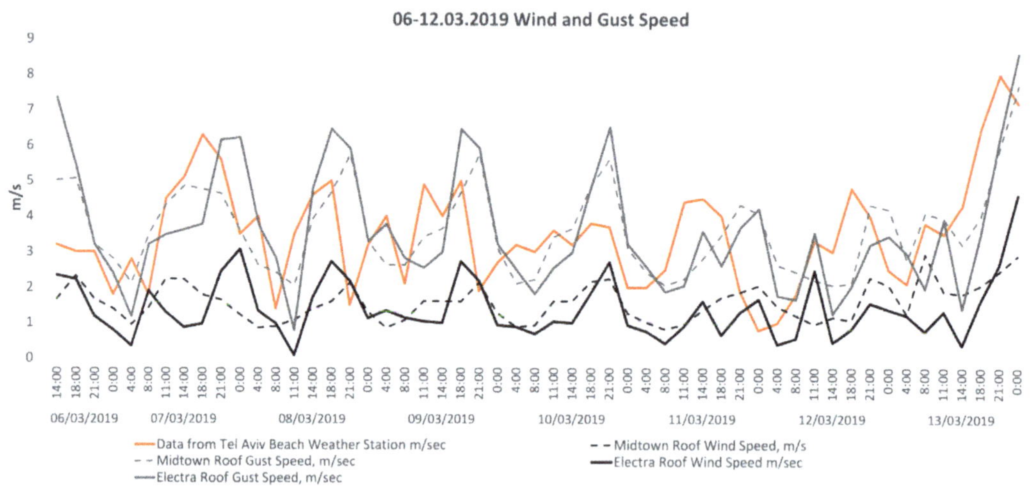

Figure 1. Wind and gust speed velocity averages between the depicted timeframes on the building tops of Electra Tower (165 m above ground) and Midtown Tower (200 m above ground). Wind speed recordings from the Tel Aviv weather station located at the beach are also included.

Figure 2. (**Left**) view of on-site monitoring station at the balcony of Midtown Tower. (**Right**) view of on-site monitoring station at the roof-top of Midtown Tower, 200 m above ground.

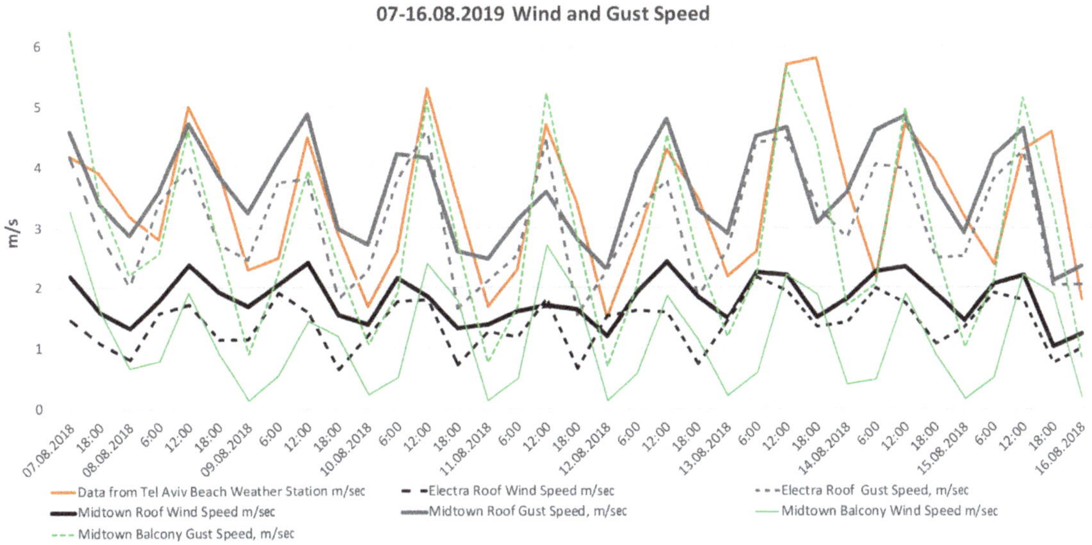

Figure 3. Wind and gust speed velocity averages between the depicted timeframes on the building tops of Electra Tower and Midtown Tower, and the balcony of Midtown Tower. Wind speed recordings from the Tel Aviv weather station located at the beach are also included.

The results of these studies prompted further research on the turbulent flow modification of wind and gust velocities per building height in the urban environment based on on-site monitoring. The focus here is on different height balconies and the effect of the wind stratifications per building height of a residential high-rise case study. Conclusions are made on the outdoor thermal comfort of the building occupants (balconies), the safety and usability of the balcony outdoor spaces design at increasing height above ground, as well as regarding missed opportunities for advancing the structure's energy efficiency.

3. Materials and Methods

3.1. The Case-Study

On-site monitoring of wind and temperature variations is conducted on different height balconies of a residential high-rise case study. The design configuration is considered a typical one, where every floor level is similar to the one above, with no considerations or detailing relating to specific heights above ground, i.e., the design of the balconies along the building envelope. Conclusions are mainly based on a critical evaluation of the residential high-rise building typology in the Mediterranean climate of Tel Aviv (Csa in the Köppen–Geiger climate classification), setting an initial understanding and delineation for a future study's roadmap in similar climates. The study covers a hot and a cold season.

The proposed case study has a West orientation, facing the prevailing W/NW wind direction in Tel Aviv, while it is distanced about 3 km downwind from the coast, with the dense urban fabric of Tel Aviv covering the intermediate space (Figure 4). The city's highway is located nearby, as are the national rail services. The residential tower is a relatively new construction approximately 100 m high and is positioned in close proximity (approximately 15 m) to its twin tower to its south (Figure 5). The towers are located within an urban plot with noticeable high-rise building activity and medium-density 2–3 floor apartment buildings. Below is a height analysis of the case-study balconies. The data are relevant to the CTBUH typical tall building characteristics, in reference to a residential tower, based on the Council's wide database of built projects [61]. The entrance, ground level floor-to-floor height is 4.65 m and a typical floor is 3.1 m high. A half floor is also estimated above the entrance level. Based on the above data, the estimated monitoring height above ground of the balconies, is:

- Floor 4 (West orientation): 4.65 m ground level floor-to-floor height + 3.1 (typical floor height) × 3 + a half floor = 4.65 + 9.3 + 1.6 = 15.55 m
- Floor 15 (South orientation): 4.65 m ground level floor-to-floor height + 3.1 (typical floor height) × 14 + a half floor = 4.65 + 43.4 + 1.6 = 49.65 m
- Floor 17 (West orientation): 4.65 m ground level floor-to-floor height + 3.1 (typical floor height) × 16 + a half floor = 4.65 + 49.6 + 1.6 = 55.85 m
- Floor 28 (West orientation): 4.65 m ground level floor-to-floor height + 3.1 (typical floor height) × 27 + a half floor = 4.65 + 83.7 + 1.6 = 89.95 m

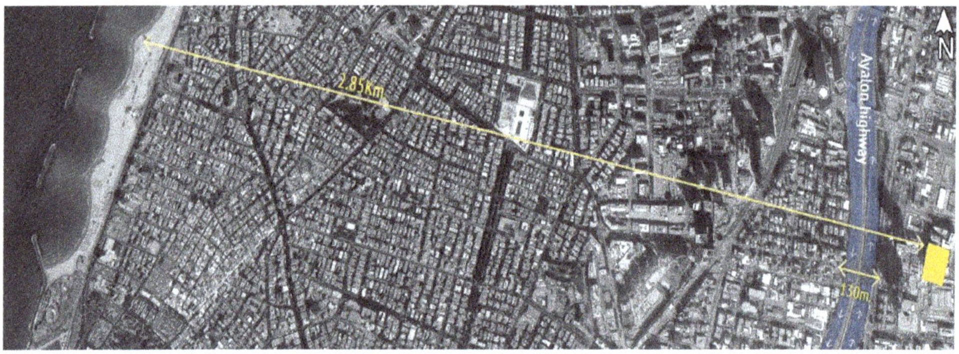

Figure 4. Location of residential tower (yellow rectangle (within the urban fabric of Tel Aviv, depicting distance from the sea promenade.

Figure 5. View of the residential tower and case-study balconies at floors 4, 17, and 28 on the West windward façade, and floor 15 on the South, side façade—between the two residential towers.

3.2. Micro-Meteorological Measurements

In situ micro-meteorological measurements were conducted at different height balconies, representing *summer* and *winter* periods. Monitoring was conducted with HOBO USB Microstations H-21 (Figure 6). Table 1 shows the specifications of the sensors used. Three monitoring locations/balconies were selected on the West windward façade at floors 4, 17, and 28, and one on the South, side façade, between the case-study tower and its neighboring one, both residential, at floor 15 (corner effect) (Figure 5). Floor 15 balcony was chosen in order to better understand the Venturi effect between the two towers, and how it affects the microclimate there. Recorded data included dry bulb temperature (°C), relative humidity (%), wind direction (degrees), wind speed (m/s), and gust speed (m/s).

The *summer* study period spans from 25 June to 1 July 2020, and the *winter* study period from 26 January to 3 February 2021. For the summer period, environmental data were recorded on floor levels 4 and 28 (West orientation—windward façade). For the winter period, initial studies showed considerably higher wind and gust speed velocities that demanded the collection of a larger set of data, as well as a higher level of analysis of the results produced. As a result, monitoring expanded to cover both the windward façades, on floor levels 4, 17, and 28 (West orientation), and the side façade on floor 15 (South orientation—between the towers). Comparisons are made between the summer and winter periods from the in situ meteorological stations, wind velocities from the weather station at the beach, and survey questionnaires administered to the respective residents.

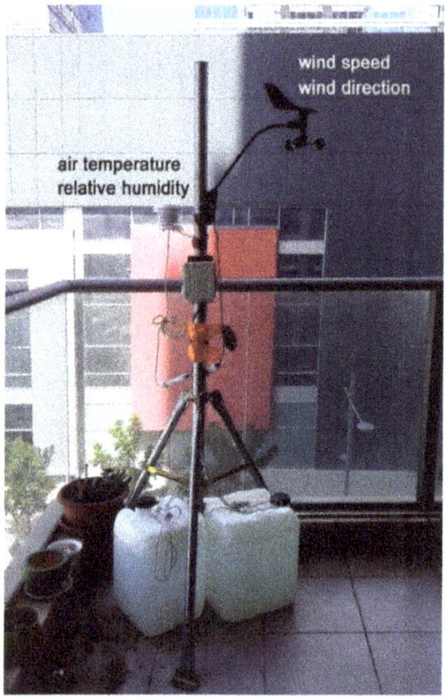

Figure 6. Installation of micro-meteorological station for data acquisition: dry bulb temperature (°C), relative humidity (%), wind speed (m/s), gust speed (m/s), and wind direction (in degrees).

Table 1. Specifications for HOBO USB Microstation H-21 used for micro-meteorological measurement.

Parameters	Sensor	Measurement Range	Accuracy	Resolution
Air temperature (°C)	HOBO S-THB-M002	−40 °C to 75 °C	±0.21 °C from 0° to 50 °C	0.02 °C at 25 °C
Relative humidity (%)	HOBO S-THB-M002	0–100%	±2.5% from 10% to 90%	0.1% RH
Wind speed (m/s)	HOBO S-WCF-M003	0 to 76 m/s	±1.1 m/s or ±5% of reading	0.5 m/s (1.1 mph)
Gust speed (m/s)	HOBO S-WCF-M003	0 to 76 m/s	±1.1 m/s or ±5% of reading	0.5 m/s (1.1 mph)
Wind direction (°)	HOBO S-WCF-M003	0 to 355°	±7 degrees	1 degree
Tripod	HSW 2 m Tripod Tower			

Figure 7 shows the micro-meteorological stations in place during the monitoring periods of the respective balconies. The photographs depict quite clearly the urban views toward the sea/west orientation. While there are a number of high-rise buildings in close proximity to the case study, these are not located directly in front of the tower. Except for floor 4 that faces a low-rise building at that level, sited at a distance of more than 15 m, and floor 15 that is located between the residential towers, again at a distance of approximately 15 m, floor 17 and floor 28 are completely exposed. Floor 17 is also exposed from above as there is no balcony directly above it, while floor 28 is shaded and covered by the balcony of the floor above, which is also the last one in the building.

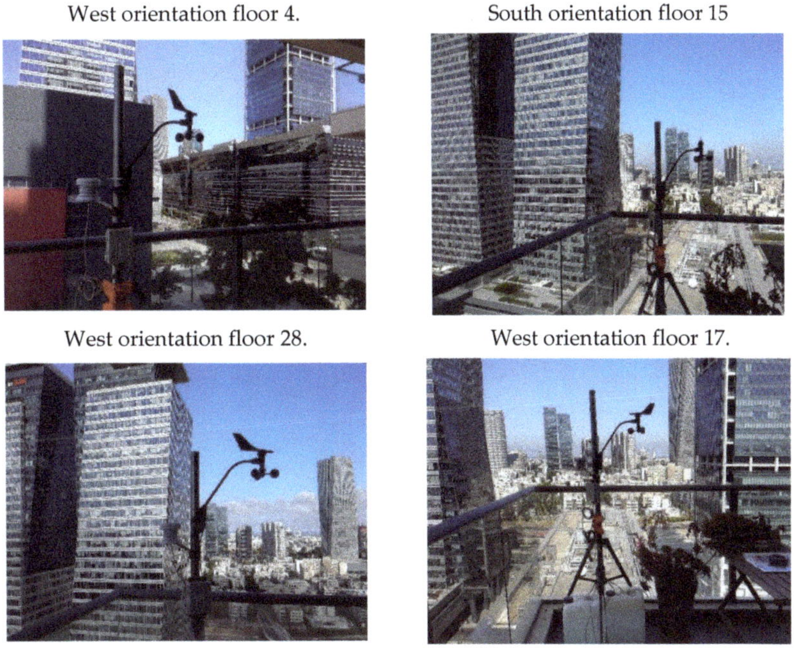

Figure 7. Micro-meteorological stations at different height balconies. Top left, clockwise: West orientation floor 4; South orientation floor 15(between the towers); West orientation floor 17; West orientation floor 28.

4. Results

4.1. Summer Period

Figure 8 graphs depict the summer timeline (25 June to 1 July 2020) in relation to wind and gust velocities, and temperature variations from the micro-mereological stations on floor 4 and floor 28. Wind speed from the weather station at the beach is also depicted as a reference point. The weather station at the beach is located 10 m above ground and is unobstructed by buildings or vegetation, as opposed to the urban location of the residential tower where the monitoring took place. However, the monitored balconies are on average much higher than the surrounding building canopy. The height of the micro-mereological station is 1.5 m, while the case study balcony heights are mentioned further up in the Case-study section. According to the above calculations, the height difference between the weather station at the beach (sea level plus 10 m) and floor 4 is 5.55 m higher, and 79.95 m higher for floor 28. The distance between floor 4 and floor 28 is about 75 m.

Data are depicted as averages between 00:00, 12:00, and 16:00 h. The specific time-frames represent best the daily fluctuations of the environmental variables under study. Monitoring results confirm that wind speed increases with height, while temperature drops. Regarding the temperature, daily values range between 23–29 °C, while the variations between floor 4 and floor 28 are approximately 1–2 °C less for floor 28. In addition, the average diurnal temperature variations are in the range of 1–2 °C less at night, suggesting an overheating of the urban environment. Wind velocities for floor 4 range between 0.3–1.7 m/s, while the increase per building height (floor 4 to floor 28) is in the range of 20% to more than 100%, with values reaching 2–3.4 m/s at floor 28. Similarly, gust speed values at floor 4 range from 1.0–5.0 m/s, while at floor 28 they accelerate to 1.7–7.0 m/s. Gust speed recordings seem to follow more closely the air velocities recorded at the beach weather station. The trend is similar, but with lower values for floor 4, while a better correlation exists with floor 28, however there, wind spikes can be higher. The highest recorded gust velocity is 7.0 m/s at floor 28, at 16:00 on 26 June 2020, with a temperature of 25.8 °C.

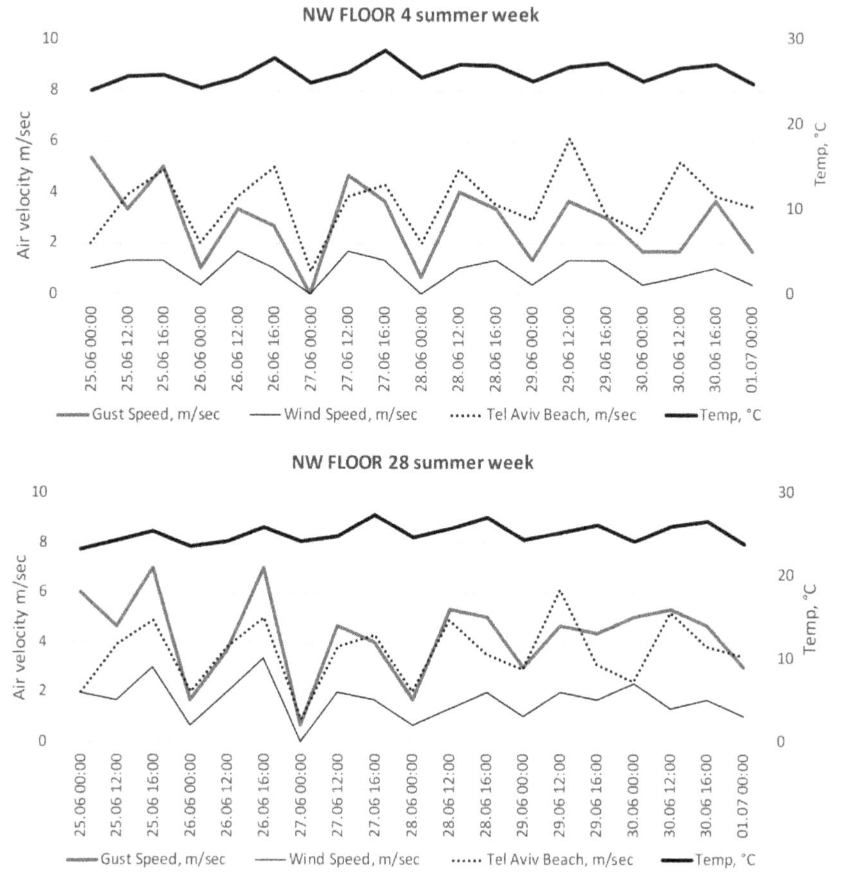

Figure 8. Summer period recordings from 25 June to 1 July 2020, depicting wind speed, gust velocity and temperature variations from the micro-meteorological stations located at floor 4 (**top**) and floor 28 (**bottom**) on the West façade, in relation to wind speed from the meteorological station at the beach at a height of 10 m, at 00:00, 12:00, and 16:00 h.

According to Lawson and Penwarden's extensive 'Land Beaufort scale' on wind effects on people at pedestrian level height [62,63], an air velocity value of 7.0 m/s refers to 'force of wind felt on body', and is at the threshold between a 'moderate breeze' and a 'strong breeze'; however, these specifications refer to wind speed at 1.75 m height. Floor 28 is at a height of almost 90 m plus the height of the standing person. In addition, the balcony's frame and railings are made of glass, possibly enhancing a feeling of discomfort to a person standing there, by exposing them to the intensified environmental variables. Figure 9 shows a view of the floor 28 balcony during summer. While the city views are astonishing the balcony is devoid of any decoration due to the high winds. The residents placed a children's pool that is hardly used according to their statements, and that is securely held in place with the weight of a metal chair. The effective use of the balcony space and the opening of the glass doors for ventilation is discussed further on in this paper through an analysis of occupants' questionnaires.

Figure 9. View of balcony at floor 28 during summer. The balcony is devoid of decoration due to the high winds, while the children's pool that is seldom used by the owners is secured in place with the weight of the metal chair.

4.2. Winter Period

Graphs in Figures 10 and 11 depict the winter period results from 26 January to 3 February 2021, in relation to wind and gust velocities, and temperature variations from the micro-meteorological stations at the residential balconies, as well as air velocities from the weather station at the beach. Due to the higher air flows during winter, the data collection expanded, covering floors 4, 17, and 28 on the windward West façade, and floor 15 on the South, side façade, between the two residential towers. The height differences between the weather station at the beach, located 10 m above ground, and the monitored balconies, are as follows: 5.55 m for floor 4, 39.65 m for floor 15, 45.85 m for floor 17, and 79.95 m for floor 28. Data are depicted as averages between 00:00, 12:00, and 14:00 h. The specific timeframes are considered to best represent the daily fluctuations of the environmental variables under study, while an hourly analysis is also conducted further down in this paper. Daily temperature variations range from about 14 °C at night to as high as 27 °C during the day, with an average of just above 18 °C for the timeframe under study, as recorded at floor 4, 15.55 m above ground. The highest temperatures are observed around 12:00. Monitoring results confirmed that temperature drops with building height between floor 4 and floors 15 and 17; however, there is a slight increase again at floor 28 during winter, e.g., 20.6 °C at 12:00 on 26 January at floor 4, drop to 18.4 °C at floor 17, and rise again to 20.1 °C at floor 28.

Further studies within the urban environment have shown higher temperatures on the roofs of the towers due to their exposure to unobstructed direct and diffuse solar radiation, as opposed to lower levels, certainly the pedestrian level, which are overshaded by the adjoining towers. A thermal comfort study on the specific climatic conditions of Tel Aviv during summers and winters of 2007–2011 reported that the city's all-year thermal comfort range is between 19–25/26 °C [64]. These results resemble the temperature stratifications of the current monitoring data. In hot climates, as per ASHRAE [65], the comfort zone shifts towards warmer conditions compared to colder climates. However, one's perception between comfort and discomfort lies between the calibration of micro-climatic conditions [66]. Outdoor thermal comfort within the urban environment is a multi-faceted subject that involves a number of environmental variables not easily quantified or controlled and is relative to peoples' specific preferences and attributes [67–70]. The impact of urban wind is pivotal. In this paper, the complexity of the outdoor urban environment is discussed in relation to building height through the presence of balconies.

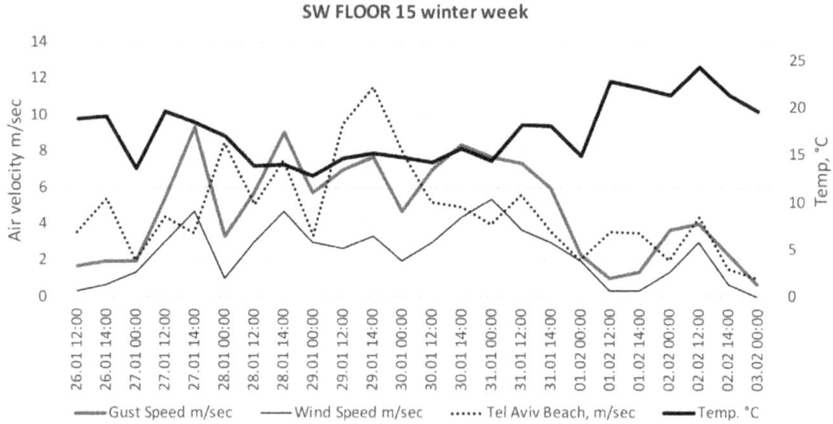

Figure 10. Winter period recordings from the 26 January to 3 February 2021, depicting wind speed, gust velocity and temperature variations from the micro-meteorological stations located at floor 15—South, side façade orientation, in relation to wind speed from the meteorological station at the beach at a height of 10 m, at 00:00, 12:00, and 16:00.

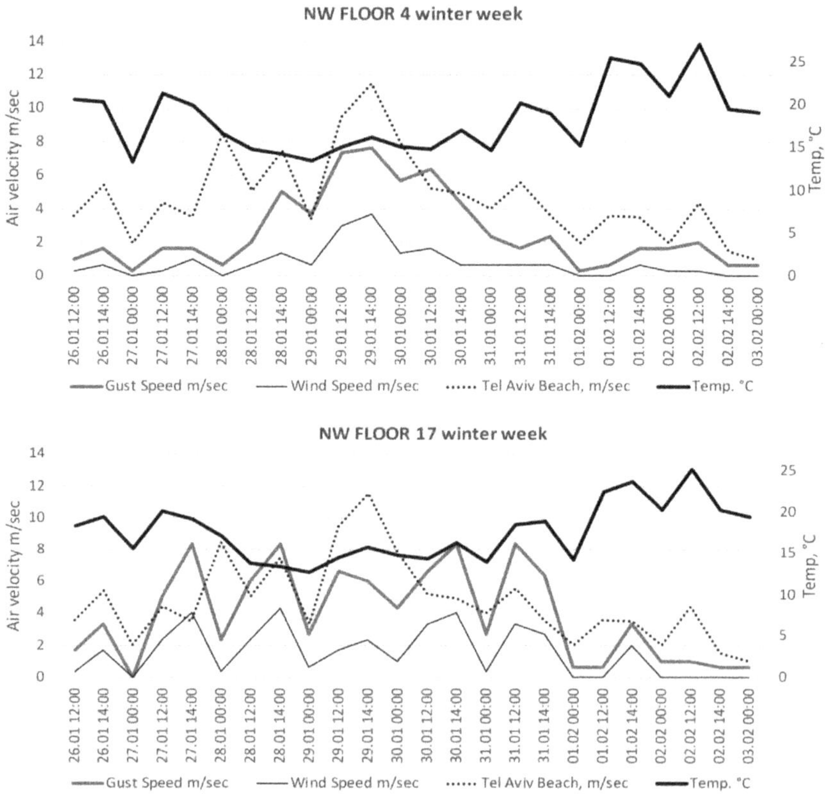

Figure 11. *Cont.*

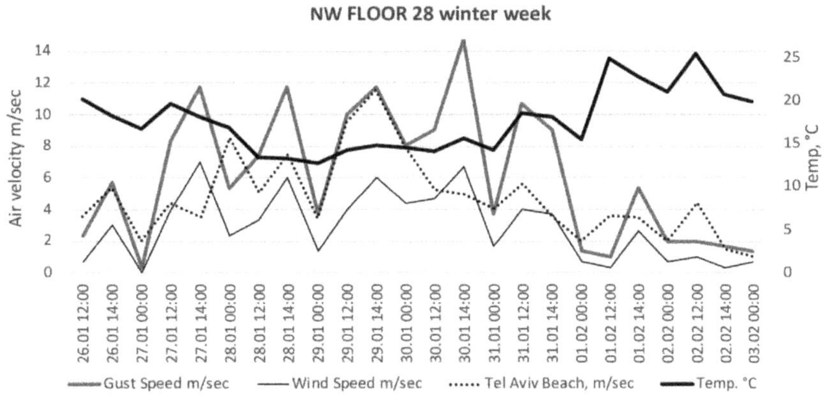

Figure 11. Winter period recordings between 26 January to 3 February 2021 depicting wind speed, gust velocity and temperature variations from the micro-meteorological stations located at floor 4 (**top**), floor 17 (**middle**) and floor 28 (**bottom**)—W façade orientation, in relation to wind speed from the meteorological station at the beach at a height of 10 m, at 00:00, 12:00, and 16:00.

During the winter period, monitored air velocities were quite high between 27 January and 2 February, with the beginning and the end of the monitoring week having weak wind days. Results confirm an increase in air velocity per building height. Wind speed averages at floor 4 are in the range of 1 m/s, with a spike to 3.5 m/s on 29 January, at 14:00. The increase, however, from floor 4 to floors 15 and 17, as a general rule, is in the range of 3, 4, and almost 5 times more, e.g., 0.7 m/s at floor 4 rises to 4.0 m/s at floor 17, 4.4 m/s at floor 15 (side elevation—corner effect), and then accelerates to 6.7 m/s at floor 28. Regarding gust speed averages, both summer and winter periods confirm that these usually reach twice the mean wind speed. In addition, similarly with the summer period, monitored gust speed velocities seem to more closely follow the air velocities at the meteorological station at the beach, especially with the increase in building height, e.g., between 29 and 30 January 2021, the trend is exactly the same. Nevertheless, with the increase in building height, both wind and gust speeds relate more to the air velocities at the beach and tend to increase considerably by the time they reach floor 28.

The correlation between air velocities at the beach and the monitoring balconies is especially prominent at the West windward façade, while the trend is not as obvious for the side balcony on floor 15. As a general rule, gust speed increases by approximately double from floor 4 to floors 15 and 17, and then another two times higher up, at floor 28. At 14:00 on 1 January, gust velocity at floor 4 was 4.4 m/s, then increased to 8.4 m/s for floors 15 and 17, and reached 14.7 m/s at floor 28, while at the same time air velocity at the beach was 4.9 m/s. Floor 28 presented considerably higher velocities throughout the windy days that span 27 January to 2 February, with gust speed averages of 8.9 m/s, while at the same time averages for floors 15 and 17 were 5.9 m/s.

In relation to the monitored wind and gust velocities, these are increased twofold and more from the ones during summer. From the graphs of Figure 7, it is quite clear that the velocities of the wind entering Tel Aviv from the beach, as recorded by the meteorological station there, can be quite high, with a maximum of 11.5 m/s on 29 January, at 14:00, and a minimum of 1.5 m/s on 2 February, at 14:00. Cohen et al.'s on-site monitoring study of outdoor pedestrian wind stratifications in three urban locations within Tel Aviv's urban fabric, recorded wind speeds between 0.9–1.9 m/s during winter, and 1.1–2.2 m/s during summer [64]. Similar values were also recorded by these authors during spot measurements, but the scope of these works is not part of this study. Nevertheless, the results highlight the considerable reductions of urban winds from the coastline inland. A further analysis on the diurnal cycle of hourly wind and gust speed velocities in relation to wind direction, and to the air velocities recorded at the beach meteorological station, is conducted in the following section for a better understanding of the wind behavior per building height during winter.

Diurnal Cycle of Hourly Air Velocities

Figure 12 illustrates a day-and-night hourly analysis on the air velocity stratifications per building height and orientation (West vs. South façade) for the monitored balconies, in relation to the maximum air velocity at the beach during the specific timeframe under study. Results confirm that wind speed increases with height; however, they also underline the importance of wind direction, depicted in degrees from north eastwards on the graphs. During the day, wind accelerates mainly from floor 4 to floor 28; however, in the early morning hours, due to the change in wind direction, floor 17 may present equal or even lower air velocities than floor 4. Nevertheless, it is quite clear than even though floor 4 could also reach high wind velocities, e.g., graph '29.01 p.m.' with daily gust speeds of 8 m/s, wind velocities at higher balconies can extend to the level of danger, making the balconies at higher altitudes dangerous and thus non-usable, redundant. An example is graph '29.01 a.m.', when at 8:30 in the morning an 18 m/s gust speed is recorded at floor 28. While the above example could be considered extreme, the graphs indicate that during high-wind winter days, wind velocities at higher floors range between 4–6 m/s, and gust speeds reach 8–10 m/s on a regular basis at floor 28.

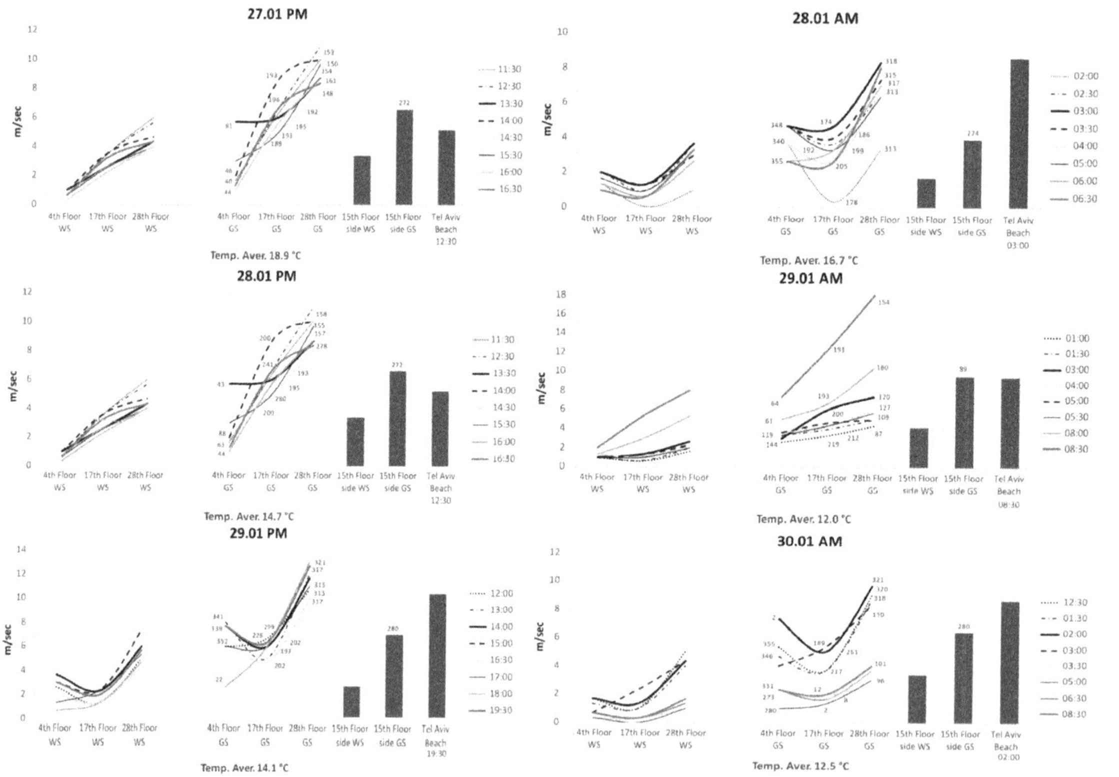

Figure 12. *Cont.*

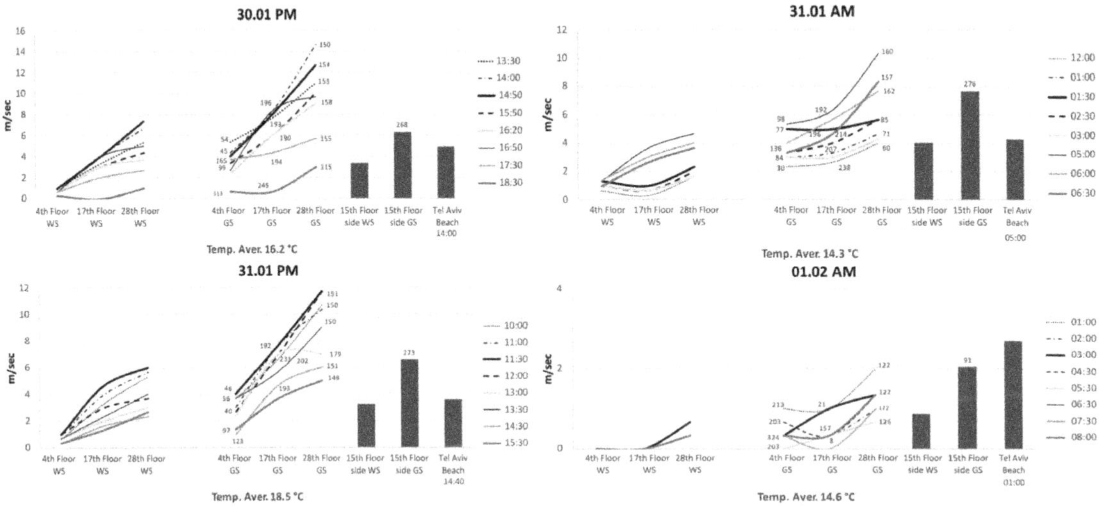

Figure 12. Day-and-night maximum wind and gust velocities between 27 January–1 February depicting floor 4, floor 17, and floor 28 (West windward façade) and floor 15 (South side façade-between towers), in relation to the maximum wind velocities at the beach. The numbers indicate wind direction in degrees from N eastwards. The continuous lines of the graph represent the wind flow from lower level to the highest, while the time frames depict the highest values recorded during the specific day.

4.3. *Wind Comfort Criteria*

Even though this case study focuses on the wind profile per building height, it was important to relate the results of these studies to 'pedestrian height urban winds' and 'wind comfort criteria', in order to provide an understanding on the effects that environmental conditions have on a person occupying the balconies discussed here, and so many others in many similar buildings. However, pedestrian wind comfort is a multifaceted matter that involves a number of parameters. Results go beyond the mechanical effects wind has on people according to the Beaufort scale, i.e., direct effect of wind force [71,72], while the thermal effects or thermal perception describe the physical, physiological and psychological parameters of peoples' comfort conditions outdoors. Physical adaptation relates to physical activity, adding or removing clothing items (reactive adaptation), or an active response to the environment, like the opening of a parasol (interactive adaptation) [73,74]. Physiological adaptation relates to changes that occur after repeated exposure to a stimulus, while psychological factors mainly relate to one's expectations from the environment, and depend on the time of exposure, control and familiarity with the environment, as well as a 'perceived naturalness' [70].

Currently, no universal wind comfort criteria for pedestrians have been agreed upon to date, while existing such are mainly based on research studies (field studies and thermal simulations) [75–77], and may differ amongst different countries and institutions. As a result, more outdoor studies are needed within the urban fabric on a global scale for establishing threshold values between tolerable and unacceptable wind conditions for pedestrians [51,78,79]. Table 2 depicts an evaluation of 'wind comfort and danger' according to Lawson and Penwarden's extended 'Land Beaufort scale' at pedestrian level height [62,63], Lawson's wind assessment in the UK [80], and NEN 8100 wind criteria according to the Netherlands Normalisation Institute (NEN) [81]. The code for pedestrian wind comfort created by the NEN is currently the only official document on a global scale. It is not a legal building requirement, but offers assistance through the provision of strict guidelines for incorporating wind comfort in building construction [72].

Table 2. Lawson's assessment in the UK [30], Land Beaufort's scale [Lawson and Penwarden], and NEN 8100 wind criteria.

The Lawson Wind Comfort Criteria	NEN 8100 Wind Comfort Criteria	Land Beaufort Scale Wind Speed at 1.75 Height	Effect	
(m/s)				
>25	>20	14.6–17.1	Dangerous to walk People blown over	Strong Gale Poor traversing
>20			Unacceptable to walk	
10–12	10	12.1–14.5	Fast walking Great difficulty with balance	Near Gale Moderate traversing
8–10		9.8–12	Pedestrian transit. Inconvenience felt when walking	
6–8		7.6–9.7	Walking, Sightseeing. Hair blown straight, difficult to walk steadily	Strong Breeze Moderate strolling
6	5	5.6–7.5	Short periods of standing/sitting Force of wind felt on body	Gentle Breeze/Moderate Breeze/Fresh Breeze Moderate: Sitting
4	2.5	3.9–5.5	Long periods of standing/sitting Raises dust, hair disarranged	
		2.4 3.8	Hair disturbed/clothing flaps	
<3.9	<2.5	1.1–2.3	Acceptable for sedentary activities. Wind felt on face	Calm/Light Air/Light Breeze Good: Walking strolling sitting
		0–1.0	No noticeable wind	

Low wind velocity (<1.0 m/s) in hot environments has a negative effect.

Nevertheless, commonly agreed findings between the different indexes see no discomfort for wind speeds up to 5 m/s, some discomfort and unpleasant feelings from 5–10 m/s, and strongly unpleasant feelings and potential danger beyond that [82,83]. However, such data, sparse as they may be, refer to pedestrian comfort and safety, whereas here we discuss comfort and safety on balconies of high-rise residential buildings. This is quite a different discussion since it implies mostly sedentary activities (sitting, dining, etc.), and involves safety discomfort (whether lack of safety is genuine or just perceived) as well as significant hazards, not least in the case of toddlers and the elderly, as well as people with a low body mass. The above monitoring results on air velocities per building height, with a focus on the winter period, range from 'uncomfortable' to 'dangerous', according to the international indexes on wind comfort (Table 2).

4.4. Resident Survey

In order to better understand the impact of the intensity of the environmental variables per building height on the thermal perception of the potential balcony users, the subjective human perspective was included in a survey using self-administered semi-structured questionnaires [84,85]. Tables 3 and 4 give an overview of the effective use of the balcony space by their occupants, as an outdoor extension of the apartment area, an understanding of their reliance on the mechanical systems of the building for heating and cooling, and their relationship to natural ventilation, as well as their perception of natural light within the premises. The questions also included the occupants' perception of noise from the highway and railway below, as well as noise from wind gusts.

The responses of the occupants are a combination of their individual preferences on comfort and satisfaction, and the effect of the environmental variables. The responses from floor 4 are the mildest. The balcony is used for recreation throughout the year, infiltration and noises are barely noticed, and natural light levels are pleasant, probably due to this floor level being shaded by a building opposite. The floor 17 balcony has a different character. While the occupants spend up to 1 h per day outdoors, both during winter and summer, they pointed out that this is because they prefer to smoke outdoors and that all balcony items are tied down due to strong winds, while windows are barely opened for ventilation. Windows and balcony doors also remain mainly closed due to noises from the highway and railway below, while noises due to wind are prominent. During winter, wind and gust velocities are mainly felt in the morning and at dusk, and during summer the sunlight is especially strong at sunset due to the balcony's west orientation.

Table 3. Occupant responses on the effective annual use of the balconies A.

South—Leeward Façade		West—Windward Façade		
	15th Floor	**4th Floor**	**17th Floor**	**28th Floor**
	Opening balcony glass doors for ventilation			
1. A few times a day	V			
2. Once day		V	V	
3. Sometimes				
4. Few times				V
5. Not at all				
	Perception of light			
1. Very strong	V		V	V
2. Strong				
3. Neutral/Pleasant		V		
4. Weak				
5. No light				
	Is the balcony in use in Winter?			
1. Daily	V		V	
2. From time to time		V		
3. Not at all				V
	Is the balcony in use in Summer?			
1. Daily	V	V	V	
2. From time to time				
3. Not at all				V
	How much time do you spend on the balcony?			
1. Less than 15 min	V			V
2. 15–30 min		V		
3. 60 min				
4. More than 60 min			V	

Table 4. Occupant responses on the effective annual use of the balconies B.

15th Floor	4th Floor	17th Floor	28th Floor
What is the purpose of the balcony for you?			
Mainly for storage	Gardening and resting	Resting and hosting	No purpose
What objects are on the porch?			
Baby carrier, equipment (tied down)	Flowerpots, a chair table, a cat litter box	Plants, chairs and table (tied down)	No objects
Are there shutters in the apartment? External, internal, both? Are these automatic?			
Internal, not automatic	No	Internal, not automatic	Automatic external shutters
Have you encountered wind gusts, whistles, infiltrations problems, noise?			
Frequent strong winds. All balcony items need to be secured. Noise and whistle noises are also present.	Barely—only on very stormy evenings in winter.	Windows are not used for ventilation, because of noise and harsh weather phenomena. A lot of noise is present from the highway and train station all day.	Everything—wind, air, noise, whistles.
Have you noticed changes between the seasons?			
Extreme weather during winter	Almost no infiltration. Slightly more in winter when it is stormy.	In the transition seasons strong winds at dusk.	Wind infiltration mainly during winter.
Additional comments			
Prefer natural ventilation over A/C. If not too hot/cold glass doors open and remain open as long as possible.	In the evening the wind is stronger than in the day	Prefer to smoke outside. The most difficult hours are in the summer from noon onwards (western sun), and in the winter the wind gusts in the morning and at dusk.	There is a lot of noise all day from the highway and train station. The windows are mostly closed because it is very cold.

The occupants on floor 15, on the other hand, prefer natural ventilation to air conditioning (A/C), so even though most of the time wind phenomena can be intense, with gust velocities ranging between 8–12m/s during the day (Table 2), these occupants try to open the glass doors to the balcony for as long as they feel comfortable. However, they do not enjoy staying on the balcony, and only use this space for storing equipment, which they tie down due to the high wind velocities. The perception of light is also strong, and occupants have added Venetian blinds to avoid glare. The most extreme responses are from the occupants of floor 28, that also relate to the pronounced environmental phenomena there. The occupants do not use the balcony at all, except very rarely in summer. They have added outdoor electric blinds and automatic opening sensors, in order to control natural light intensity and introduce some natural ventilation within the premises. Air infiltration due to high wind pressures increases during winter, while whistling noise due to the wind and infiltration, as well as noise from the adjacent highway and railway, are prominent throughout the day.

This section may be divided by subheadings. It should provide a concise and precise description of the experimental results, their interpretation, as well as the experimental conclusions that can be drawn.

The behavioral patterns of the occupants are comparable with the intensity of the recorded environmental variables per building height. Except for floor 4, the character of the other, higher balconies is mainly problematic, and they are not used, thus redundant. On floor 17, the residents' need and will to enjoy their outdoor space overcame their negative feelings stemming from increased wind velocities and noise levels, while on floor 15 the balcony is no different from a storage room. Residents on floor 28 seem to suffer the most and take the least pleasure of this extra space, as they are unable to use it either for pleasure or storage. The higher wind velocities are also counterproductive in terms of natural ventilation and cooling within the premises. These could have been beneficial during the high summer temperatures, while introducing fresh air into the premises, which is an important consideration all year round. In addition, regarding high-rise buildings located near highways and railways, most residents noted that noise levels were quite high, as the sound is not filtered by the surrounding built environment.

5. Discussion

This study is part of a wider research on the sustainability of high-rise buildings and their successful integration within the urban environment, given their increasing numbers worldwide. Results emphasize the complications of the varying microclimate in tall building design, and by doing so, emphasize the importance of a different detailing of the building envelope with height, in order to accommodate the changing environmental variables, improve the energy efficiency of the structure, and secure safe and pleasant conditions for the occupants. The building envelope's importance as the interface between indoors and outdoors is paramount [86]. This relationship is relevant to the specific climatic conditions of the building's location, and in high-rise buildings also becomes relevant to the changing microclimate with height [45,46]. Previous studies have confirmed the importance of the thermal properties of the materials of the building envelope on energy loads, with a focus on the high-rise office and residential typologies, based on thermal simulations [49,87,88]. This study draws conclusions on the appropriateness of a typical residential tower design with exposed balconies through on-site monitoring. Monitoring data are juxtaposed with the occupants' perception of comfort in relation to the height of the balconies.

The results in fact revealed that the higher the apartment the less comfortable were its owners to occupy and enjoy the outdoor balcony space (ref. Table 4). These findings stand in contrast with evidence provided of a rent gradient within tall buildings, where rent premiums increase with building height, while the addition of balcony space is considered a further premium [89] the cost of which is both reflected in the price of the apartment and the municipal monthly taxes calculated per floor area. In high-rise Hong Kong, the provision of a balcony is also considered a 'green feature', as it provides sun-shading, offers plant space, and with height above ground mitigates street pollution and noise, while an apartment without a balcony is expected to sell for considerably less, depending on the surrounding views and location [90]. Conroy et al.'s survey on the subject of 'higher-floor premium' of high-rise buildings in San Diego confirmed a 2.2% increase in sales prices with the addition of every floor level; however, the increase rate decreases with height above ground [91]. Nevertheless, while sale prices per floor level are dependent on a number of variables, e.g., additional building materials with height, size of apartment, provision of car-parking, etc., there is also the perception that units on higher floors are generally more 'desirable' due to extended views, noise and pollution reduction [92–95].

The current studies were conducted in the Mediterranean climate of Tel Aviv, yet it is quite clear, that contrary to the mild climatic conditions, wind and gust velocities tend to intensify considerably with height, reaching the point of danger, especially during winter. Such relationships are especially prominent in the graphs of Figure 12 that depict minimum–maximum winter values between day and night. Regarding the occupants, the main issues that were stressed involved increased air velocities and light intensity, increased infiltration, and wind whistling, as well as noises from the highway and railway at ground level. The intensity of these phenomena relates to the density of the urban environment and the number of high-rise buildings that exist in close proximity to one another. As height increases the structure is exposed both to higher wind and light intensities. Failure to accommodate these into building design results in malfunctioning of the building itself. With regard to the high-rise buildings being located near railroad and railway stations and highways, this is in accordance with the decisions of Tel Aviv's Planning and Building Committee for urban development. However, noise level studies that would provide an understanding on the impact such relationships have on the high-rise occupants are presently lacking.

The current case study rejected the notion of reduced noise from the ground level as height increases, with residents of the higher-up apartments being more aware of noises from the city's highway and railway, than the apartment closer to the street level (ref. Table 4, floor 4). The density of the urban fabric is definitely a parameter here, with the residents of high-rise apartments, in Hong Kong and Singapore for example, where high-rise building density is increased, having a different view on the subject. Three-dimensional simulations of an urban district in Singapore confirm that the changes in noise levels according to height in high-rise buildings are relative to the effects of obstruction, distance attenuation, ground absorption and noise barrier screen [96]. A study on noise level profiles at different levels of three high-rise buildings in Milan confirmed that these are relevant to the characteristics of the urban environment where the specific building is located, and that the results sometimes contravene the general rule that noise levels at higher floors are reduced [97]. As a result, a noise level assessment on high-rise structures becomes important in order to guide mitigation measures.

In terms of design optimization, opportunities are missed for introducing, for instance, a second façade layer enclosing the balcony space. A double-skin façade DSF spanning the width of the balcony could potentially solve the issues noted above, i.e., increased wind and gust velocities, infiltration and noise, as well as improve the effective use of this outdoor space at higher altitudes and advance the energy efficiency of the structure. The detailing of the DSF façade, however, becomes very important [88,98]. The exterior layer of glass doors could have the possibility of opening and closing subject to the use patterns of the occupants and the changes in environmental variables, while considerably reducing annual energy loads [49,88]. The balcony's outdoor space could act as a greenhouse during winter, warming passively the interior, while during summer the strategic opening of the glass doors alongside dynamic shading could allow for the introduction of natural ventilation within the premises, irrespective of the floor height above ground, reducing cooling loads. However, examples of opening doors—partitions may be problematic, even at the ground floor level, let alone in high-rise buildings. Better yet, new technologies could make such relationships easier to accomplish. The above strategy could result in annual energy load reductions and a more sociable and friendly outdoor space at greater heights.

The introduction of a double-skin façade layer into the high-rise balcony space could be explored further, while other strategies could also be appropriate in order to accommodate more socially acceptable, energy-efficient design proposals in the high-rise building typology [99–101]. Nevertheless, similar monitoring studies on high-rise buildings are lacking, while the inclusion of the occupants' perception through surveys appears to be vital towards an understanding of indoor/outdoor human comfort conditions in tall buildings. Additionally, further research, to micro-model air flow conditions using powerful CFD tools, could also prove valuable in the design of high-rise buildings. The goals and results of this study underline the complexity of the high-rise typology and the need for further studies on the relationship between design and function, especially in relation to the changing microclimate with height.

6. Conclusions

The growing numbers of high-rise buildings globally, a result of urbanization and population growth, highlight the need for in-depth studies relating to their economic, environmental, and social characteristics. The outcome will be the implementation of design strategies relating to the tall building typology, in order for it to successfully contribute towards the establishment of sustainable urban environments. This case study questions the appropriateness of the design of exposed balconies along the high-rise building envelope in the Mediterranean climate of Tel Aviv. On-site monitoring at different height balconies recorded the changing environmental variables with height and the increased wind flow fields around the tower. The results were juxtaposed with the perception of comfort of the respective occupants.

The conclusions of this case study highlight that while there are different parameters governing tall buildings relative to their height above ground, such as environmental, social, economic, etc., current design practices do not reflect upon such, with the use of a typical floorplan design all along the structure's height being one example. The introduction of exposed balconies along the high-rise building envelope, a design solution which is currently gaining popularity, especially in warmer climates, does not seem to yield the success expected. The balconies' copy/paste approach from one floor level to the next is not appropriate as height increases, and may also reach a level of danger like, for example, the environmental conditions faced in this case study by the residents of floor 28 during winter.

The tall building typology has seen a dramatic evolution from its birth to date. The main design parameters relate to function, architectural style, height and structural strategy, and accordingly affect the structure's energy performance over time [102–104]. The new generation of tall buildings, however, from the lower-height range to the tallest, will have to address issues of energy efficiency, an increasing level of climate change mitigation strategies, as well as the provision of a safe and comfortable environment for the people occupying them. Planning authorities should become stricter and require a higher level of analysis of the proposed designs, e.g., a detailed environmental impact assessment report addressing the building's impact on its surroundings, as well as the impact of the changing climate with height. An ideal scenario would be the establishment of green design guidelines on high-rise development according to climate, as well as incentives for 'excellence' in design, as part of the planning procedure.

In addition, tall buildings and their relationship to the urban environment should be studied from a holistic perspective and that can only be achieved through the partnership and commitment of all parties involved in city planning, urban and building design. In this process, a successful dialogue between government agencies, policy makers, urban designers, city planners, academics, architects and engineers, and their clients, becomes vital and should form a continuous process of information and expertise exchange. Such procedures should become the norm towards a sustainable urban future, where high-rise buildings could be seen as assets, instead of problematic urban additions.

Author Contributions: Conceptualization, ideas, formulation, and evolution of research goals and aims took place by all the authors present in this paper. Preparation, creation (writing the initial draft) and presentation of the published work conducted by I.A.M. Critical review, commentary or revision by I.A.M. and H.I.-B.-S. Visualization/data presentation by S.S. with the assistance of H.I.-B.-S. All the results/experiments and other research outputs were conducted by S.S. and H.I.-B.-S. Oversight and leadership responsibility for the research activity planning and execution, including mentorship was by I.A.M. Instrumentation and other analysis tools were supplied by I.A.M. and H.I.-B.-S. Experiments and data/evidence collection were completed by S.S. and H.I.-B.-S. Statistical, mathematical, computational, and other formal techniques to analyze or synthesize study data were analyzed by S.S. All authors have read and agreed to the published version of the manuscript.

Funding: This research received no external funding.

Data Availability Statement: Data available on request from the authors.

Acknowledgments: This research is partly supported by the Department of Civil and Environmental Engineering, Faculty of Engineering Sciences, Ben-Gurion University of the Negev (BGU); and BGU's Center for Energy and Sustainability. Access to Midtown Project and Electra Tower was facilitated by the buildings' CEOs and managers, I. Hamamy and M. Bar Lev, respectively, and their staff. We kindly acknowledge their cooperation and kind assistance, as well as those of the residential tower's residents whose balconies were studied.

Conflicts of Interest: The authors declare no conflict of interest.

References

1. Cook, R.; Browning, B.; Garvin, C. Sustainability and energy considerations. In *The Tall Buildings Reference Book*; Routledge: London, UK, 2013; pp. 145–155.
2. Donnolo, M.; Galatro, V.; Janes, L. Bank of America Tower at One Bryant Park: New York City, NY: Ventilation in Wonderland. *High Perform. Build.* 2014, pp. 51–58. Available online: http://www.hpbmagazine.org/attachments/article/11791/14Su-Bank-of-America-Tower-at-One-Bryant-Park-New-York-City-NY.pdf (accessed on 16 March 2019).
3. Simmonds, P. *American Society of Heating, Refrigerating and Air-Conditioning Engineers. ASHRAE Design Guide for Tall, Supertall, and Megatall Building Systems*; American Society of Heating, Refrigerating and Air-Conditioning Engineers: Peachtree Corners, GA, USA, 2015.
4. EU. Directive 2010/31/EU of the European Parliament and of the Council of 19 May 2010 on the Energy Performance of Buildings. *Off. J. Eur. Union* **2010**, *153*, 13–35.
5. GBEE. About PlaNYC Green Buildings & Energy Efficiency. 2015. Available online: https://www1.nyc.gov/html/gbee/html/about/about.shtml (accessed on 10 February 2019).
6. Voss, K.; Musall, E.; Lichtmeß, M. From Low-Energy to Net Zero-Energy Buildings: Status and Perspectives. *J. Green Build.* **2011**, *6*, 46–57. [CrossRef]
7. Al-Kodmany, K. Sustainability and the 21st Century Vertical City: A Review of Design Approaches of Tall Buildings. *Buildings* **2018**, *8*, 102. [CrossRef]
8. Ali, M.; Armstrong, P. Overview of Sustainable Design Factors in High-Rise Buildings. In Proceedings of the CTBUH 8th World Congress, Dubai, United Arab Emirates, 3–5 March 2008. Available online: http://grundfos.kernelglobal.dk/BLUEPRINT/BP_03_2011/GJP/3_DOWNLOAD_FILES/3_SPREAD3/B4/Future-high-rise-technology-water.pdf (accessed on 19 December 2019).
9. Cangelli, E.; Fais, L. Energy and environmental performance of tall buildings: State of the art. *Adv. Build. Energy Res.* **2012**, *6*, 36–60. [CrossRef]
10. Gültekin, A.B.; Yavaşbatmaz, S. Sustainable design of tall buildings. *J. Croat. Assoc. Civ. Eng.* **2013**, *65*, 449–461. [CrossRef]
11. Holtermans, R.; Kok, N.; Pogue, D. International Green Building Adoption Index 2018. CBRE, 2018. Available online: https://static1.squarespace.com/static/57fa8eb0b3db2b529d44fa66/t/5ade1f3188251b9506c7b346/1524506429700/IGBAI2018.pdf (accessed on 13 March 2018).
12. Long, M. *Green Building Accelerates around the World, Poised for Strong Growth by 2021*; USGBC: Washington, DC, USA, 2018. Available online: https://www.usgbc.org/articles/green-building-accelerates-around-world-poised-strong-growth-2021 (accessed on 8 March 2020).
13. Wood, A.; Salib, R. *Natural Ventilation in High-Rise Office Buildings*; CTBUH Techical Guide; Routledge: London, UK, 2013. Available online: https://store.ctbuh.org/PDF_Previews/Reports/2012_CTBUHNaturalVentilationGuide_Preview.pdf (accessed on 16 February 2019).
14. Calhoun, K.; Torbert, R. The Bank of America Tower—What Is and What Could Have Been. RMI OUTLET: Plug into New Ideas. 2013. Available online: http://blog.rmi.org/blog_2013_09_05_the_bank_of_america_tower (accessed on 15 March 2020).
15. Short, C.A.; Song, J.; Mottet, L.; Chen, S.; Wu, J.; Ge, J. Challenges in the low-carbon adaptation of China's apartment towers. *Build. Res. Inf.* **2018**, *46*, 899–930. [CrossRef]

16. Tong, Z.; Chen, Y.; Malkawi, A. Estimating natural ventilation potential for high-rise buildings considering boundary layer meteorology. *Appl. Energy* **2017**, *193*, 276–286. [CrossRef]
17. Roaf, S.; Crichton, D.; Nicol, F. Natural Ventilation in High-Rise Office Buildings. In *Adapting Buildings and Cities for Climate Change—A 21st Century Survival Guide*, 2nd ed.; Elsevier Ltd.: Amsterdam, The Netherlands, 2009; pp. 205–236. Available online: http://library.uniteddiversity.coop/Ecological_Building/Adapting_Buildings_and_Cities_for_Climate_Change.pdf (accessed on 15 June 2019).
18. Ahlfeldt, G.; McMillen, D.P. Tall Buildings and Land Values: Height and Construction Cost Elasticities in Chicago, 1870–2010. *Rev. Econ. Stat.* **2018**, *100*, 861–875. [CrossRef]
19. Choi, H.S.; Ho, G.; Joseph, L.; Mathias, N. CTBUH Height Criteria. In *Outrigger Design for High-Rise Buildings*, 1st ed.; Routledge: Abingdon, UK, 2012; pp. 80–82. [CrossRef]
20. Lotfabadi, P. Analyzing passive solar strategies in the case of high-rise building. *Renew. Sustain. Energy Rev.* **2015**, *52*, 1340–1353. [CrossRef]
21. Lotfabadi, P. Solar considerations in high-rise buildings. *Energy Build.* **2015**, *89*, 183–195. [CrossRef]
22. Torcellini, P.; Pless, S.; Deru, M.; Crawley, D. Zero Energy Buildings: A Critical Look at the Definition. In Proceedings of the ACEEE Summer Study, Pacific Grove, CA, USA, 14–18 August 2006. Available online: http://www.nrel.gov/docs/fy06osti/39833.pdf (accessed on 20 April 2019).
23. Acero, J.A.; Koh, E.J.; Ruefenacht, L.A.; Norford, L.K. Modelling the influence of high-rise urban geometry on outdoor thermal comfort in Singapore. *Urban Clim.* **2021**, *36*, 100775. [CrossRef]
24. Allegrini, J.; Carmeliet, J. Coupled CFD and building energy simulations for studying the impacts of building height topology and buoyancy on local urban microclimates. *Urban Clim.* **2017**, *21*, 278–305. [CrossRef]
25. Forouzandeh, A. Numerical modeling validation for the microclimate thermal condition of semi-closed courtyard spaces between buildings. *Sustain. Cities Soc.* **2018**, *36*, 327–345. [CrossRef]
26. Kubota, T.; Miura, M.; Tominaga, Y.; Mochida, A. Wind tunnel tests on the relationship between building density and pedestrian-level wind velocity: Development of guidelines for realizing acceptable wind environment in residential neighborhoods. *Build. Environ.* **2008**, *43*, 1699–1708. [CrossRef]
27. Stavrakakis, G.; Tzanaki, E.; Genetzaki, V.; Anagnostakis, G.; Galetakis, G.; Grigorakis, E. A computational methodology for effective bioclimatic-design applications in the urban environment. *Sustain. Cities Soc.* **2012**, *4*, 41–57. [CrossRef]
28. Anon. Minimum Design Loads for Buildings and Other Structures. In *ANSI/ASCE Standard*; (Issue 7 98); American Society of Civil Engineers: Reston, VA, USA, 2000. [CrossRef]
29. Avini, R.; Kumar, P.; Hughes, S.J. Wind loading on high-rise buildings and the comfort effects on the occupants. *Sustain. Cities Soc.* **2018**, *45*, 378–394. [CrossRef]
30. Dimoudi, A.; Kantzioura, A.; Zoras, S.; Pallas, C.; Kosmopoulos, P. Investigation of urban microclimate parameters in an urban center. *Energy Build.* **2013**, *64*, 1–9. [CrossRef]
31. *EN 1991-1-4*; 2005 Wind Actions. EUROCODES: Brussels, Belgium, 2008.
32. Okafor, C.V.; Okolie, K.C.; Cyril, M.E.; Okafor, C.P. Analysis of Wind Effect on High-Rise Building for Different Terrain Category. *Eur. J. Eng. Technol. Res.* **2017**, *2*, 23–30. [CrossRef]
33. Robertson, L.E.; Naka, T. *Tall Building Criteria and Loading*; ASCE Publications: Reston, VA, USA, 1980.
34. Taniike, Y. Interference mechanism for enhanced wind forces on neighboring tall buildings. *J. Wind. Eng. Ind. Aerodyn.* **1992**, *42*, 1073–1083. [CrossRef]
35. Yang, F. High-Rise Urban Form and Environmental Performance—An Overview on Integrated Approaches to Urban Design for a Sustainable High-Rise Urban Future. *Int. J. High-Rise Build.* **2016**, *5*, 87–94. [CrossRef]
36. Elshaer, A.; Gairola, A.; Adamek, K.; Bitsuamlak, G. Variations in wind load on tall buildings due to urban development. *Sustain. Cities Soc.* **2017**, *34*, 264–277. [CrossRef]
37. Liu, X.; Huang, B.; Li, R.; Zhang, J.; Gou, Q.; Zhou, T.; Huang, Z. Wind environment assessment and planning of urban natural ventilation corridors using GIS: Shenzhen as a case study. *Urban Clim.* **2022**, *42*, 101091. [CrossRef]
38. Liu, Y.; Xu, Y.; Zhang, F.; Shu, W. A preliminary study on the influence of Beijing urban spatial morphology on near-surface wind speed. *Urban Clim.* **2020**, *34*, 100703. [CrossRef]
39. Wang, W.; Wang, D.; Chen, H.; Wang, B.; Chen, X. Identifying urban ventilation corridors through quantitative analysis of ventilation potential and wind characteristics. *Build. Environ.* **2022**, *214*, 108943. [CrossRef]
40. Chauhan, B.S.; Chakrabarti, A.; Ahuja, A.K. Study of wind loads on rectangular plan tall building under interference condition. *Structures* **2022**, *43*, 105–130. [CrossRef]
41. Kim, J.-W.; Baik, J.-J.; Han, B.-S.; Lee, J.; Jin, H.-G.; Park, K.; Yang, H.; Park, S.-B. Tall-building effects on pedestrian-level flow and pollutant dispersion: Large-eddy simulations. *Atmos. Pollut. Res.* **2022**, *13*, 101500. [CrossRef]
42. Boiné, K.; Demers, C.M.; Potvin, A. Spatio-temporal promenades as representations of urban atmospheres. *Sustain. Cities Soc.* **2018**, *42*, 674–687. [CrossRef]
43. Deng, J.-Y.; Wong, N.H. Impact of urban canyon geometries on outdoor thermal comfort in central business districts. *Sustain. Cities Soc.* **2019**, *53*, 101966. [CrossRef]
44. Zhao, H.; Xu, G.; Shi, Y.; Li, J.; Zhang, Y. The characteristics of dynamic and non-uniform thermal radiation experienced by pedestrians in a street canyon. *Build. Environ.* **2022**, *222*, 109361. [CrossRef]

45. ASHRAE. Airflow around Buildings. In *ASHARE Fundamentals*; American Society of Heating, Refrigeration, and Air Conditioning Engineers: Peachtree Corners, GA, USA, 2009.
46. NOAA; NASA; USAI. *U.S. Standard Atmosphere 1976 (Vol. 1)*; National Oceanic and Atmospheric Administration: Washington, DC, USA; National Aeronautics and Space Administration: Washington, DC, USA; United States Air Force: Washington, DC, USA, 1976. [CrossRef]
47. Ellis, P.G.; Torcellini, P.A. Simulating tall buildings using EnergyPlus. In Proceedings of the Ninth International Building Performance Simulation Association (IBPSA) Conference and Exhibition (Building Simulation 2005), Montréal, QC, Canada, 15–18 August 2005.
48. Lotfabadi, P. High-rise buildings and environmental factors. *Renew. Sustain. Energy Rev.* **2014**, *38*, 285–295. [CrossRef]
49. Saroglou, T.; Theodosiou, T.; Givoni, B.; Meir, I.A. Studies on the optimum double-skin curtain wall design for high-rise buildings in the Mediterranean climate. *Energy Build.* **2020**, *208*, 109641. [CrossRef]
50. Saroglou, T.; Theodosiou, T.; Meir, I.A. Improving the Energy Efficiency of a Mediterranean High-Rise Envelope. *CTBUH J. Int. J. Tall Build. Urban Habitat II.* 2020, pp. 20–27. Available online: https://global.ctbuh.org/resources/papers/download/4301-improving-the-energy-efficiency-of-a-mediterranean-high-rise-envelope.pdf (accessed on 10 September 2020).
51. Saroglou, T.; Itzhak-Ben-Shalom, H.; Meir, I.A. Pedestrian thermal perception: Studies around two high-rise buildings in the Mediterranean climate. *Build. Res. Inf.* **2021**, *50*, 171–191. [CrossRef]
52. Feng, X.; Myint, S.W. Exploring the effect of neighboring land cover pattern on land surface temperature of central building objects. *Build. Environ.* **2016**, *95*, 346–354. [CrossRef]
53. Leung, L.; Ray, S.D. New York City Energy Benchmarking Data Paper Type: Low-energy Tall Buildings? Room for Improvement as Demonstrated by New York City Energy Benchmarking Data. *Int. J. High-Rise Build.* **2013**, *2*, 285–291. Available online: http://global.ctbuh.org/resources/papers/download/2276-low-energy-tall-buildings-room-for-improvement-as-demonstrated-by-new-york-city-energy-benchmarking-data.pdf (accessed on 15 December 2018).
54. Zhang, J.; Li, Z.; Wei, Y.; Hu, D. The impact of the building morphology on microclimate and thermal comfort-a case study in Beijing. *Build. Environ.* **2022**, *223*, 109469. [CrossRef]
55. Fox, J. Tel Aviv's Skyline in 2025. Sustainable City Blog. 2011. Available online: http://www.sustainablecityblog.com/2011/03/tel-avivs-skyline-in-2025/ (accessed on 25 April 2019).
56. SI 1045. Thermal Insulation of Buildings. The Standards Institution of Israel: Tel Aviv, Israel, 2019.
57. Bitan, A.; Rubin, S. *Climatic Atlas of Israel for Physical and Environmental Planning and Design*; Ramot Publishing: Tel Aviv, Israel, 1991.
58. Hunt, J.; Poulton, E.; Mumford, J. The effects of wind on people; New criteria based on wind tunnel experiments. *Build. Environ.* **1976**, *11*, 15–28. [CrossRef]
59. Melbourne, W. Criteria for environmental wind conditions. *J. Wind Eng. Ind. Aerodyn.* **1978**, *3*, 241–249. [CrossRef]
60. Liu, J.; Niu, J.; Mak, C.M.; Xia, Q. Detached eddy simulation of pedestrian-level wind and gust around an elevated building. *Build. Environ.* **2017**, *125*, 168–179. [CrossRef]
61. CTBUH. Calculating the Height of a Tall Building Where Only the Number of Stories Is Known. 2015. Available online: http://www.ctbuh.org/TallBuildings/HeightStatistics/HeightCalculator/tabid/1007/language/en-GB/Default.aspx (accessed on 10 April 2019).
62. Blocken, B.B.; Carmeliet, J. Pedestrian Wind Environment around Buildings: Literature Review and Practical Examples. *J. Therm. Envel. Build. Sci.* **2004**, *28*, 107–159. [CrossRef]
63. Lawson, T.V.; Penwarden, A.D. The effetcs of wind on people in the vicinity of buildings. In Proceedings of the 4th International Conference on Wind Effects on Buildings and Structures, London, UK, 10–13 September 1975; pp. 605–622.
64. Cohen, P.; Potchter, O.; Matzarakis, A. Human thermal perception of Coastal Mediterranean outdoor urban environments. *Appl. Geogr.* **2013**, *37*, 1–10. [CrossRef]
65. ASHRAE. *Thermal Environmental Conditions for Human Occupancy*; American Society of Heating, Refrigerating and Air-Conditioning Engineers (ANSI/ASHRAE) Standards: Peachtree Corners, GA, USA, 1992.
66. Spagnolo, J.; de Dear, R. A field study of thermal comfort in outdoor and semi-outdoor environments in subtropical Sydney Australia. *Build. Environ.* **2003**, *38*, 721–738. [CrossRef]
67. ASHRAE 2004–ANSI/ASHRAE Standard 55-2004. Thermal Environmental Conditions for Human Occupancy. American Society of Heating, Refrigerating and Air-Conditioning Engineers: Peachtree Corners, GA, USA, 2004; 2004.
68. Coccolo, S.; Kämpf, J.; Scartezzini, J.-L.; Pearlmutter, D. Outdoor human comfort and thermal stress: A comprehensive review on models and standards. *Urban Clim.* **2016**, *18*, 33–57. [CrossRef]
69. Olwig, K.R. Life between Buildings: Using Public Space. *Landsc. J.* **1989**, *8*, 54–55. [CrossRef]
70. Szűcs, Á. Wind comfort in a public urban space—Case study within Dublin Docklands. *Front. Arch. Res.* **2013**, *2*, 50–66. [CrossRef]
71. Stathopoulos, T.; Wu, H.; Zacharias, J. Outdoor human comfort in an urban climate. *Build. Environ.* **2004**, *39*, 297–305. [CrossRef]
72. Willemsen, E.; Wisse, J.A. Design for wind comfort in The Netherlands: Procedures, criteria and open research issues. *J. Wind Eng. Ind. Aerodyn.* **2007**, *95*, 1541–1550. [CrossRef]
73. Nikolopoulou, M.; Baker, N.; Steemers, K. Thermal comfort in outdoor urban spaces: Understanding the human parameter. *Sol. Energy* **2001**, *70*, 227–235. [CrossRef]

74. Nikolopoulou, M.; Steemers, K. Thermal comfort and psychological adaptation as a guide for designing urban spaces. *Energy Build.* **2003**, *35*, 95–101. [CrossRef]
75. Blocken, B.; Stathopoulos, T.; van Beeck, J. Pedestrian-level wind conditions around buildings: Review of wind-tunnel and CFD techniques and their accuracy for wind comfort assessment. *Build. Environ.* **2016**, *100*, 50–81. [CrossRef]
76. Melbourne, W.H.; Joubert, P.N. Problems of wind flow at the base of tall buildings. In Proceedings of the 3rd International Conference on Wind Effects on Building Structures, Tokyo, Japan; 1971; pp. 105–114.
77. Yoshie, R.; Mochida, A.; Tominaga, Y.; Kataoka, H.; Harimoto, K.; Nozu, T.; Shirasawa, T. Cooperative project for CFD prediction of pedestrian wind environment in the Architectural Institute of Japan. *J. Wind Eng. Ind. Aerodyn.* **2007**, *95*, 1551–1578. [CrossRef]
78. Adamek, K.; Vasan, N.; Elshaer, A.; English, E.; Bitsuamlak, G. Pedestrian level wind assessment through city development: A study of the financial district in Toronto. *Sustain. Cities Soc.* **2017**, *35*, 178–190. [CrossRef]
79. Du, Y.; Mak, C.M. Improving pedestrian level low wind velocity environment in high-density cities: A general framework and case study. *Sustain. Cities Soc.* **2018**, *42*, 314–324. [CrossRef] [PubMed]
80. Lawson, T.V. *The Determination of the Wind Environment of a Building Complex before Construction*; Report Number TVL 9025; Department of Aerospace Engineering, University of Bristol: Bristol, UK, 1990.
81. *NEN 8100*; Netherlands Normalisation Institute, Wind Comfort en Wind Danger in the Built Environment. NEN: Delft, The Netherlands, 2006.
82. Lawson, T.V. The Wind Environment of Buildings: A Logical Approach to the Establishment of Criteria. *J. Wind. Eng. Ind. Aerodyn.* **1973**, *3*, 93–105. [CrossRef]
83. Penwarden, A. Acceptable wind speeds in towns. *Build. Sci.* **1973**, *8*, 259–267. [CrossRef]
84. Nikolopoulou, M.; Lykoudis, S. Thermal comfort in outdoor urban spaces: Analysis across different European countries. *Build. Environ.* **2006**, *41*, 1455–1470. [CrossRef]
85. Pearlmutter, D.; Jiao, D.; Garb, Y. The relationship between bioclimatic thermal stress and subjective thermal sensation in pedestrian spaces. *Int. J. Biometeorol.* **2014**, *58*, 2111–2127. [CrossRef]
86. Givoni, B. *Man, Climate and Architecture*; Van Nostrand Reinhold: Washington, DC, USA, 1981.
87. Saroglou, T.; Meir, I.A.; Theodosiou, T.; Givoni, B. Towards energy efficient skyscrapers. *Energy Build.* **2017**, *149*, 437–449. [CrossRef]
88. Saroglou, T.; Theodosiou, T.; Givoni, B.; Meir, I.A. A study of different envelope scenarios towards low carbon high-rise buildings in the Mediterranean climate—Can DSF be part of the solution? *Renew. Sustain. Energy Rev.* **2019**, *113*, 109237. [CrossRef]
89. Baxter, R.; Hastings, N.; Law, A.; Glass, E.J. The Vertical City: Rent Gradients and Spatial Structure Crocker. *Anim. Genet.* **2008**, *39*, 561–563. [CrossRef]
90. Chau, K.W.; Wong, S.K.; Yiu, C.Y. The value of the provision of a balcony in apartments in Hong Kong. *Prop. Manag.* **2004**, *22*, 250–264. [CrossRef]
91. Conroy, S.; Narwold, A.; Sandy, J. The value of a floor: Valuing floor level in high-rise condominiums in San Diego. *Int. J. Hous. Mark. Anal.* **2013**, *6*, 197–208. [CrossRef]
92. Chin, T.-L.; Chau, K.; Ng, F. The Impact of the Asian Financial Crisis on the Pricing of Condominiums in Malaysia. *J. Real Estate Lit.* **2004**, *12*, 33–50. [CrossRef]
93. Ong, S.E. Prepayment risk and holding period for residential mortgages in Singapore—Evidence from condominium transactions data. *J. Prop. Investig. Finance* **2000**, *18*, 586–602. [CrossRef]
94. So, H.; Tse, R.; Ganesan, S. Estimating the influence of transport on house prices: Evidence from Hong Kong. *J. Prop. Valuat. Investig.* **1997**, *15*, 40–47. [CrossRef]
95. Tyrväinen, L.; Miettinen, A. Property Prices and Urban Forest Amenities. *J. Environ. Econ. Manag.* **2000**, *39*, 205–223. [CrossRef]
96. Zhao, W.-J.; Liu, E.-X.; Poh, H.J.; Wang, B.; Gao, S.-P.; Png, C.E.; Li, K.W.; Chong, S.H. 3D traffic noise mapping using unstructured surface mesh representation of buildings and roads. *Appl. Acoust.* **2017**, *127*, 297–304. [CrossRef]
97. Benocci, R.; Bisceglie, A.; Angelini, F.; Zambon, G. Influence of traffic noise from local and surrounding areas on high-rise buildings. *Appl. Acoust.* **2020**, *166*, 107362. [CrossRef]
98. Christian, S.; Richard, H.; Herwig, H.; Thomas, M.; Wolfgang, S.; Matheos, S.; Ifigenia, F.; Hans, E.; Heike, E.-K.; De Matos Mario Duarte Rogerio, B.Å. BESTFACADE—Best Practice for Double Skin Façades. 2008. Available online: http://scholar.google.com/scholar?hl=en&btnG=Search&q=intitle:Best+practice+for+double+skin+facades#1 (accessed on 11 April 2019).
99. Attia, S. Evaluation of adaptive facades: The case study of Al Bahr Towers in the UAE. *Qsci. Proc.* **2017**, *2016*, 8. [CrossRef]
100. Capeluto, G. Adaptability in envelope energy retrofits through addition of intelligence features. *Arch. Sci. Rev.* **2019**, *62*, 216–229. [CrossRef]
101. Niu, J. Some significant environmental issues in high-rise residential building design in urban areas. *Energy Build.* **2004**, *36*, 1259–1263. [CrossRef]
102. Ali, M.M.; Armstrong, P.J. Strategies for Integrated Design of Sustainable Tall Buildings. University of Illinois at Urbana-Champaign School of Architecture, 2006; pp. 1–22. Available online: http://www.aia.org/aiaucmp/groups/ek_public/documents/pdf/aiap080038.pdf (accessed on 20 February 2019).

103. Ali, M.M.; Moon, K.S. Structural Developments in Tall Buildings: Current Trends and Future Prospects. *Arch. Sci. Rev.* **2007**, *50*, 205–223. [CrossRef]
104. Oldfield, P.; Trabucco, D.; Wood, A. Five energy generations of tall buildings: An historical analysis of energy consumption in high-rise buildings. *J. Arch.* **2009**, *14*, 591–613. [CrossRef]

buildings

MDPI

Article

Health Risk Assessment of Inhalable Dust Exposure during the Welding and Grinding Process of Subway Aluminum Alloy Components

Can Li *, Duanjun Han, Xiaoqing Wei *, Jinlin Yang and Chunlong Wu

College of Civil Engineering, Hunan University of Technology, Zhuzhou 412007, China; m21081400008@stu.hut.edu.cn (D.H.); m22085900009@stu.hut.edu.cn (J.Y.); m22081400006@stu.hut.edu.cn (C.W.)
* Correspondence: canli@hut.edu.cn (C.L.); weixiaoqing@hut.edu.cn (X.W.)

Abstract: The subway factory industry is developing rapidly in China, but there are some occupational health risk assessments of inhalable dust in this industry. Therefore, this study aimed to explore the contamination level and health risks of dust in an aluminum alloy body workshop of subway factories in Hunan Province, China. A total of 160 dust samples were collected from the welding and grinding areas. The main elements of PM_{10} were analyzed. The health risks of aluminum dust and $PM_{2.5}$ were evaluated. The Monte Carlo method was adopted to compare the sensitivity of the Hazard Quota (HQ) of aluminum dust and carcinogenic risk (CR) of $PM_{2.5}$ to the exposure parameters for workers. The results showed that the PM_{10} concentration in the grinding area was higher, while the $PM_{2.5}$ concentration in the welding area was higher. The metal element with a mass fraction of 27.7% was aluminum. In both areas, the probability of the aluminum dust HQ exceeding 1 was approximately 17% and 68%, respectively. The $PM_{2.5}$ CR exceeded the acceptable upper limit value (1.0×10^{-4}). The main risk factor of aluminum dust HQ was concentration, while the main risk factors of $PM_{2.5}$ CR were concentration and exposure duration. These findings provide basic data for enhancing health risk management in the subway industry.

Keywords: inhalable dust; health risk assessment; quantitative evaluation; sensitivity analysis

Citation: Li, C.; Han, D.; Wei, X.; Yang, J.; Wu, C. Health Risk Assessment of Inhalable Dust Exposure during the Welding and Grinding Process of Subway Aluminum Alloy Components. *Buildings* **2023**, *13*, 2469. https://doi.org/10.3390/buildings13102469

Academic Editor: Andrea Petrella

Received: 20 July 2023
Revised: 23 September 2023
Accepted: 25 September 2023
Published: 28 September 2023

1. Introduction

With the acceleration of China's urbanization process, the subway has become an important part of urban public transportation. In the subway factory industry, aluminum products are widely used for welding because of their unique physical and chemical properties [1], and the performance and lifespan of welded products, such as the subway aluminum alloy car body, are directly affected by welding quality [2]. During the welding and grinding processing of aluminum subway components, a large quantity of dust particles is inevitably generated. In the post-pandemic era, the human health risks of particulate matter have received widespread attention. These dusts contain harmful metallic elements and metal oxides, such as Aluminum, Manganese, etc. [3] Long-term exposure to low doses of these dusts, especially $PM_{2.5}$, is very detrimental to worker health [4,5]. Fine particles entering the human body will harm human organs such as the respiratory tract, heart, lungs and blood vessels [6]. For example, the heart rate variability of shipyard welders is mainly caused by their exposure to welding fumes [7]. The metal oxide of aluminum in welding smoke is the main cause of metal fume fever, which may trigger respiratory problems [8]. Therefore, it is necessary to monitor the dust concentration and accurately assess the health risks of dust to enhance the management of the subway factory working environment.

Over the past few decades, extensive research has been conducted in particle properties and their interactions with human health [9]. Numerous reports on the health risk assessment of dust exposure have also been published. Huang et al. [10] investigated the characteristics and health effects of the fine particulate matter ($PM_{2.5}$) at the subway sites in

Chengdu, China. The results showed that arsenic was the predominant element affecting the potential non-carcinogenic and carcinogenic risks of subway workers. Alias et al. [11] identify the composition and health risk of $PM_{2.5}$ in naturally ventilated schools located in the state of Selangor and Kuala Lumpur in Peninsular Malaysia. The authors found that for all stations, the Hazard Quota (HQ) value was lower than the acceptable limits, while the excess lifetime carcinogenic risk (CR) value was slightly higher than the acceptable level (1.0×10^{-6}) for Cr and Ni. Wang et al. [12] conducted a health risk assessment on exposure to particulate matter and other air pollutants in the underground parking garage environment under natural ventilation conditions in Xi'an, China. The results indicated that the entropy values of cardiac and non-cardiac risks of staff were the highest. Hamzah et al. [13] investigated the association between metal dust exposure and respiratory health in male factory workers. Dueck et al. [14] found that compared to welder apprentices, professional welders should pay more attention to the health risks of the exposure to manganese in inhalable particles. Tong et al. [15] analyzed contamination levels and health effects of automobile factory dust. The results showed that reducing the average time spent in the factory could effectively mitigate the health risks of dust to workers. In automotive plants, the method of local exhaust ventilation could also reduce worker particle exposure [16]. However, few studies have been conducted on the health risk assessment of dust in the welding and grinding environment of subway factory industries.

Furthermore, the Monte Carlo simulation method is a very effective method for the human health risk assessment of environmental contaminants. Dehghani et al. [17] adopted the Monte Carlo simulation technique to assess the health risks of heavy metal exposure in a steel plant, and found that further improved control measures should be proposed for reducing the occupational exposure levels. In addition, quantitative health risk analysis based on the Monte Carlo simulation can provide an important scientific basis for evaluating potential human risk assessment [15,18]. Therefore, this probabilistic simulation method was employed to analyze the uncertainty of exposure parameters in this study.

We selected an aluminum alloy body workshop within subway manufacturing plants located in Hunan Province, China as a case study. First, dust samples in the welding and grinding areas were collected at fixed points. In addition, the dust mass concentrations were determined via the membrane-weighing method. Then, the probabilistic health risk evaluation model was adopted to evaluate the non-carcinogenic Hazard Quota (HQ) and carcinogenic risk (CR) of inhalable dust. Finally, the sensitivities of the exposure parameters influencing the HQ and CR of dust were investigated. The research results could provide reference for managers to make health risk management decisions in the subway factory industry.

2. Materials and Methods

2.1. A Subway Aluminum Alloy Body Workshop

At present, China has taken the global lead in terms of the speed and scale of subway system planning and construction. With the increase in Gross Domestic Product (GDP), major cities in China tend to develop longer subway systems [19]. Thus, an aluminum alloy body workshop within subway factory industries in Hunan Province was selected in this study. In the workshop, the subway aluminum alloy bodies are welded and grinded. Their sizes were 67 m long × 18 m wide × 15 m high. The dust generated mainly included welding fumes and grinding particles. A displacement ventilation system with 14 cylindrical air supply outlets (2 m height) and 14 louvers return air outlets (12.5 m height) was adopted to meet the indoor air quality of the Chinese mandatory national occupational health standard "GBZ 2.1—2019 Occupational exposure limits for hazardous factors in the workplace Part 1: Chemical Hazardous Factors" [20].

2.2. Sample Collection and Analysis

A total of 160 dust samples were collected from 8 to 21 June 2020 under normal welding and grinding conditions. The fixed-point sampling process follows the national standard "Specification for Air Sampling for Monitoring Hazardous Substances in the Workplace" [21]. As shown in Figure 1, two sampling points (S1–S2) are set up in the welding and grinding area (height 1.5 m) of the workshop, where S1 is located in the welding area and S2 is located in the grinding area. In this study, the mass concentrations of total suspended particulate matter (TSP), PM_{10} and $PM_{2.5}$ were monitored and recorded. The surface charge of particles is a common phenomenon. For example, laser printers emit negatively charged particles during operation [22]. The maximum surface charge density of large spherical particles in a normal atmosphere is 27 $\mu C/m^2$, and decreases inversely with the increase in the square root of particle size [23]. When charged colloidal particles interact with oppositely charged ions, they can form relatively large aggregates that are stable for days or weeks [24]. In order to avoid the influence of particle surface charge on sample collection, a neutral adsorption film, namely carbon support film (effective diameter of the membrane Φ40 mm, China), was selected in this study. An intelligent small-flow $TSP/PM_{10}/PM_{2.5}$ sampler (Laoying 2030D, Qingdao, China) equipped with the carbon support membrane was used for dust sampling with a flow rate of 16.67 L/min. During the two-week period, samples were collected at each sampling point at 9:30–9:45, 11:00–11:15, 14:30–14:45, and 16:00–16:15 on each weekday for 15 min.

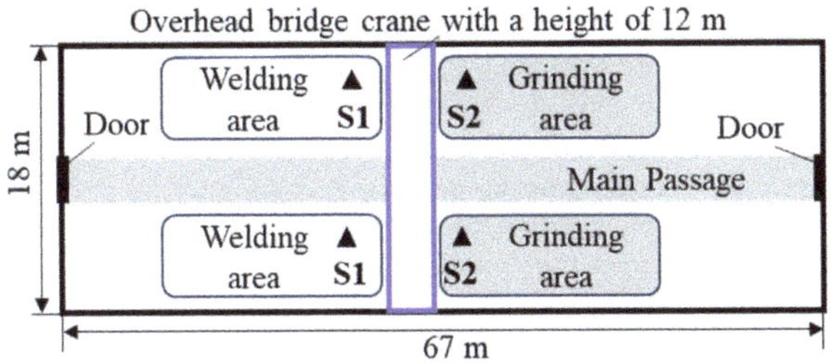

Figure 1. Location of the sampling points in the welding and grinding area.

The dust concentration was calculated using the membrane increment method. All membranes were placed in a desiccator for 2.5 h prior to sampling. Then, the weight and number of these membranes were recorded. At each sampling point, a dust sampler equipped with a filter membrane collected a 15 min air sample at the height of the breathing belt of the welding and grinding workers. After dust sampling, the membrane was accurately weighed using a microbalance with a sensitivity of 0.01 mg. When the diameter of the filter membrane was 40 mm, the increase in the total dust of the filter membrane should have been controlled within the range of [0.1 mg, 10 mg] to avoid dust falling off due to overload. The dust concentration in the working environment in the workshop was equal to the dust increment of the filter membrane divided by the volume of the air sample. The main elements and their weight percent of dust sample micro-area in the welding and grinding area were analyzed using the method of scanning electron microscopy combined with energy spectroscopy method.

2.3. Health Risk Assessment

This study aimed to evaluate the dust health risks of welding and grinding workers in an aluminum alloy body workshop of subway factory factories. Therefore, the purpose of this study was to evaluate the health risks of inhalable exposure to aluminum dust and $PM_{2.5}$. A non-carcinogenic HQ greater than 1 indicates that the worker is at risk of developing adverse health effects, and the opposite means that there is less health risk. The HQ of respirable dust is calculated as follows [25]:

$$HQ = ADD/RfC \tag{1}$$

$$ADD = \frac{C \times IR \times ET \times EF \times ED}{BW \times AT} \quad (2)$$

where RfC is the reference concentration, mg/m^3; RfCs for $PM_{2.5}$ [25] and aluminum dust [26] are 0.005 mg/m^3, respectively; ADD is the average daily dose (ADD) of indoor air pollutants inhaled through the respiratory tract, mg/(kg·d); C is the mass concentration of air pollutants inhaled through the respiratory tract, mg/m^3; BW is the body weight, kg; IR is the inhalation rate, m^3/d. Based on data from Ref. [27], the linear fitting relationship between IR and BW is: IR = 0.0325 × BW + 0.6045 (R^2 = 0.9985).ET is the exposure time, h/d; EF is the exposure frequency, d/a; ED is the exposure duration, a;AT is the average exposure time, h. AT for non-carcinogenic effects: ED × 365 × 24 (h). For carcinogenic effects fixed as: 70 × 365 × 24 = 613,200 (h).

The CR values were divided into three scales: an insignificant CR less than 10^{-6} indicated a negligible CR, an acceptable CR value between 10^{-6} and 10^{-4} indicated that the CR should be considered, and a significant CR value greater than 10^{-4} indicated a strong CR must be considered [28–30]. The CR value of $PM_{2.5}$ through the inhalation route was calculated as follows:

$$CR = ADD \times IUR \times CF \quad (3)$$

where IUR is the inhalation unit risk, IUR for $PM_{2.5}$ is 0.008 $(\mu g/m^3)^{-1}$; CF is the conversion factor, 1000 μg/mg.

2.4. Sensitivity Analysis

As a very important class of methods, sensitivity analysis has been widely applied in many research fields, such as the occupational health risk assessment. It could quantitatively describe the importance of risk assessment model input variables to output variables. Monte Carlo simulation, which is a technique used to perform sensitivity analysis, could be used to quantitatively assess the health risks of air pollutant exposure and deal with the uncertainties associated with it [31,32]. Based on the Monte Carlo simulation, sensitivity analysis was adopted to evaluate the influence of exposure parameters on health risk evaluation indicators. In the simulation results, when the sensitivity is greater than 0, the variable has a positive impact on the prediction results, and when the sensitivity is less than 0, it indicates a negative impact. The impact of variables on health risk increases as the absolute value of sensitivity increases. Previous studies have shown that based on Oracle Crystal Ball software (Version: 11.1.3.0.0), 10,000 iterations are sufficient to ensure the stability and accuracy of simulation results [33]. Therefore, the number of iterations was set to 10,000.

3. Results and Discussion

3.1. Monitoring Results and Analysis

3.1.1. Dust Mass Concentration

The statistical analysis results of the dust mass concentration in the working environment of this workshop are shown in Table 1. Based on the mass concentration data of dust samples, the probability distribution law of dust concentration was obtained by using Crystal Ball software. The results of the Anderson–Darling test show that the dust concentration distribution in the workshop working environment is normally distributed. The results of the existing studies show that the same is true of the distribution of dust concentration in foundries [15]. The mean mass concentration (MMC) ± standard deviations (SD) of total suspended particulate matter (TSP) in the welding and grinding areas were (1052 ± 509 $\mu g/m^3$) and (2562 ± 1528 $\mu g/m^3$), respectively, with PM10 fraction accounting for more than 75%. To date, the particulate matter with the greatest impact on human health has been recognized as less than 10 μm in diameter [34]. Adverse health effects may increase with decreasing particle size, especially for dusts less than 2.5 μm in diameter [35]. As shown in Table 1, the mass concentration of $PM_{2.5}$ (237 ± 116 $\mu g/m^3$) in the welding area is higher, whereas those of TSP (2562 ± 1528 $\mu g/m^3$) and PM10 (1937 ± 1487 $\mu g/m^3$)

in the grinding area are higher. The dust concentration is lower than the occupational exposure limits for welding fumes of 4 mg/m^3. The average mass concentration of $PM_{2.5}$ (237 μg/m^3) in the welding area is more than that (88 μg/m^3) in the grinding area. PM_{10} and $PM_{2.5}$ can deposit deeply in the upper and lower respiratory tracts, respectively, while ultrafine particles (<0.1μm) are mainly deposited in the alveoli [36]. Therefore, the occupational health risk management of the working environment, especially the welding area, should be enhanced.

Table 1. Dust mass concentration (μg/m^3) in the workshop environments.

Working Area	Dust	Distribution	Mean	SD	Min	Max
	TSP	Normal	1052	509	475	1714
Welding area	PM10	Normal	838	518	260	1517
	PM2.5	Normal	237	116	117	394
	TSP	Normal	2562	1528	756	4283
Grinding area	PM10	Normal	1937	1487	304	3565
	PM2.5	Normal	88	16	67	108

3.1.2. Main Elements of PM_{10}

The deposition characteristics of inhaled particles depend on flow velocity and particle size, and their chemical composition is very complex [36]. As shown in Figure 2, the main elements and their weight percentages of respirable particles PM_{10} in the microregion of dust samples were analyzed using scanning electron microscopy combined with energy spectroscopy. The trajectory of submicron particles is similar to that of tracer gases [37], and it is easy to aggregate during airborne propagation, making the particle size larger and becoming a botryoidal aggregate. Among these dusts, the mass fractions of the elements O, Al and Si are 46.0%, 27.7% and 7.5%, respectively. The metal element with a mass fraction of more than 10% is aluminum, followed by Mg (4.6%), Fe (3.6%) and Ca (2.5%). Mg, Fe, Ca and Na in $PM_{2.5}$ were not harmful to the human body [38]. The Mn element in $PM_{2.5}$ might pose a non-carcinogenic risk to adults and children through respiratory pathways [26,39]. In this study, the impact of manganese on workers' health risks was ignored due to its low mass fraction (0.7%). The mass fraction of non-metallic elements C is 2.1%. Elemental carbon can cause lung cancer or inflammation [40], while the adverse effects of organic carbon, such as Polycyclic aromatic hydrocarbons (PAHs) in an office room with intense printing activity, on respiratory health are more significant [29]. In addition, the lung deposition and the toxicity of airborne carbon nanotubes are both dependent on the nanotubes' shape and form [41]. Therefore, the health risk assessment of respirable aluminum dust exposure is mainly considered.

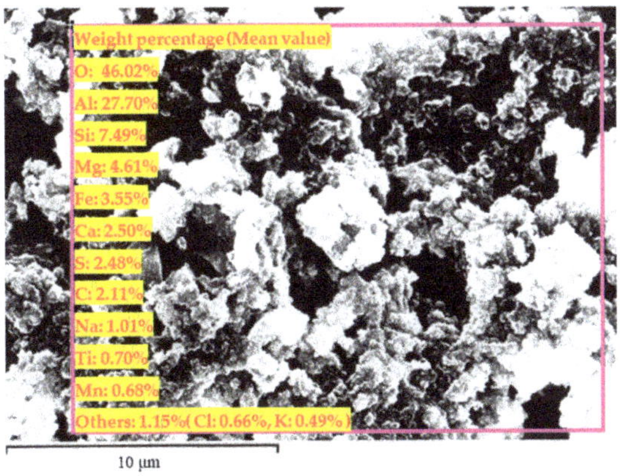

Figure 2. Sample particles observed under scanning electron microscopy and their main elemental components obtained by energy spectrum analysis.

The statistical analysis results of the mass concentration of respirable aluminum dust in the working environment are shown in Table 2. The mass concentrations of aluminum dust in the welding and grinding areas are normally distributed, and their MMC ± SD are 235 ± 145 μg/m^3 and 484 ± 372 μg/m^3, respectively. The concentration of aluminum dust is lower than the occupational exposure limits for aluminum metal and aluminum alloy dust of 3 mg/m^3 and aluminum oxide dust of 4 mg/m^3.

Table 2. Inhalable aluminum dust concentration (μg/m^3) in the working environments.

Working Area	Distribution	Mean	SD	Min	Max
Welding area	Normal	235	145	73	425
Grinding area	Normal	484	372	76	891

3.2. Exposure Parameters

Exposure parameters play a vital role in human health risk assessment. We interviewed 18 welders and 10 grinding workers (all men) under their normal working conditions. Data such as workers' body weight (BW), exposure time (ET), exposure frequency (EF) and exposure duration (ED) were collected through the method of site survey. Crystal Ball software was used to test the statistical values (Chi square, Kolmogorov–Smirnov, and Anderson–Darling) to analyze the distribution of these exposure parameters. In this study, BW and IR were normally distributed based on the Anderson–Darling test. Based on data from Ref. [27], the linear fitting relationship between IR and BW was obtained. Triangular distributions of ET, EF and ED fit best. These exposure parameters are shown in Table 3.

Table 3. Exposure parameters of male workers engaged in welding and grinding.

Exposure Parameter	Unit	Distribution	Probable Value	Min	Max	SD
IR	m^3/h	Normal	2.37	2.12	2.97	0.81
ED	a	Triangular	26	5	35	
EF	d/a	Triangular	296	272	318	
ET	h/d	Triangular	9.50	8	10.25	
BW	kg	Normal	54.3	46.5	72.8	6.2

3.3. Health Risk Assessment

The soluble aluminum compounds in inhalable particles can enter the bloodstream from the lungs of workers exposed to aluminum welding fumes. This can cause disturbances in their cognitive processes, memory and concentration, and changes in mood and electroencephalogram [42]. The results of the simulation of non-carcinogenic health risks of inhalable exposure to aluminum dust are shown in Figure 3. The geometric mean ± SD of the aluminum dust HQ in the welding and grinding areas was 0.697 ± 0.286 and 1.37 ± 0.663, respectively. The probability of the HQ of aluminum dust in both areas exceeding 1 was approximately 17% and 68%, respectively. The aluminum dust generated during the grinding process far exceeded that generated during the welding process, which led to a greater probability that the grinding workers could inhale aluminum dust and exposed to a non-carcinogenic risk value of more than 1. HQ > 1 indicates that aluminum dust exposure exceeded threshold and has a high non-carcinogenic risk, which should be a cause for concern. Therefore, workers should pay more attention to protection against the non-carcinogenic risks of inhalable aluminum dust, especially grinding workers.

$PM_{2.5}$ can induce lung cancer and chronic airway inflammatory diseases [43]. The results of the simulation of the health risks of $PM_{2.5}$ inhalable exposure to carcinogenesis are shown in Figure 4. The geometric mean ± SD of CR for the $PM_{2.5}$ in the welding and grinding areas was $8.90 \times 10^{-3} \pm 3.98 \times 10^{-3}$ and $3.11 \times 10^{-3} \pm 1.00 \times 10^{-4}$, respectively. The $PM_{2.5}$ CR in the both areas exceeded 1.0×10^{-4}, which meant that $PM_{2.5}$ in the workshop working environment had a significant CR to human health. Existing research results showed that the probability of the inhalation silica dust CR in an automobile foundry was below 1.0×10^{-5} [15], while that of $PM_{2.5}$ CR for male and female technicians in an academic metallurgy workshop had exceeded 1×10^{-5} [25]. The comparison results indicate that the health risks of all elements in inhalable dust should be evaluated to avoid the impact of a single element on effective risk management in terms of health risk assessment. The $PM_{2.5}$ generated during welding far exceeds that generated during the grinding process. Therefore, workers in an aluminum alloy body workshop of subway factory industries, especially welding workers, should focus on the $PM_{2.5}$ CR.

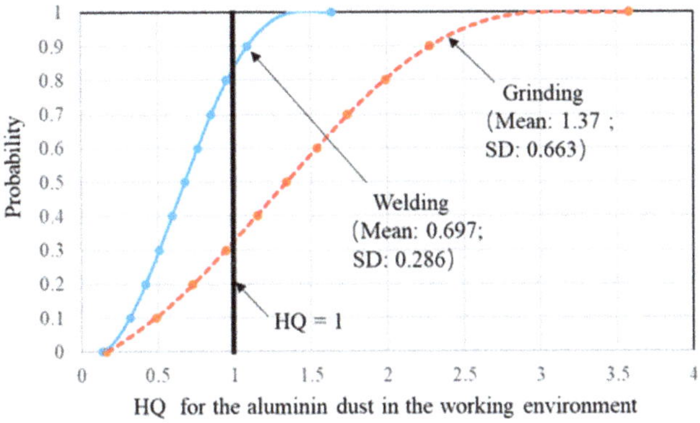

Figure 3. Evaluation of non-carcinogenic HQ of inhalable exposure to aluminum dust.

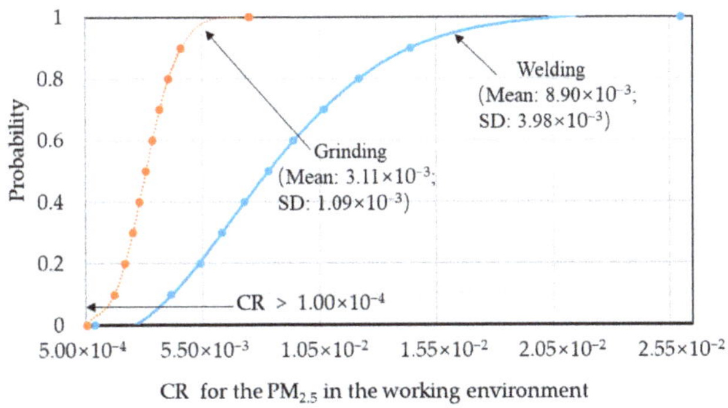

Figure 4. Health risk evaluation of carcinogenic health risk of $PM_{2.5}$ exposure.

3.4. Sensitivity Analysis

The sensitivity analysis results of aluminum dust HQ (HQ-Al) and $PM_{2.5}$ CR (CR-$PM_{2.5}$) at two types of work area (S1 for welding and S2 for grinding) are shown in Figure 5. Aluminum dust concentration with a sensitivity of 88% is the most important cause of non-carcinogenic health risks of aluminum dust in welding (HQ-Al-S1) and grinding areas (HQ-Al-S2), while the remaining exposure parameters have relatively little sensitivity. This finding is consistent with previous findings [15]. This shows that the mass concentration of inhalable aluminum dust in the working area should be reduced by rationally designing and optimizing the form of indoor ventilation, so as to reduce the health risk of aluminum dust to workers.

Dust concentration and exposure duration have a considerable impact on the results of $PM_{2.5}$ carcinogenic risk assessment of welder (CR-PM2.5-S1) and grinding workers (CR-PM2.5-S2). Dust concentration with a sensitivity of 45.6% and exposure duration with a sensitivity of 45.9% were the two main factors leading to the health risk of $PM_{2.5}$ carcinogenesis in the welded area, while dust concentration with a sensitivity of 12.2% and duration of exposure with a sensitivity of 71.7% were the two main factors leading to the CR of $PM_{2.5}$ in the grinding area. Therefore, managers need to explore different measures and effective methods of reducing the dust health risks involved. For example, the use of more advanced welding and grinding equipment could reduce dust emissions from pollution sources, and the method of optimizing indoor airflow organization could effectively control the concentration field of dust in the working environment. These are two very effective prevention strategies for improving the health risk management of dust exposure in the work environment.

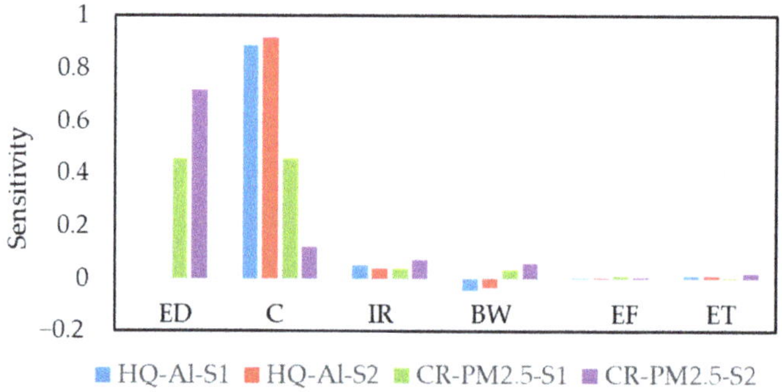

Figure 5. Sensitivity analysis of non-carcinogenic health risk (HQ-Al) and $PM_{2.5}$ carcinogenic health risk (CR-$PM_{2.5}$) of aluminum dust.

3.5. Limitations and Future Research Directions

The above research results indicated the contamination levels and health risks of welding fumes and grinding particles in the aluminum alloy body workshop of subway manufacturing plants, and had theoretical and practical significance in the health risk management for workers. However, some limitations were found in the research process of health risk assessment of inhalable exposure to dust in the workshop. Firstly, the health risk assessment results of welding and grinding workers had certain limitations because of the lack of female workers' participation in this study. In addition, some exposure parameters, such as the respiratory rate of workers, were mainly indirectly selected from existing research results due to the lack of standard reference values. These uncertain factors of exposure parameters would inevitably have adverse effects on the results of health risk assessment. In addition, the toxicity values of some pollutants, such as the Reference Concentration (RfC) standard value of PM_{10}, had not been found as suitable reference values, which affected the subsequent research on health risk assessment. Furthermore, the health risk protection measures taken by workers, such as wearing a KN100 dust mask, were not taken into account in this study, which could result in the calculated values of workers' health risk indicators being higher than the actual values. Finally, the content of this study is only a small fraction of health issues in welding and grinding, viz. respiratory problems in welding and grinding, skin cancer issues in welding, metal fume fever in welders and mill grinding noise. For example, harmful gases such as ozone and NO_x are inevitably generated during the welding process of aluminum alloys [44], but the concentration of these gases was not monitored in this study.

Based on the above analysis results, the following research is recommended as follows: firstly, from the perspective of basic research, extensive investigation and research on exposure parameters of welding and grinding workers should be conducted, and the corresponding databases should be established. Extensive research should be conducted on the physicochemical properties of aerosols, the anatomical structure of the respiratory tract, and the physiology of the respiratory tract in order to further investigate the deposition patterns of particles in the respiratory tract [36]. Then, from the perspective of the research object, similar research work should be conducted on the health risk assessment of different subway manufacturing plants in China (the world's factory). Furthermore, from the perspective of research methods, it is necessary to explore the application of artificial intelligence modeling methods in human health risk assessment, such as the artificial-intelligence-based risk assessment of older adults [45]. Finally, from the perspective of health risk assessment, a comprehensive evaluation index system for health issues in welding and grinding should be established based on existing research results.

4. Conclusions

The quantitative assessment of the health risks of inhalable exposure to welding fumes and grinding particles in an aluminum alloy body workshop of subway factory industries is an indispensable part of the occupational health and safety management system. In this study, the dust mass concentration, exposure parameters for workers and health risk assessment of inhalable aluminum dust and $PM_{2.5}$ were investigated and studied. The research results showed that the mass concentration of $PM_{2.5}$ in the welding area was higher, while the mass concentrations of PM_{10} and total suspended particulate matter in the grinding area were higher. The geometric means of the non-carcinogenic Hazard Quota (HQ) of aluminum dust in the welding and grinding areas were 0.697 and 1.37, respectively, and the probability of their HQ exceeding 1 was approximately 17% and 68%, respectively. Therefore, workers should pay more attention to the protection against the non-carcinogenic risks of inhalable aluminum dust, especially grinding workers. The geometric mean carcinogenic risk (CR) of $PM_{2.5}$ in the welding and grinding areas was 8.90×10^{-3} and 3.11×10^{-3}, respectively. Both exceeded the acceptable upper limit value of $PM_{2.5}$ CR (1.0×10^{-4}), which meant that $PM_{2.5}$ in the working environment has a significant CR to workers, especially welders. Aluminum dust concentration with a sensitivity of more than 89% is the most significant cause of the non-carcinogenic health risks of aluminum dust in welding and grinding areas. Dust concentration and exposure duration have a considerable impact on the $PM_{2.5}$ CRs of workers. The remaining exposure parameters have relatively little sensitivity. Therefore, managers need to develop different measures and effective health risk management methods to reduce the dust health risks based on the sensitivity analysis results. These findings may provide valuable information for a better understanding of the contamination level and health risk assessment of inhalable welding fumes and grinding particles in subway factory industries.

Author Contributions: Investigation, C.L., D.H., J.Y. and C.W.; Writing—original draft, X.W.; Writing—review and editing, C.L. These authors contributed equally to this work. All authors have read and agreed to the published version of the manuscript.

Funding: This research was funded by the National Natural Science Foundation of China (No. 51246008), Hunan Provincial Natural Science Foundation of China (No. 2021JJ50038, 2022JJ50075).

Data Availability Statement: All data generated or analyzed during this study are included in this article.

Acknowledgments: The authors would like to thank all participants of the case study.

Conflicts of Interest: The authors declare no conflict of interest.

References

1. Sun, J.; Zong, Z. Service life evaluation method of metro vehicle aluminum alloy carbody. *Urban Mass Transit* **2023**, *S1*, 46–50. [CrossRef]
2. Li, X.; Wang, S.; Liu, C.; Zhang, J.; Chen, Y. Fatigue life assess-ment of metro carbody based on submodel method. *J. Southwest Jiaotong Univ.* **2022**, *57*, 295–300+330.
3. Zimmer, A.T.; Biswas, P. Characterization of the aerosols resulting from arc welding processes. *J. Aerosol Sci.* **2001**, *32*, 993–1008. [CrossRef]
4. Wang, Y.; Kuo, Y.; Wang, L. Long-term metal fume exposure assessment of workers in a shipbuilding factory. *Sci. Rep.* **2022**, *12*, 790. [CrossRef] [PubMed]
5. Chen, R.; Hu, B.; Liu, Y.; Xu, J.; Yang, G.; Xu, D.; Chen, C. Beyond PM2.5: The role of ultrafine particles on adverse health effects of air pollution. *Biochim. Biophys. Acta (BBA) Gen. Subj.* **2016**, *1860*, 2844–2855. [CrossRef]
6. Chen, W.; Liu, Y.; Wang, H.; Hnizdo, E.; Sun, Y.; Su, L.; Zhang, X.; Weng, S.; Bochmann, F.; Hearl, F.J.; et al. Long-term exposure to silica dust and risk of total and cause-specific mortality in Chinese workers: A cohort study. *PLoS Med.* **2012**, *9*, e1001206. [CrossRef]

7. Han, B.-C.; Liu, I.J.; Chuang, H.-C.; Pan, C.-H.; Chuang, K.-J. Effect of welding fume on heart rate variability among workers with respirators in a shipyard. *Sci. Rep.* **2016**, *6*, 34158. [CrossRef]
8. Rahul, M.; Sivapirakasam, S.P.; Vishnu, B.R.; Balasubramanian, K.R.; Mohan, S. Health issue owing to exposure with welding fumes and their control strategies at the source—A review. *Mater. Today Proc.* **2021**, *46*, 9239–9245. [CrossRef]
9. Riediker, M.; Zink, D.; Kreyling, W.; Oberdörster, G.; Elder, A.; Graham, U.; Lynch, I.; Duschl, A.; Ichihara, G.; Ichihara, S.; et al. Particle toxicology and health—Where are we? *Part. Fibre Toxicol.* **2019**, *16*, 19. [CrossRef]
10. Huang, S.; Ma, R.; Zhang, P.; Hu, C.; Wu, D.; Sun, L.; Deng, M. Characteristics and health risk assessments of fine particulate matter at the overground and underground subway sites in Chengdu. *Build. Environ.* **2023**, *242*, 110577. [CrossRef]
11. Alias, A.; Latif, M.T.; Othman, M.; Azhari, A.; Abd Wahid, N.B.; Aiyub, K.; Khan, M.F. Compositions, source apportionment and health risks assessment of fine particulate matter in naturally-ventilated schools. *Atmos. Pollut. Res.* **2021**, *12*, 101190. [CrossRef]
12. Wang, J.; Huang, Y.; Bo, Y.; Zhang, Y.; Zhang, L. Air pollution characteristics and human health risk assessment of underground parking garages in Xi'an, China. *Indoor Built Environ.* **2022**, *32*, 632–651. [CrossRef]
13. Hamzah, N.A.; Tamrin, S.B.M.; Ismail, N.H. Metal Dust Exposure and Respiratory Health of Male Steel Workers in Terengganu, Malaysia. *Iran. J. Public Health* **2014**, *43*, 154–166.
14. Dueck, M.; Rafiee, A.; Mino, J.; Nair, S.; Kamravaei, S.; Pei, L.; Quemerais, B. Welding Fume Exposure and Health Risk Assessment in a Cohort of Apprentice Welders. *Ann. Work Expo. Health* **2021**, *65*, 775–788. [CrossRef]
15. Tong, R.; Cheng, M.; Ma, X.; Yang, Y.; Liu, Y.; Li, J. Quantitative health risk assessment of inhalation exposure to automobile foundry dust. *Environ. Geochem. Health* **2019**, *41*, 2179–2193. [CrossRef]
16. Buonanno, G.; Morawska, L.; Stabile, L. Exposure to welding particles in automotive plants. *J. Aerosol Sci.* **2011**, *42*, 295–304. [CrossRef]
17. Dehghani, F.; Omidi, F.; Fallahzadeh, R.A.; Pourhassan, B. Health risk assessment of occupational exposure to heavy metals in a steel casting unit of a steelmaking plant using Monte-Carlo simulation technique. *Toxicol. Ind. Health* **2021**, *37*, 431–440. [CrossRef]
18. Soltanpour, Z.; Rasoulzadeh, Y.; Mohammadian, Y. Occupational Exposure to Metal Fumes Among Iranian Welders: Systematic Review and Simulation-Based Health Risk Assessment. *Biol. Trace Elem. Res.* **2023**, *201*, 1090–1100. [CrossRef]
19. Lin, D.; Nelson, J.D.; Beecroft, M.; Cui, J. An overview of recent developments in China's metro systems. *Tunn. Undergr. Space Technol.* **2021**, *111*, 103783. [CrossRef]
20. *GBZ 2.1-2019*; Occupational Exposure Limits for Hazardous Agents in the Workplace in Part 1: Chemical Hazardous Agents. NHC: Beijing, China, 2019.
21. *GBZ 159-2019*; Specifications of Air Sampling for Hazardous Substances Monitoring in the Workplace. NHC: Beijing, China, 2019.
22. Jiang, H.; Lu, L. Measurement of the surface charge of ultrafine particles from laser printers and analysis of their electrostatic force. *Atmos. Environ.* **2010**, *44*, 3347–3351. [CrossRef]
23. Hamamoto, N.; Nakajima, Y.; Sato, T. Experimental discussion on maximum surface charge density of fine particles sustainable in normal atmosphere. *J. Electrost.* **1992**, *28*, 161–173. [CrossRef]
24. Sennato, S.; Truzzolillo, D.; Bordi, F.; Sciortino, F.; Cametti, C. Colloidal particle aggregates induced by particle surface charge heterogeneity. *Colloids Surf. A Physicochem. Eng. Asp.* **2009**, *343*, 34–42. [CrossRef]
25. Mbazima, S.J. Health risk assessment of particulate matter 2.5 in an academic metallurgy workshop. *Indoor Air* **2022**, *32*, e13111. [CrossRef] [PubMed]
26. Di Vaio, P.; Magli, E.; Caliendo, G.; Corvino, A.; Fiorino, F.; Frecentese, F.; Saccone, I.; Santagada, V.; Severino, B.; Onorati, G.; et al. Heavy Metals Size Distribution in PM10 and Environmental-Sanitary Risk Analysis in Acerra (Italy). *Atmosphere* **2018**, *9*, 58. [CrossRef]
27. Guo, M.; Wang, Z.; Jian, N.; Gu, Y.; Xu, S.; Tan, Y.; Duan, X. Transformation Coefficient between Occupational Exposure Limits and Ambient Air Pollutant Limits Based on the Exposure Characteristics of Chinese Population. *Res. Environ. Sci.* **2022**, *35*, 2518–2526. [CrossRef]
28. Castro, A.; Götschi, T.; Achermann, B.; Baltensperger, U.; Buchmann, B.; Felber Dietrich, D.; Flückiger, A.; Geiser, M.; Gälli Purghart, B.; Gygax, H.; et al. Comparing the lung cancer burden of ambient particulate matter using scenarios of air quality standards versus acceptable risk levels. *Int. J. Public Health* **2020**, *65*, 139–148. [CrossRef]
29. Arı, A. A comprehensive study on gas and particle emissions from laser printers: Chemical composition and health risk assessment. *Atmos. Pollut. Res.* **2020**, *11*, 269–282. [CrossRef]
30. Candeias, C.; Ávila, P.F.; Ferreira da Silva, E.A.; Rocha, F. Metal(Loids) Bioaccessibility in Road Dust from the Surrounding Villages of an Active Mine. *Atmosphere* **2021**, *12*, 685. [CrossRef]
31. Qu, C.; Li, B.; Wu, H.; Wang, S.; Giesy, J.P. Multi-pathway assessment of human health risk posed by polycyclic aromatic hydrocarbons. *Environ. Geochem. Health* **2015**, *37*, 587–601. [CrossRef]
32. Othman, M.; Latif, M.T.; Mohamed, A.F. Health impact assessment from building life cycles and trace metals in coarse particulate matter in urban office environments. *Ecotoxicol. Environ. Saf.* **2018**, *148*, 293–302. [CrossRef]
33. Chiang, K.C.; Chio, C.P.; Chiang, Y.H.; Liao, C.M. Assessing hazardous risks of human exposure to temple airborne polycyclic aromatic hydrocarbons. *J. Hazard. Mater.* **2009**, *166*, 676–685. [CrossRef]
34. Zhao, J.; Feng, Y.; Bezerra, M.; Wang, J.; Sperry, T. Numerical simulation of welding fume lung dosimetry. *J. Aerosol Sci.* **2019**, *135*, 113–129. [CrossRef]

35. Gou, G.; Zhang, M.; Chen, H.; Chen, J.; Li, P.; Yang, Y.P. Effect of humidity on porosity, microstructure, and fatigue strength of A7N01S-T5 aluminum alloy welded joints in high-speed trains. *Mater. Des.* **2015**, *85*, 309–317. [CrossRef]
36. Morawska, L.; Buonanno, G. The physics of particle formation and deposition during breathing. *Nat. Rev. Phys.* **2021**, *3*, 300–301. [CrossRef] [PubMed]
37. Yang, S.; Duque Mahecha, S.; Aguacil Moreno, S.; Licina, D. Integration of Indoor Air Quality Prediction into Healthy Building Design. *Sustainability* **2022**, *14*, 7890. [CrossRef]
38. Liu, X.; Wang, Z.; Bai, H.; Zhang, S.; Mu, L.; Peng, L. Characteristics and health risk assessments of heavy metals in PM2.5 in Taiyuan and Yuci college town, China. *Air Qual. Atmos. Health* **2020**, *13*, 909–919. [CrossRef]
39. Guo, Q.; Li, L.; Zhao, X.; Yin, B.; Liu, Y.; Wang, X.; Yang, W.; Geng, C.; Wang, X.; Bai, Z. Source Apportionment and Health Risk Assessment of Metal Elements in PM2.5 in Central Liaoning's Urban Agglomeration. *Atmosphere* **2021**, *12*, 667. [CrossRef]
40. Kim, W.-G.; Yong, S.-D.; Yook, S.-J.; Ji, J.H.; Kim, K.-H.; Bae, G.-N.; Chung, E.-K.; Kim, J.H. Comparison of black carbon concentration and particle mass concentration with elemental carbon concentration for multi-walled carbon nanotube emission assessment purpose. *Carbon* **2017**, *122*, 228–236. [CrossRef]
41. Hammer, T.; Sachinidou, P.; He, X.; Pan, Z.; Bahk, Y.K. Advanced filtration and lung deposition models of airborne multi-walled carbon nanotubes for inhalation exposure assessment. *NanoImpact* **2020**, *19*, 100240. [CrossRef]
42. Riihimäki, V.; Aitio, A. Occupational exposure to aluminum and its biomonitoring in perspective. *Crit. Rev. Toxicol.* **2012**, *42*, 827–853. [CrossRef]
43. Li, R.; Zhou, R.; Zhang, J. Function of PM2.5 in the pathogenesis of lung cancer and chronic airway inflammatory diseases. *Oncol. Lett.* **2018**, *15*, 7506–7514. [CrossRef] [PubMed]
44. Cole, H.; Epstein, S.; Peace, J. Particulate and Gaseous Emissions When Welding Aluminum Alloys. *J. Occup. Environ. Hyg.* **2007**, *4*, 678–687. [CrossRef] [PubMed]
45. Wiil, U.K. Important steps for artificial intelligence-based risk assessment of older adults. *Lancet Digit. Health* **2023**, *5*, E635–E636. [CrossRef] [PubMed]

buildings

MDPI

Article

Exploring Information and Comfort Expectations Related to the Use of a Personal Ceiling Fan

Romina Rissetto [1,*] and Marcel Schweiker [2]

[1] Karlsruhe Institute of Technology, Faculty of Architecture, Building Science and Technology, 76131 Karlsruhe, Germany
[2] Healthy Living Spaces Lab, Institute for Occupational, Social and Environmental Medicine, Medical Faculty, RWTH Aachen University, 52074 Aachen, Germany; mschweiker@ukaachen.de
* Correspondence: romina.rissetto@kit.edu; Tel.: +49-721-608-46982

Abstract: Rising global temperatures have increased the need for research into human adaptability and comfort in buildings. To reduce comfort-related energy demands, low-energy-consumption alternatives for space cooling, such as personal environmental control systems (PECS), are being investigated. The implementation of PECS in office buildings is still underway, and little is known about how occupants' expectations can influence their satisfaction with PECS and indoor environmental quality. This study examines the influence of tailored information and occupants' comfort expectations on their thermal perceptions and satisfaction with a personal ceiling fan. Seventy-six participants completed an online questionnaire and attended a half-day session at 30 °C in a climate chamber in Germany. A manipulation technique to activate personal norms was used to test the influence of information on expectations. Results indicated higher reported thermal comfort in participants with more positive thermal expectations, regardless of their expectations of the building systems. These effects were largely moderated by personal norms, indicating the importance of activating normative motivations to increase thermal comfort. Occupants with negative expectations improved their perceptions of the fan when making personal adjustments to stay comfortable. However, this effect was not moderated by personal norms. Practical implications focus on manipulating occupants' comfort expectations, e.g., by providing occupants with normative messages and individual control, to achieve greater comfort and acceptance of personal building controls in naturally ventilated buildings.

Keywords: psychological adaptation; adaptive behaviors; personal ceiling fan; personal norms; test chamber; thermal perception; thermal comfort

Citation: Rissetto, R.; Schweiker, M. Exploring Information and Comfort Expectations Related to the Use of a Personal Ceiling Fan. *Buildings* **2024**, *14*, 262. https://doi.org/10.3390/buildings14010262

Academic Editors: Yingdong He and Nianping Li

Received: 7 December 2023
Revised: 11 January 2024
Accepted: 15 January 2024
Published: 17 January 2024

1. Introduction

The global climate emergency has led to a push to deliver habitable indoor spaces, resulting in a growing demand for space cooling. A compounding increase in the use of air conditioning is expected, which will sharply escalate global carbon dioxide emissions. By better understanding how humans perceive and adapt to their thermal built environment, it may be possible to reduce the comfort-related energy demands of buildings. The literature on adaptive thermal comfort has gained particular attention over the past twenty years [1]. According to the theory of adaptive thermal comfort [2], three mechanisms take place in the adaptive processes of occupants in buildings—namely behavioral, physiological, and psychological mechanisms. Although many efforts have been made to understand the different factors that influence human adaptation, there is still a gap between predicted and actual occupant comfort and behavior observed in field studies [1,3].

The concept of comfort expectations has been studied as a relevant dimension of psychological adaptation to the environment [2]. According to the expectation hypothesis, an expectation (or anticipatory attitude) affects people's attitude towards thermal comfort attainment. Thus, the expectation of specific thermal conditions is certainly a major aspect of subjective assessment and satisfaction [4,5]. Some empirical evidence from China [6] suggested that long-term thermal experiences can raise thermal comfort expectations and that it is easier and quicker to enhance an individual's thermal expectations but harder to lower them. Accordingly, occupants in air-conditioned buildings are quicker to complain whenever the indoor temperature slightly strays from the usual set point because they have come to expect thermal constancy [7].

Relaxing comfort expectations could be an alternative path to promote resilience in buildings. A strategy to transform expectations could be achieved by widening occupants' thermal acceptability through adaptive behaviors, especially in free-running and green buildings [8,9]. Adding adaptive capacity in buildings, that is, the ability to implement effective adaptation strategies, is strongly related to control strategies [10]. Luo et al. [11] suggested the implementation of personal environmental control systems (PECS) as an adaptive strategy. PECS have the advantage of controlling the localized environment at the occupant's workstation according to their preferences rather than conditioning an entire room. Thus, PECS have the potential not only to save energy but also to improve comfort by addressing intra- and interpersonal differences among occupants [12,13]. Personal fans have been widely implemented as a type of PECS, as the cooling effect of air movement increases the thermal comfort and acceptability range of occupants in moderately warm thermal conditions [14–16].

By giving occupants the responsibility of managing certain aspects of the building, more information needs to be provided related to the passive features and building control systems in order to pursue an energy-efficient approach [17]. On the one hand, this might reduce the gap between how designers expect occupants to use a building and how they actually do. On the other hand, information feedback has been shown to help occupants save energy. For example, Schweiker et al. [18] found that participants receiving training and information about passive strategies were more likely to apply such methods and to reduce high-energy-consumption devices, such as AC-units. Day et al. [19] found that individuals who reported effective training and therefore understood how to operate the building controls were significantly more likely to be satisfied with their office environment when compared to individuals who did not receive any kind of training. Brown et al. [20] found a positive relationship between knowledge of a building's systems and higher use of personal control.

Research Gap and Scientific Contribution

Although several experimental and field studies have shown the potential of providing effective information and increasing occupants' knowledge to promote energy-saving behaviors [21,22] and increase occupant satisfaction [23,24], little work has examined how tailored information may influence the interaction between comfort expectations and satisfaction with PECS in naturally ventilated buildings. Thus, this study aims to understand whether information and knowledge can manipulate occupants with different positive or negative expectations about PECS and some aspects of the indoor environmental quality (IEQ), as such expectations could, in turn, influence occupants' satisfaction with the building controls and their perception of the indoor environment. The following research questions will be examined:

- To what extent do occupants' different expectations of the indoor environment and adaptive possibilities influence their a) thermal and indoor air quality perception and b) their satisfaction with a type of PECS?
- To what extent can tailored information to activate normative motivations be used to manipulate thermal and indoor air quality perception and satisfaction with a type of PECS of occupants with different expectations?

To address the identified research questions, this study investigates the relationship between occupants with different expectations of their built environment and their satisfaction with their indoor environment, as well as how expectations of a type of PECS can be manipulated to achieve greater satisfaction with the device. The existing definitions, relevant studies in the literature, and the research hypotheses are presented in Section 2. An experimental study and an online survey were conducted to test the research hypotheses. The study design and methods for data collection and analysis are presented in Section 3. The results and related hypotheses are discussed in Section 4.

This work contributes to the research on the acceptance of a type of PECS to increase its prominence and implementation in buildings and adds knowledge to the adaptive comfort literature by deepening the concept of comfort expectations in naturally ventilated buildings. The application of a theory-based definition of comfort expectation in a case study and the relationship between occupant expectations and their acceptance of a personal control device constitute the novelty of this paper.

2. Literature Review, Definitions and Hypotheses

2.1. Thermal and Behavioral Expectations

Evidence indicates occupants' expectations of indoor building environments influence their perceptions of climatic conditions, and unmet expectations of building performance can lead to dissatisfaction with indoor conditions. To better understand the mismatch between the occupants' predicted thermal perceptions of indoor environments and reported satisfaction, expectations in prior studies have been conceptualized in different ways. Fanger et al. [25] introduced an expectancy factor that relates expectations to past experiences, such as habituation to warm environments and exposure to air-conditioned buildings. Schweiker et al. [26] investigated how observed expectations affect occupants' thermal comfort levels and found a significant influence of thermal memory on expected comfort. Comfort expectations have mainly been analyzed in relation to perceived control [6,8,27], thermal experience and exposure [28–31], and thermal memory [32].

Despite the mentioned efforts in the literature, there is a lack of evidence-based and theory-driven characterization of the psychological adaptive concept of expectation [1]. In a recent study, the authors of reference [33] proposed a framework to operationalize expectations through cognitive mechanisms from well-established psychological and comfort theories (i.e., self-efficacy, perceived control, thermal history, and personal norms and attitudes). The model was tested through a nationwide survey, and it was concluded that expectations are key drivers of comfort and comfort-related behaviors. Based on the psycho-physiological model of Auliciems [34] and the adaptive theory, the framework established that expectations can be distinguished by thermal and behavioral expectations, which can be defined as follows:

- Thermal expectations: the thermal experience foreseen by occupants; the anticipated result, their perception of what will occur.
- Behavioral expectations: the likelihood of engaging in a specific behavior to adapt to the thermal environment to improve their comfort.

The results of the study [33] showed that the more the positive thermal expectations of the indoor environment were, the greater the associated reported thermal comfort was. Similarly, a positive relationship was found between more positive behavioral expectations toward a specific behavior and the probability of performing that action. A negative correlation was found between thermal and behavioral expectations, supporting the adaptive principle. The theoretical framework was empirically tested by asking the survey respondents to envision a working space with defined adaptive opportunities, but participants' actual comfort votes and adaptive actions were not captured in real-time and under the actual thermal conditions and building settings. Although the relationship between thermal and behavioral expectations was evaluated, the combined effect of positive–negative thermal and behavioral expectations on participants' thermal comfort and behavior responses was not investigated. It would be meaningful to classify different types of thermal and behavioral expectations for groups of participants with similar cognitive mechanisms, as this may give insights into how different groups of occupants may be targeted according to their shared expectations.

2.2. Provided Information and Building Interactions

In addition to individual differences in preferences and expectations for thermal comfort, variations in occupants' behaviors may result from their inadequate understandings of the building controls and purpose design of the building [35] or from missing knowledge or feedback regarding the effect of occupant actions (e.g., [36,37]). Studies on feedback and feedforward information revealed that the decisions made by occupants can be manipulated by providing feedback about the consequences of their previous actions [38] or feedforward information advising occupants prior to their actions [21,39,40]. Only a few studies have analyzed the impact of feedback and feedforward information on the decision process of occupants with respect to their building interactions and their change in comfort level after such decisions.

Meinke et al. [41] concluded that participants tended to interact more rationally with their built environment when receiving information about the consequences of different cooling strategies on comfort and energy consumption. They also found that when occupants were more aware of their control options, it led to increased perceived control and, consequently, higher comfort. Brown et al. [20] found that occupants' knowledge of the building, i.e., awareness and understanding of the building's environmental features and control systems, was positively related to the use of personal control in green buildings. More recently, Arpan et al. [42] investigated the effect of information on building occupants' expectations of sustainable buildings. They concluded that potential building occupants who are informed about the common features of sustainable buildings and how they function may have more positive a priori expectations about the thermal and indoor air quality conditions in the building. Accordingly, it could be hypothesized that providing information about the benefits and operation of PECS could create positive expectations towards the device and, consequently, satisfaction with it.

2.3. Normative Motivations

Additional results from the above-mentioned study [33] showed that behavioral expectations were partially explained by personal norms: participants with greater motivations towards passive cooling strategies (stronger personal norms) will express higher expectations to successfully modify their indoor environment. Changes in user expectations may be reflected in expectations of building systems and occupant behavior [43]. Research conducted in intervention studies suggests that normative motivations, i.e., people who prioritize collective interests over their personal ones, have a significant impact on anticipating and designing interventions to encourage energy-saving behavior [39]. Accordingly, when activating personal norms, occupants' behavior is driven by feelings of moral obligation to act in a norm-concordant manner. In this sense, occupants with strong personal norms suggest that they are intrinsically motivated to act pro-socially—following normative considerations—and increase their sense of self-worth. For example, Hameed et al. [44] studied patterns of adoption of low-carbon practices and concluded that normative motivations were key drivers for the purchase of energy-saving air conditioners in Pakistan. Similarly, Gerhardsson et al. [45] found that lighting behaviors, such as improving lighting technology, were driven by normative goals, while Wall et al. [46] found that environmentally motivated participants who were motivated to save energy were more tolerant of the poor performance of energy-efficient energy lamps than less environmentally motivated participants. By activating personal norms, individuals tend to act according to those norms and are more willing to make concessions to meet their standards of behavior, especially those who are more environmentally engaged [47].

A theoretical foundation prominently used in psychology to analyze behavioral change and promote pro-environmental behaviors is the goal-framing theory [48]. According to this theory, goals determine or "frame" what people pay attention to, what knowledge and attitudes become most cognitively accessible, how people evaluate different aspects of the situation, and what alternatives are being considered. A "goal-frame" is the way in which people process information and act on it. If people change their goals, they will also perceive the situation differently. When it is activated or "focalized", a goal is a combination of a motive and an activated knowledge structure. There are three types of frames: gain, hedonistic, and normative frames. The latter two will be considered for the present study. Hedonistic frames activate subgoals that promise to improve how one feels in a particular situation. Their time horizon is very short, and people in this frame are sensitive to what changes their pleasure and mood. For example, feeling warm in a room may decrease a person's thermal comfort. The normative frames of "act appropriately" activate goals related to what is appropriate, and people in this frame are sensitive to what should be done according to their self or others, including, for example, turning off the thermostat when the windows are open even if the person does not pay the bill simply because it is the right thing to do.

In an experimental study, Li et al. [24] used social normative messages to investigate intended occupant interactions with a PECS, showing a positive influence of feedforward information on the intended use of a personal desk fan. Thus, informing individuals with strong pro-environmental norms about PECS features designed to protect the environment should activate their normative goals and subsequently motivate them to act—or perceive the situation—in a manner that is congruent with those personal norms. Accordingly, it might be expected that some occupants would have more positive expectations of PECS and be more tolerant of indoor conditions when these overarching personal goals are activated.

2.4. Hypotheses

The review of the state of the art has shown that there is a lack of studies assessing the effect of occupants' expectations and normative motivations on their satisfaction with thermal and indoor air quality conditions and personal controls in buildings. Based on the state of the art in combination with the definitions presented, a preliminary framework for the assessment of expectancy was developed (Figure 1), which summarizes results from a previous study on thermal and behavioral expectations [33] and proposes a new investigation to assess the effect of expectancy groups on thermal satisfactions and satisfaction with PECS tailored by normative motivations.

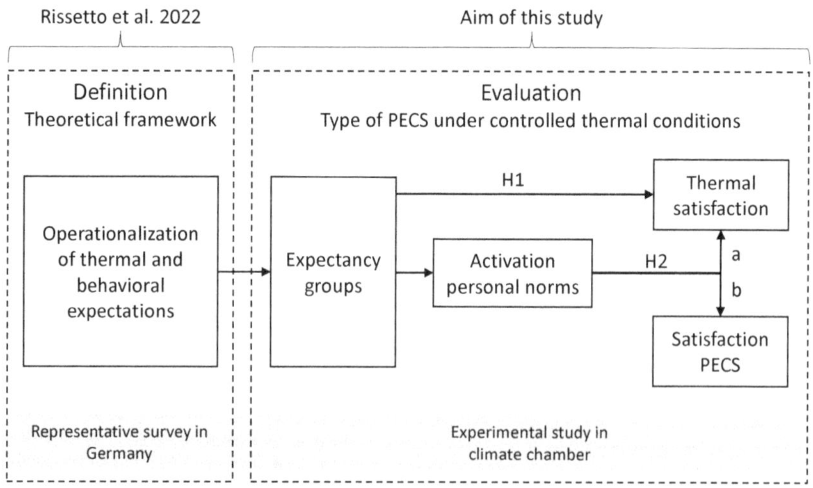

Figure 1. Proposed framework to assess occupants' expectancy, personal norms, and satisfaction with thermal conditions and a type of PECS, together with an existing theoretical framework [33]. H1, H2a and H2b are the investigated hypotheses.

Based on the above-stated research questions, the following hypotheses will be investigated:

- H1: A person with more positive expectations about the thermal conditions in the room and towards a type of PECS will find the climatic conditions more acceptable, expressing higher thermal satisfaction than a person with more negative expectations.
- H2: By activating normative motivations through tailored information, expectations can be influenced in a positive direction so that (a) participants with more positive expectations will express higher thermal satisfaction and (b) participants with more negative expectations will show a change in expectations after using the PECS.

3. Methods

In order to assess the proposed hypotheses, an experimental study in a laboratory setting and an online survey were conducted. First, participants were asked to complete an online questionnaire prior to attending a half-day session at the LOBSTER test chamber in Karlsruhe, Germany [49]. The latter consists of two identical office rooms, each with two operable windows and blinds facing north. The surface of the test facility (except for the glass facade) is activated with a capillary tube system, which allows set point temperature of each surface to be changed individually. For this study, each room was equipped with a personal ceiling fan. The sessions took place over 15 working days in August 2021. All procedures were approved by the data protection officer and the ethics committee of the Karlsruhe Institute of Technology and were conducted in accordance with the Declaration of Helsinki. The study is described as follows.

3.1. Recruitment and Participation

Participants were recruited primarily through the local newspaper and university websites. Participants had to be non-smokers and be German or have a good command of the German language to ensure that they understood and were capable of answering the provided questionnaires. They received monetary compensation for participating in the survey and the test chamber session. A total of 76 participants (35 male and 41 female), aged 18–34 and 50–70 years, took part in the half-day experiment and completed the online questionnaire. The aim of including those age groups was to increase the probability of participation and control the sample, as there was a higher probability that individuals of those groups were able to participate in the experiment during working hours and have higher motivation to receive a monetary compensation (e.g., students or retired participants). For the session in the LOBSTER, participants were asked to wear long pants, a shirt, and closed shoes. Clothing data were collected in the initial questionnaire, and the clothing level was estimated based on self-reported clothing items in the questionnaire and converted to clo values based on ISO 7730 [50]. An average value of 0.44 clo (SD = 0.12) was calculated with an additional value of 0.10 clo to account for the insulation provided by the desk chair. The participants were not allowed to change their clothing level (e.g., by taking off their sweater or shoes) during the test.

3.2. Pre-Test: Online Questionnaire

Participants completed an online background questionnaire one week before the LOBSTER session. The focus of the questionnaire was to assess participants' psychological constructs that represent expectations about the indoor environment and PECS, as well as related topics, such as sustainability or passive climate control strategies in buildings. The questions were based on the expectancy framework proposed by Rissetto et al. [33]. The questionnaires were sent via Limesurvey [51]. The purpose of this pre-test was to obtain the long-term attitudes of the participants without the possible influence of the controlled environment and the experience with the personal ceiling fan in the test chamber.

The survey consisted of three parts. The first section included an anonymous ID code to allow a comparison with the results of the session in the test chamber (see Section 3.3) and the measures of control variables, mainly current mood, experience with and evaluation of fans, experience working in an office environment (e.g., use of air conditioning and the use of building controls to adjust to climatic conditions). The second section included the main measures of this study: (1) measures of thermal expectations and behavioral expectations, (2) self-efficacy, perceived control, personal norms, thermal history, and attitudes, and (3) reported comfort and behavior. Finally, the third section included temperature type and sensitivity to indoor air quality and humidity, as well as expectations of ceiling fans. The last item was included to analyze the effect of information on a possible change in fan expectations (related to H2b).

3.3. Session in the LOBSTER

The same participants participated in a half-day session (either morning or afternoon) in the test chamber. Each session lasted three and a half hours, and a single participant occupied each room. Figure 2 describes the complete schedule before and during one session in the chamber. For the first 10 min, the study and the schedule were explained to the participants in the hallway. During the first half hour (acclimation phase), they entered the respective room and adapted to the climatic conditions. Both groups experienced warm indoor thermal conditions, so the walls' surface temperature was set to 30 °C. Participants were not able to modify the indoor environmental conditions of the rooms. During the next three hours, they engaged in personal activities, such as reading their own material or working on the computers provided. Meanwhile, they had the opportunity to perform different adaptive measures to restore their comfort with the thermal environment: (1) turning on the ceiling fan, (2) tilting the window(s), or (3) drinking water or another beverage.

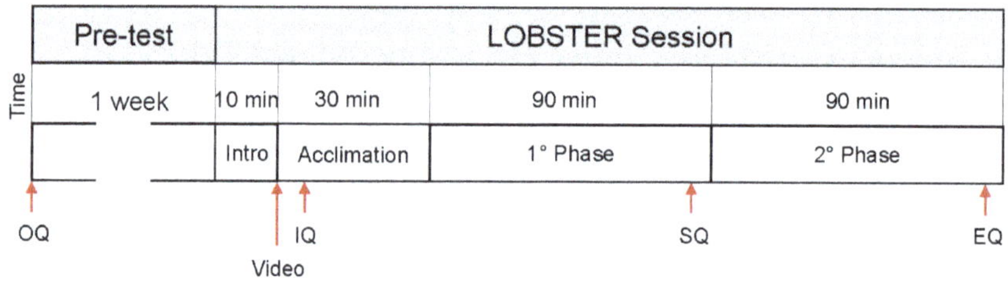

Figure 2. Timeline of surveys and experimental conditions before and during the session in the LOBSTER. OQ: online questionnaire. IQ: initial questionnaire; SQ: status questionnaire; EQ: end questionnaire.

Figure 3 shows the workstation, the personal ceiling fan, and the corresponding sensor equipment. The participants were seated 50 cm away from the center of the personal ceiling fan and 1.50 m from the windows. The personal ceiling fan corresponds to a type of PECS as it is workstation-related, i.e., each occupant owns a device, and can be individually controlled by the occupant. The axial fan had a small rotating area, and it was integrated into an acoustic panel to improve the acoustics in the room. The integrated ceiling fan had an adjustable grille to direct the airflow to the head of the participants, which in this study was directed towards the side of the participant's head. The influence of different airflow directions was previously tested for this personal fan [52], and no significant difference was found between top, back, frontal and side airflow. The air velocity of the ceiling fan could be adjusted by the participants using a remote control. Further descriptions of the ceiling fan can be found in Rissetto et al. [52].

Participants completed various questionnaires during their stay via a web interface based on pre-set schedules (Figure 2). The focus was to collect information mainly on their perception of and satisfaction with the IEQ and the personal ceiling fan. As the questions were asked in the German language, most of the questions and corresponding scales were based on the German index "INKA: Instrument für Nutzerbefragungen zum Komfort am Arbeitsplatz" to assess comfort in office buildings [53], which is based on the questionnaire of ASHRAE 55 [54]. The questionnaires were divided into three blocks according to different experimental phases: an initial questionnaire (IQ) at the beginning of the acclimation phase (first 30 min of the experiment), an intermediate questionnaire

(SQ) at the end of the first hour and a half after the acclimation phase (phase 1), and a final questionnaire (EQ) asked 10 min prior the end of the second phase of the experimental part of the session (phase 2). Participants were exposed to the same thermal conditions in phases 1 and 2, but each phase indicated the appearance of the comfort questionnaires at different points in time (SQ and EQ) to evaluate the comfort votes during the length of the study. To analyze a possible change in fan expectations, participants were asked about their experiences with fans and their expectations and preferences with the personal ceiling fan and PECS in general to examine whether the expectations reported in the background questionnaire (Section 3.2) changed after using the personal device and having received the targeted information (see Section 3.3.2). Table 1 summarizes the key variables collected on the questionnaires relevant to this paper.

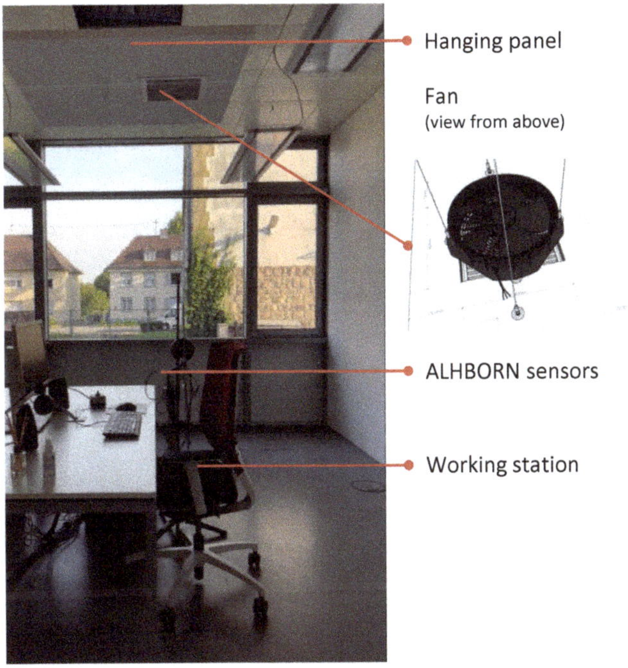

Figure 3. Setup of ceiling fan, sitting position, and sensors in the office room in the test chamber.

Indoor and outdoor parameters were collected from sensors through the building management system (BMS). Air temperature (Mean = 29.7 °C, SD = 0.6), globe temperature (Mean = 29.6 °C, SD = 0.6), relative humidity (Mean = 41.9%, SD = 4.5), and air velocity (Mean = 0.13 m/s, SD = 0.1) were collected with AHLBORN comfort meters placed at the height of 1.10 m and 0.25 m away from the participant's head. The corresponding resolutions are 0.01 °C, 0.01 °C, 0.1%, and 0.001 m/s; the accuracies are ±0.2 K, ±(0.30 K + 0.005 × T), ±2.0%, and ±(3% reading + 0.01), respectively. Interactions with the remote control and with the windows were recorded by the BMS. The fan speed chosen by participants through the remote control was recorded as a continuous variable between 0 and 100%. At the end of the sessions, participants were asked about their drink consumption. Physiological data were also collected, including heart rate (EcgMove 4: r = 12 bit, input range CM = 560 mV, DM = +/−5 mV) and skin temperature (iButton DS1921H: r = 0.125 °C, a = +/− 1 °C). The resulting analysis of the physiological data was not included in this paper. All data were recorded at 1 min intervals.

Table 1. The main information obtained by the questionnaires. All answers are integer values. Note: the provided questionnaires were in German; the English translations in the table were not used in the study and are presented only for understanding purposes. The German version is available from the authors per request.

Measure	Description of Item	Response Categories	Mean (SD)
Thermal sensation [a]	"Wie fühlen Sie sich jetzt gerade?" (*How do you feel right now?*)	−3 (cold) to +3 (hot)	4.79 (0.55)
Thermal comfort [a]	"Empfinden Sie dies als…" (*Right now, do you find this environment...?*)	1 (extremely uncomfortable) to 5 (comfortable)	3.87 (0.55)
Thermal preference [a]	"Wie hätten Sie es jetzt gerade lieber?" (*Right now, would you prefer to be...?*)	1 (much cooler) to 7 (much warmer)	3.29 (0.54)
Thermal acceptability [a]	"Wie empfinden Sie diese Temperaturbedingungen jetzt gerade?" (*Right now, do you find the thermal environment...?*)	1 (clearly unacceptable) to 4 (clearly acceptable)	3.47 (0.55)
Indoor air quality perception [a]	"Wie nehmen Sie die Raumluftqualität im Büro wahr?" (*How do you perceive the indoor air quality in the office?*)	1 (very good) to 7 (very bad)	4.26 (1.02)
Fan satisfaction [b]	To maintain comfortable indoor temperatures, the ceiling fan is more effective than I expected; To maintain comfortable indoor temperatures, the ceiling fan is more effective than I expected; If I could choose, I would rather use a ceiling fan than open the windows; I have control over the personal ceiling fan; The ceiling fan is easy to operate; The ceiling fan fits well with the floor plan and furnishings of the office; I can understand the advantages of the ceiling fan; The ceiling fan is quiet; Being able to adjust the air velocity myself is an advantage of the ceiling fan; Improving the indoor climate is a benefit of using the ceiling fan; If I could choose, I would use the fan as an energy-saving cooling strategy; If I could choose, I would use a ceiling fan instead of turning on an air conditioner; I consider myself capable of operating the personal ceiling fan; I should avoid opening the window when it is very warm outside.	1 (strongly disagree) to 7 (strongly agree)	8.24 (0.76) [6.05, 9.40] [d]
Fan expectations [c]	Same as before, but slightly modified and adapted in the form of *"I expect that ..."*	1 (strongly disagree) to 7 (strongly agree)	4.53 (2.11) [1.52, 8.83] [d]

[a] Measured in IQ, SQ, and EQ during the LOBSTER session. [b] Measured in EQ during the LOBSTER session. Scale reliability: 0.70. [c] Measured in background questionnaire (pre-test). Scale reliability: 0.93. [d] Unstandardized values resulting from principal component analysis (PCA) conducted with all presented questions (see Section 3.4.1).

3.3.1. Classification of Expectancy Groups

Participants were divided into groups to investigate the influence of different "levels" of expectancy on occupant satisfaction (related to H1). The clustering process was adapted from a previous study [55] following these steps:

- Using a training dataset, the cluster structure was calculated to explain a selected threshold of 80% of the variance using the k-means method [56].
- As the k-means method requires the number of clusters as an input, the elbow method was applied to calculate the optimal number of clusters.
- A test dataset was fitted to the obtained cluster structure using a support vector machine (SVM) method [57], which is a class of supervised learning algorithms that train the classifier function using labeled data.

A pre-analysis of the data from the nationwide survey to assess comfort expectations [33], explained in Section 2.1, was used as the training dataset to define the cluster structure. An expectancy value was obtained for each participant by assigning two scores: one for thermal expectations and one for behavioral expectations. The scores were obtained by principal component analysis (PCA). The obtained expectancy value was used to define the cluster centers using the k-means algorithm. The results from the elbow method showed an optimal number of three clusters. The corresponding label (cluster) was assigned to each point of the training dataset.

Prior to the LOBSTER session, the new scores for expectancy values were obtained for each participant using the data from the online survey explained in Section 3.2 (test data). With the labeled data, the SVM linear classifier was used to fit the participants' scores from the test data into the defined cluster structure. Figure 4 shows the results of the SVM. The different colors represent the three clusters. We can interpret the cluster classification as follows: participants in cluster 1 had positive fan expectations and negative thermal expectations; participants in cluster 2 had positive thermal and fan expectations; participants in cluster 3 had near-neutral thermal expectations and negative fan expectations.

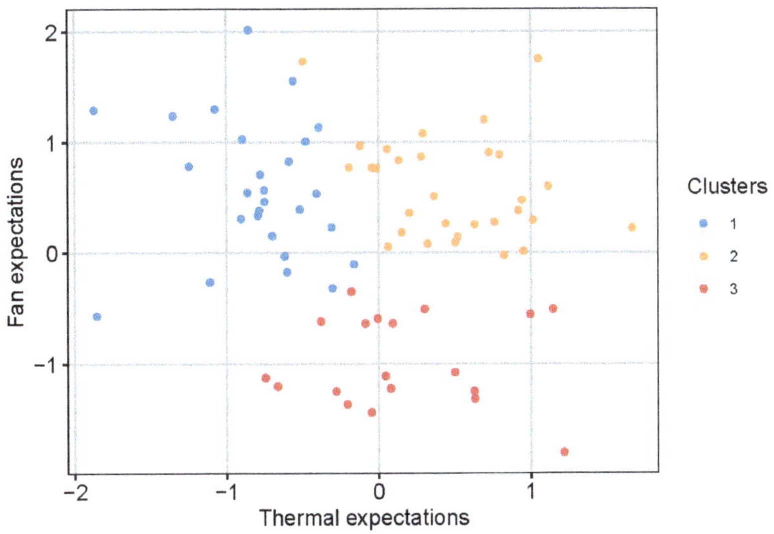

Figure 4. Classification groups for thermal and behavioral expectations based on the SVM method.

This clustering process was carried out before the session in the LOBSTER to similarly distribute participants according to daytime (morning/afternoon) and information groups (Section 3.3.2).

3.3.2. Manipulation Technique

The experimental study used a manipulation technique to test the effect of information on occupants' expectations and satisfaction (H2). The main goal of the manipulation technique is to activate hedonistic frames in all participants and to test whether normative frames predominate over hedonistic frames according to the different information provided. To activate the hedonistic frames, the office rooms were set to warm conditions, which can act as a stimulus for subjects to perform an action to restore thermal comfort (hedonistic motivation). Previous studies investigating the cooling effect of air movement under controlled conditions in test rooms [58,59] found that thermal comfort can be achieved at an indoor temperature set point of 30 °C if a personally controlled fan was provided. For this study, a setpoint of 30 °C was selected to trigger warm discomfort and encourage the use of the personal fans to achieve thermal comfort without compromising health and productivity issues that may affect occupants' satisfaction in the room. To fulfill the hedonistic frames, i.e., to restore thermal comfort, participants were provided with adjustment options, such as turning on the ceiling fan, opening the window, and drinking a beverage. Questions about the fulfillment of hedonistic frames were asked in the final questionnaire (EQ).

Participants watched a video (see Figure 2 in Section 3.3) that provided information about sustainability and energy efficiency in buildings to activate the normative frames. Two different videos were created. The control group was shown a shorter video containing general information about sustainability, climate change, and political energy targets in Germany, as well as the aim of the study. The experimental group was shown a longer video (https://www.youtube.com/watch?v=JJdQMij2kT0, accessed on 14 January 2024) that had the same initial content as the control group but included additional information about benefits and scientific explanations on how ceiling fans work.

The inclusion of general information is to set a "baseline" of information for all participants. The distinction between videos (additional information on personal ceiling fans) is intended to increase motivation to use the low-energy-consumption device in opposition to other non-energy-efficient strategies, such as opening the windows when it is too warm outside. Accordingly, participants were divided into the experimental group (long video) and the control group (short video). Both groups received instructions with a standardized text on how to operate the adaptive strategies: turning on and adjusting the air velocity of the ceiling fan, opening the tilt windows, and recording beverage intake in liters. Different adaptive opportunities to counteract thermal discomfort were given based on the work from Meinke et al. [41] to evaluate the influence of the provided information about the potential change in comfort and energy consumption of the personal ceiling fan on the experimental group. Participants were similarly distributed according to their cluster group of expectations described in the previous section.

During the session, participants were also asked to rate the educational video. All questions had a 7-point Likert scale ranging from strongly disagree (1) to strongly agree (7).

3.4. Data Analysis

All data preparation and analysis were performed in the software environment R (Version 4.1.3) [60]. The following subsections describe the assumptions and methods used for data analysis.

3.4.1. Sample Size and Checks on Random Assignment

The sample size was calculated using G*Power 3.1.9.7 [61]. Since the sample size was less than the required to achieve a small effect size, a large effect size was necessary (>0.8). For a t-test between two independent group means with an α value of 0.05, a power $(1 - \beta)$ of 0.95, and an effect size of 0.8, the required sample size was 74 participants.

Before testing the hypotheses, we verified the equivalence of the participant groups in the two research conditions using t-tests and Chi-square analyses (see Appendix A for results of these equivalence tests). Table A1 shows the distribution of participants from the different clusters according to their demographics and other characteristics, as well as the experimental conditions. Body mass index was categorized into two groups according to the WHO classification [62]: BMI < 25 kg/m^2 = normal and BMI > 25 kg/m^2 = overweight. The results showed that BMI and previous experience in working in an office were significantly different between clusters. Accordingly, we controlled for those variables by entering them as covariates in the tests of H1–H2. Table A2 shows the distribution of participants from the different clusters according to their actual mood, video rating, and fan use (air velocity and duration of fan turned on). None of the variables were significantly different between groups.

Additionally, we verified differences in indoor climate perception between the expectation clusters. To capture changes in the reported thermal comfort between the acclimation phase and the rest of the experimental phase, a mean value for comfort votes was taken for the whole test. Results of the Kruskal–Wallis tests showed that participants in cluster 2 were significantly more comfortable with the thermal conditions during the whole test compared to the other two groups ($H(2) = 6.65$, $p < 0.05$, $\eta^2 = 0.06$). A post hoc analysis was performed using the Dunn test to determine which levels of the independent variable differed from each other. The pairwise comparison test showed that cluster 2 is significantly different from cluster 1 ($p < 0.05$) but not from cluster 3 ($p = 0.089$). In addition, no differences were found for thermal sensation, preference, and acceptability and indoor air quality perception between groups. Therefore, only thermal comfort was kept for further analysis as the dependent variable to test the proposed hypotheses.

To evaluate changes in participants' fan expectations and evaluation, questions related to the expectations of personal ceiling fans from the background questionnaire (Section 3.2) and the last questionnaire from the LOBSTER session were analyzed. Firstly, a principal component analysis (PCA) was conducted on 13 questions from the background questionnaire. The weights from the background questionnaire were calculated to obtain the scores for the equivalent questions in the LOBSTER session. A single component was obtained for fan expectations (pre-test) and fulfilled expectations (LOBSTER session). To obtain a value representing the change between fan expectations (before the session) and evaluation (after the session), the difference between the two variables was calculated. The resulting variable was called "fan evaluation" (M = 3.71, SD = 2.25).

3.4.2. Hypotheses Testing: Statistical Tests

To test the hypothesis that groups of occupants with different types of thermal and behavioral expectations will express different thermal satisfaction (H1), a regression analysis was conducted. The single-answer options for measuring participants' evaluation of the temperature could not be assumed to be equidistant but needed to be considered as ordered categorical data [63]. Therefore, an ordinal model was selected to test the relationship between these ordinal response variables and one or more independent variables using the *clm* (cumulative link model) function from the R package ordinal [64]. The independent variable was the expectancy group (cluster), which was treated as categorical (1, 2, or 3). The hypothesis that the effect of information on participants' thermal satisfaction would be particularly strong among participants with more positive expectations of the indoor air quality and thermal conditions and the use of the personal ceiling fan (H2a) was tested with a conditional process analysis [65] using Hayes' PROCESS model 1 of moderation for R with cluster as the multicategorical variable. To test for possible changes in the expectations of participants with negative expectations after providing information (H2b), an additional process analysis was conducted with fan evaluation as the dependent variable. Similar to the evaluation approach for thermal comfort, fan evaluation was considered ordered categorical data.

4. Results

A series of predictive models were run to examine the above-mentioned hypotheses. H1 predicted that greater reported thermal comfort would be reported by participants with more positive thermal and behavioral expectations. To test this hypothesis, the total effect model was examined by testing the simple effect of the independent variable and control variables on the outcome variable. H1 was supported, as belonging to cluster 2 (the group with more positive thermal and behavioral expectations) was associated with significantly greater reported thermal comfort (Table 2). The coefficient in the model indicates a positive relationship: the more positive the thermal and behavioral expectations, the higher comfort participants in this group reported. A likelihood ratio test was performed with an ANOVA test. The results showed that the model that includes the expectation groups as a variable is significantly better than an intercept-only baseline model ($\chi^2 < 0.05$). Control variables of BMI and previous experience in working in an office did not significantly influence thermal comfort.

Table 2. Results of the ordinal regression analysis to test the effect of expectancy cluster on thermal comfort.

	Estimate	Std. Error	z-Value	*p*-Value
Cluster 2 [a]	1.65	0.67	2.45	0.015 *
Cluster 3 [a]	0.076	0.66	0.11	0.909
BMI (overweight)	−0.42	0.56	−0.76	0.449
Experience (yes)	−0.91	0.60	−1.50	0.133

* $p < 0.05$; [a] Results against cluster 1.

H2a predicted that the effect of the expectancy cluster on thermal comfort would be especially strong among participants with greater existing personal norms to protect the environment and save energy, as activated by tailored information (long video). This hypothesis was supported (see Figure 5 for coefficients and *p*-values), as the moderation model was significant ($F(5, 70) = 3.08$, $p < 0.05$, $R^2 = 0.18$). Tailored information to activate personal norms (the long video) seems to have prompted higher reported thermal comfort in participants from cluster 2 compared to those from clusters 1 and 3. Those participants who did not receive tailored information (the short video) expressed similar reported thermal comfort regardless of their expectancy cluster, indicating no effect of video on the relationship between expectancy and thermal comfort.

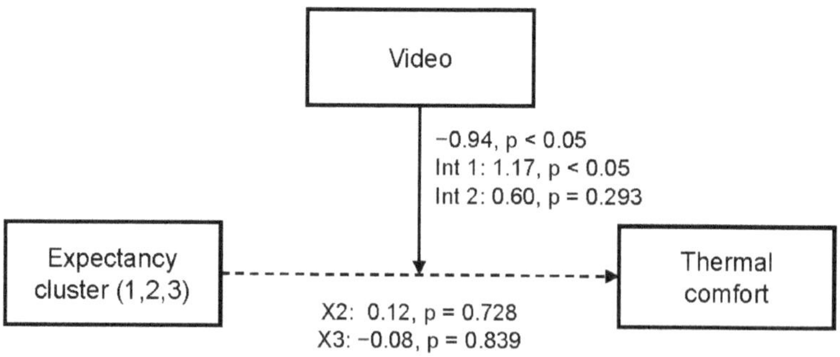

Figure 5. Model of moderating effects of video on thermal satisfaction. Unstandardized coefficients are shown. Dotted lines indicate nonsignificant relationships. Short video condition coded as 0; long video condition coded as 1. Expectancy cluster coded as dummy variables for multicategorical variables. Video significantly moderates the effect of cluster 2 on thermal comfort (Int 1).

Although data from the test of the total effect model for H1 (Table 2) identified a significant effect of expectancy cluster on reported thermal comfort, this effect was nonsignificant in the moderation model that included video (tailored information). Note that tests of direct effects (the path from expectancy cluster to thermal comfort shown in

Figure 5) reflect the influence of a predictor variable on an outcome variable while holding any moderation variables constant; this is in contrast to the total effect model, which only estimates the effect of expectancy cluster and the control variables on thermal comfort. Such findings indicate that the effect of expectancy cluster on reported thermal comfort is significant depending on the value of video. Additionally, the moderation model that included the effect of video explained higher variance ($R^2 = 0.18$) than the total effect model ($R^2 = 0.09$).

H2b predicted that by activating personal norms (the long video), the change in reported satisfaction with the personal fan would be greater among those participants with more negative expectations. This hypothesis was not supported (see Figure 6 for coefficients and *p*-values), as the moderation effect was not significant for any of the expectancy clusters. However, the expectancy cluster had a significant effect on reported fan satisfaction, and the model was significant ($F(5, 70) = 2.78$, $p < 0.05$, $R^2 = 0.17$). The negative coefficients indicate that those participants from clusters 2 and 3 may express a lower change in reported fan satisfaction compared to those from cluster 1.

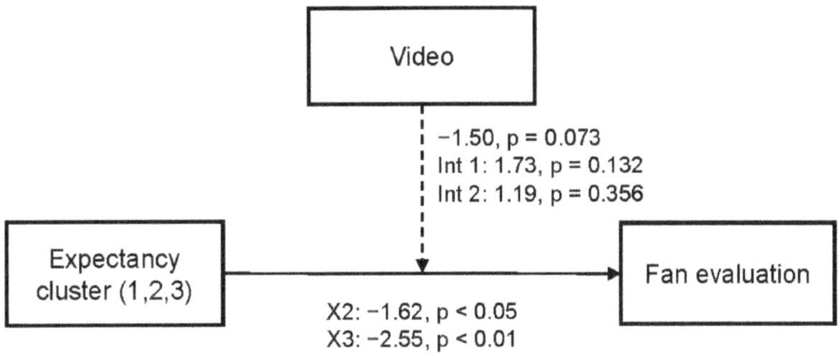

Figure 6. Model of moderating effects of video on changes in fan satisfaction. Unstandardized coefficients shown. Dotted lines indicate non-significant relationships. Short video condition coded as 0; long video condition coded as 1. Expectancy cluster coded as dummy variables for multicategorical variables. Video did not significantly moderate the effect of expectancy cluster on fan satisfaction.

5. Discussion

Previous studies have examined the effect of expectations on occupants' thermal and overall satisfaction [2,66,67], indicating cultural, geographical, and building-type differences [6]. By combining occupants' expectations of indoor thermal conditions and expectations towards building control opportunities, the current study proposed to distinguish occupants according to their expectancy levels. Thus, the relationship between participants' comfort expectations, described as thermal and behavioral expectations, and their thermal comfort in a simulated work environment was tested (H1). The study found that reported thermal comfort was greater among those participants with more positive thermal and behavioral expectations (cluster 2) and significantly differed from participants with negative thermal expectations (cluster 1) but not from the cluster expecting neutral thermal conditions and having negative behavioral expectations (cluster 3). These results may reflect the assimilation effect given by the coherence between expected and experienced indoor conditions that lead to greater thermal satisfaction [7,68]. Additionally, these findings reflect the higher importance of thermal expectations in predicting comfort compared to the effect of behavioral expectations. This could be associated with the modest expectations of occupants towards building controls in naturally ventilated buildings, which is the building type mostly found in the city where this study took place. Usually, occupants in naturally ventilated buildings do not associate their discomfort with the thermal environment provided by the building, as they may be more in contact with the outdoor conditions (e.g., by opening the window), and therefore do not expect their comfort to change due to the building's performance but through their actions [31].

Due to significant BMI differences between cluster groups, this variable was included in the model. However, BMI did not significantly influence participants' thermal perception. These results may contradict the general tendency in the literature that BMI differences exist [13,69]. However, BMI classification has recently been criticized as inaccurate and misleading [70]. Because BMI is based only on height and weight and does not take into account other body characteristics such as body fat content, muscle mass, and body composition, it is possible that some of the participants were misclassified without taking into account factors that affect human thermoregulation. Further research on the thermoregulatory process considering actual measurements of body composition should be carried out.

These first results reinforce Brown and Coles' [20] statement that expectations play an important role in shaping occupant comfort and indoor environmental behavior. However, this main effect seems best explained by the moderating role of normative motivations. The activation of personal norms was found to significantly moderate the influence of expectancy on reporter thermal comfort (H2a). Those who watched a video with detailed information about sustainable buildings and the benefits of the personal fan reported greater thermal comfort than those who watched a video with general information about the study. Although the test of H1 identified a significant influence of expectancy cluster on reporter thermal comfort, when the variable video was added to the model, that effect became non-significant. This finding suggests that the positive association between the expectancy and thermal comfort identified in the test of H1 could be largely explained by the activation of personal norms elicited among participants with more positive expectations. Additionally, greater variance in thermal comfort was explained by the moderation model that included video as compared to that explained by the total effect model, which isolated the effect of comfort expectations. Accordingly, we suggest that future studies examine the potential influence of other social-psychological constructs, such as personal norms, on perceptions of IEQ, along with additional attempts to identify which types of occupants are likely to feel more comfortable based on their social–psychological characteristics to shape their comfort expectations.

We anticipated, but did not find, a moderation effect of active personal norms on the influence of expectancy on changes in fan evaluation (H2b). An explanation for this lack of influence could be that hedonic goals were a priority for all participants rather than their normative motivations [48]. Given the moderately warm indoor temperatures, participants' comfort needs (i.e., the need to restore comfort due to the warm thermal sensation) may have become more relevant, and the potential influence of the video may not have been strong enough to rate the fan according to normative principles but rather according to its effectiveness to restore comfort (prioritizing hedonic goals). Although the moderation effect of the video was not significant in the model, there was a significant effect of expectancy on changes in fan evaluation. Greater changes in fan evaluation after participation in the experimental session (i.e., fulfilled expectations) were observed for participants with negative thermal expectations compared to participants with positive thermal expectations. These findings indicated that individually controlling the fan to increase thermal comfort may have effectively induced a change towards a more positive fan evaluation, especially in participants with lower comfort expectations. However, these results do not eliminate the possible effect of tailored information on fan evaluations and behavioral interactions, which may vary depending on the way the information is delivered. For instance, Schweiker et al. [18] found that participants who participated in a workshop were more likely to change their behavior than those who only received an information brochure. Future studies could investigate other ways of providing information to investigate whether the association of occupants' different expectations with actual normative behaviors, specifically with PECS, could be moderated by personal norms. As studied by Li et al. [24], normative messaging in personal environmental control systems could not only enhance thermal comfort but induce a higher probability of using personal devices, such as personal fans, to restore comfort.

5.1. Practical Implications

The findings of this study suggest that it may be useful to address and attempt to influence occupants' expectations of indoor thermal conditions and building operations. This is particularly relevant to the implementation of PECS in buildings as positive expectations of the indoor environment and the use of PECS may have implications for reducing energy consumption while increasing occupant satisfaction in buildings. The positive effect of information on higher tolerance of the expected indoor environment conditions, together with the provision of personal, low-intensive cooling strategies, could support the acceptance and use of PECS, such as personal ceiling fans, to ensure occupant satisfaction with the thermal environment in naturally ventilated buildings.

5.2. Limitations

This study was conducted in a laboratory setting, an unfamiliar environment to the participants. We therefore could not measure the extent to which expectations influence on-site perceptions of the thermal environment in a familiar environment, where occupants may have different expectations of the climatic conditions, as suggested by Schweiker et al. [26]. We suggest that future studies investigate such a relationship. In the present study, normative messaging was tested on the evaluation of and satisfaction with one adaptive strategy that was available for all participants. The possible effect of personal norms may be different if (1) multiple adaptive strategies with different normative impacts (e.g., low-energy-consumption strategies vs. the use of air conditioning) have been tested simultaneously, giving participants multiple adaptive possibilities, and (2) the actual behaviors have been tested in addition to the adaptive strategy's evaluation. Furthermore, the influence of information and expectancy group was examined for the personal ceiling fan for a constant temperature condition and a German sample. We suggest that additional studies be conducted with other types of PECS, different thermal conditions, and a variety of samples to examine whether the type of adaptive strategy, climatic conditions, or relevant cultural differences influence the effect of information on thermal comfort. Finally, we suggest that future studies examine the extent to which more information about the features of PECS and other types of manipulation techniques influence real-time, on-site IEQ perceptions and behaviors.

6. Conclusions

This study investigated the effect of occupants' expectations on their satisfaction with the thermal environment and a personal ceiling fan as influenced by the activation of normative goals. Our results indicate that building occupants who have more positive expectations about indoor thermal conditions may express higher levels of thermal comfort than those with more negative comfort expectations, regardless of their expectations of the building systems. Our findings also indicate that comfort expectations can be influenced by the activation of personal norms. By activating normative motivations, occupants may perceive indoor conditions as more comfortable. Those expectations should be associated with the expected satisfaction and fulfilled expectations of adaptive actions in order to stay comfortable in a building. To the extent that thermal expectations are negative, occupants might improve their perceptions of personal building controls (such as a personal ceiling fan) when making personal adjustments in order to stay comfortable. Our findings suggest that building designers could focus and manipulate occupants' comfort expectations, e.g., by providing occupants with normative messages and individual control, to achieve greater comfort and acceptance of personal building controls, such as PECS, in naturally ventilated buildings.

Author Contributions: Conceptualization, R.R. and M.S.; methodology, R.R. and M.S.; software, R.R.; validation, R.R.; formal analysis, R.R.; investigation, R.R.; resources, R.R.; writing—original draft preparation, R.R.; writing—review and editing, R.R. and M.S.; visualization, R.R.; supervision, M.S.; project administration, R.R.; funding acquisition, M.S. All authors have read and agreed to the published version of the manuscript.

Funding: The data collection and analysis were conducted within Project ID 03ET1563A, funded by the German Federal Ministry for Economics and Climate Action (BMWK). Schweiker's reviewing, supervision, and editing work was supported by a research grant (21055) from VILLUM FONDEN. The KIT-Publication Fund of the Karlsruhe Institute of Technology funded the APC.

Institutional Review Board Statement: This study was conducted in accordance with the Declaration of Helsinki and approved by the data protection officer of the Karlsruhe Institute of Technology (date of approval: 10 June 2021). Note: the Ethics Committee of the Karlsruhe Institute of Technology began assigning application numbers for applications in 2023. Therefore, there is no approval number for this study. However, the approval letter is available from the authors upon request.

Informed Consent Statement: Informed consent was obtained from all subjects involved in the study.

Data Availability Statement: The data presented in this study are available upon request from the corresponding author. The data are not publicly available due to privacy restrictions.

Acknowledgments: Special thanks to Nicolas Carbonare for his support during the experimental phase and fruitful discussions on the formal analysis of the work and to Laura Arpan for her guidance on the statistical analysis of the data. We acknowledge support by the KIT-Publication Fund of the Karlsruhe Institute of Technology.

Conflicts of Interest: The authors declare no conflicts of interest. The funders had no role in the design of the study; in the collection, analyses, or interpretation of data; in the writing of the manuscript; or in the decision to publish the results.

Abbreviations

The following abbreviations are used in this manuscript:

BMI	Body mass index
BMS	Building management system
EQ	End questionnaire
IEQ	Indoor environmental quality
Int	Intercept
IQ	Initial questionnaire
LOBSTER	Laboratory for Occupant Behavior, Satisfaction, Thermal comfort, and Environmental Research
OQ	Online questionnaire
PCA	Principal component analysis
PECS	Personal environmental control system
SD	Standard deviation
SQ	Start questionnaire
SVM	Support vector machine
WHO	World Health Organization

Appendix A. Additional Tables

Table A1. Participant demographics and other characteristics, as well as experimental conditions according to expectation clusters and results of tests of equivalence of research conditions.

	Cluster 1	Cluster 2	Cluster 3	Full Sample	Test of Independence		
	N	*N*	*N*	*N*	χ^2	df	*p*-Value
Sex					0.15	2	0.928
Female	13	13	9	35			
Female	14	17	10	41			

Table A1. *Cont.*

	Cluster 1	Cluster 2	Cluster 3	Full Sample	Test of Independence		
	N	*N*	*N*	*N*	χ^2	df	*p*-Value
Age					0.15	2	0.929
Young	18	19	13	50			
Elderly	9	11	6	26			
BMI					6.12 *	2	0.047
Normal	12	20	15	47			
Overweight	15	10	4	29			
Daytime					0.47	2	0.079
Morning	12	16	9	37			
Afternoon	15	14	10	39			
Office					4.67	2	0.097
1	9	18	11	38			
2	18	12	8	38			
Video					0.52	2	0.771
Short	15	14	9	38			
Long	12	16	10	38			
Experience with fans					5.13	2	0.077
Yes	2	4	6	12			
No	25	26	13	64			
Experience with ceiling fans					2.05	2	0.359
Yes	7	8	2	17			
No	29	22	17	59			
Previous worked in office					7.25 *	2	0.027
Yes	9	3	8	20			
No	18	27	11	56			

* $p < 0.05$.

Table A2. Participants' votes and fan use according to expectation clusters and results of tests of equivalence of research conditions.

	Cluster 1 M (SD)	Cluster 2 M (SD)	Cluster 3 M (SD)	Test of Independence χ^2	df	*p*-Value
Actual mood [a]	3.04 (1.02)	2.57 (1.14)	3.21 (1.13)	4.53	2	0.104
Air velocity level [%]	55.48 (21.44)	46.52 (29.62)	48.92 (25.10)	2.29	2	0.318
Duration fan on [min]	127.99 (5.69)	123.35 (19.50)	124.00 (24.87)	2.90	2	0.235
Video rating 1 [b]	0.21 (0.89)	−0.21 (1.14)	0.04 (1.23)	3.28	2	0.194
Video rating 2 [b]	0.03 (0.99)	−0.04 (1.02)	0.02 (1.04)	0.12	2	0.940

[a] Integer values. Scale ranged from 1 (very bad) to 7 (very good): *"How is your mood right now?"*. [b] Unstandarized values resulting from PCA conducted with seven questions. Two components resulted from the analysis.

References

1. de Dear, R.; Xiong, J.; Kim, J.; Cao, B. A review of adaptive thermal comfort research since 1998. *Energy Build.* **2020**, *214*, 109893. [CrossRef]
2. de Dear, R.J.; Brager, G.S. Developing an Adaptive Model of Thermal Comfort and Preference. *ASHRAE Trans.* **1998**, *104*, 145.
3. Mahdavi, A.; Berger, C. Predicting Buildings' Energy Use: Is the Occupant-Centric "Performance Gap" Research Program Ill-Advised? *Front. Energy Res.* **2019**, *7*, 124. [CrossRef]
4. Höppe, P. Different aspects of assessing indoor and outdoor thermal comfort. *Energy Build.* **2002**, *34*, 661–665. [CrossRef]
5. Keeling, T.P.; Roesch, E.B.; Clements-Croome, D. Cognitive Appraisals Affect Both Embodiment of Thermal Sensation and Its Mapping to Thermal Evaluation. *Front. Psychol.* **2016**, *7*, 800. [CrossRef]
6. Luo, M.; de Dear, R.; Ji, W.; Bin, C.; Lin, B.; Ouyang, Q.; Zhu, Y. The dynamics of thermal comfort expectations: The problem, challenge and impication. *Build. Environ.* **2016**, *95*, 322–329. [CrossRef]
7. de Dear, R. Adaptive comfort applications in Australia and Impacts on Building Energy Consumption. In Proceedings of the 6th International Conference on Indoor Air Quality, Ventilation & Energy Conservation in Buildings, Sendai, Japan, 28–31 October 2007; pp. 17–24.
8. Deuble, M.P.; de Dear, R.J. Green occupants for green buildings: The missing link? *Build. Environ.* **2012**, *56*, 21–27. [CrossRef]
9. Leaman, A.; Bordass, B. Are users more tolerant of 'green' buildings? *Build. Res. Inf.* **2007**, *35*, 662–673. [CrossRef]

10. Kwok, A.G.; Rajkovich, N.B. Addressing climate change in comfort standards. *Build. Environ.* **2010**, *45*, 18–22. [CrossRef]
11. Luo, M.; Cao, B.; Ji, W.; Ouyang, Q.; Lin, B.; Zhu, Y. The underlying linkage between personal control and thermal comfort: Psychological or physical effects? *Energy Build.* **2016**, *111*, 56–63. [CrossRef]
12. Rawal, R.; Schweiker, M.; Kazanci, O.B.; Vardhan, V.; Jin, Q.; Duanmu, L. Personal comfort systems: A review on comfort, energy, and economics. *Energy Build.* **2020**, *214*, 109858. [CrossRef]
13. Wang, Z.; de Dear, R.; Luo, M.; Lin, B.; He, Y.; Ghahramani, A.; Zhu, Y. Individual difference in thermal comfort: A literature review. *Build. Environ.* **2018**, *138*, 181–193. [CrossRef]
14. Schiavon, S.; Yang, B.; Donner, Y.; Chang, V.W.C.; Nazaroff, W.W. Thermal comfort, perceived air quality, and cognitive performance when personally controlled air movement is used by tropically acclimatized persons. *Indoor Air* **2017**, *27*, 690–702. [CrossRef]
15. Pasut, W.; Arens, E.; Zhang, H.; Zhai, Y. Enabling energy-efficient approaches to thermal comfort using room air motion. *Build. Environ.* **2014**, *79*, 13–19. [CrossRef]
16. Zhai, Y.; Zhang, H.; Zhang, Y.; Pasut, W.; Arens, E.; Meng, Q. Comfort under personally controlled air movement in warm and humid environments. *Build. Environ.* **2013**, *65*, 109–117. [CrossRef]
17. Arens, E. *Behavior and Buildings—Leveraging Occupants to Improve Energy and Comfort*; Center for the Built Environment: Berkeley, CA, USA, 2010.
18. Schweiker, M.; Shukuya, M. Investigation on the effectiveness of various methods of information dissemination aiming at a change of occupant behaviour related to thermal comfort and exergy consumption. *Energy Policy* **2011**, *39*, 395–407. [CrossRef]
19. Day, J.K.; Gunderson, D.E. Understanding high performance buildings: The link between occupant knowledge of passive design systems, corresponding behaviors, occupant comfort and environmental satisfaction. *Build. Environ.* **2015**, *84*, 114–124. [CrossRef]
20. Brown, Z.; Cole, R.J. Influence of occupants' knowledge on comfort expectations and behaviour. *Build. Res. Inf.* **2009**, *37*, 227–245. [CrossRef]
21. Fischer, C. Feedback on household electricity consumption: A tool for saving energy? *Energy Effic.* **2008**, *1*, 79–104. [CrossRef]
22. Carrico, A.R.; Riemer, M. Motivating energy conservation in the workplace: An evaluation of the use of group-level feedback and peer education. *J. Environ. Psychol.* **2011**, *31*, 1–13. [CrossRef]
23. Vellei, M.; Natarajan, S.; Biri, B.; Padget, J.; Walker, I. The effect of real-time context-aware feedback on occupants' heating behaviour and thermal adaptation. *Energy Build.* **2016**, *123*, 179–191. [CrossRef]
24. Li, Z.; Loveday, D.; Demian, P. Feedback messaging, thermal comfort and usage of office-based personal comfort systems. *Energy Build.* **2019**, *205*, 109514. [CrossRef]
25. Fanger, P.; Toftum, J. Extension of the PMV model to non-air-conditioned buildings in warm climates. *Energy Build.* **2002**, *34*, 533–536. [CrossRef]
26. Schweiker, M.; Rissetto, R.; Wagner, A. Thermal expectation: Influencing factors and its effect on thermal perception. *Energy Build.* **2020**, *210*, 109729. [CrossRef]
27. Zhou, X.; Ouyang, Q.; Zhu, Y.; Feng, C.; Zhang, X. Experimental study of the influence of anticipated control on human thermal sensation and thermal comfort. *Indoor Air* **2014**, *24*, 171–177. [CrossRef]
28. Luo, M.; Wang, Z.; Brager, G.; Cao, B.; Zhu, Y. Indoor climate experience, migration, and thermal comfort expectation in buildings. *Build. Environ.* **2018**, *141*, 262–272. [CrossRef]
29. Chun, C.; Kwok, A.; Mitamura, T.; Miwa, N.; Tamura, A. Thermal diary: Connecting temperature history to indoor comfort. *Build. Environ.* **2008**, *43*, 877–885. [CrossRef]
30. Cândido, C.; de Dear, R.; Lamberts, R.; Bittencourt, L. Cooling exposure in hot humid climates: Are occupants 'addicted'? *Archit. Sci. Rev.* **2010**, *53*, 59–64. [CrossRef]
31. Kim, J.; de Dear, R. Impact of different building ventilation modes on occupant expectations of the main IEQ factors. *Build. Environ.* **2012**, *57*, 184–193. [CrossRef]
32. Rajkovich, N.; Kwok, A. Thermal Comfort in Urban Arcades: A Study of Prediction, Expectation, and Memory. In Proceedings of the Proceedings of ARCC 2003 Spring Research Conference, Tempe, AZ, USA, 10–12 April 2003.
33. Rissetto, R.; Rambow, R.; Schweiker, M. Assessing comfort in the workplace: A unified theory of behavioral and thermal expectations. *Build. Environ.* **2022**, *216*, 109015. [CrossRef]
34. Auliciems, A. Towards a psycho-physiological model of thermal perception. *Int. J. Biometeorol.* **1981**, *25*, 109–122. [CrossRef]
35. Kempton, W.; Feuermann, D.; McGarity, A.E. "I always turn it on super" user decisions about when and how to operate room air conditioners. *Energy Build.* **1992**, *18*, 177–191. [CrossRef]
36. Karjalainen, S. *Why It Is Difficult to Use a Simple Device: An Analysis of a Room Thermostat*; Springer: Berlin/Heidelberg, Germany, 2007.
37. Vastamäki, R.; Sinkkonen, I.; Leinonen, C. A behavioural model of temperature controller usage and energy saving. *Pers. Ubiquitous Comput.* **2005**, *9*, 250–259. [CrossRef]
38. Froehlich, J. Promoting energy efficient behaviors in the home through feedback: The role of human-computer interaction. In Proceedings of the Human Computer Interaction Consortium, HCIC Workshop, Fraser, CO, USA, 1–5 September 2009.
39. Matthies, E.; Kastner, I.; Klesse, A.; Wagner, H.J. High reduction potentials for energy user behavior in public buildings: How much can psychology-based interventions achieve? *J. Environ. Stud. Sci.* **2011**, *1*, 241–255. [CrossRef]

40. Jain, R.K.; Taylor, J.E.; Culligan, P.J. Investigating the impact eco-feedback information representation has on building occupant energy consumption behavior and savings. *Energy Build.* **2013**, *64*, 408–414. [CrossRef]
41. Meinke, A.; Hawighorst, M.; Wagner, A.; Trojan, J.; Schweiker, M. Comfort-related feedforward information: Occupants' choice of cooling strategy and perceived comfort. *Build. Res. Inf.* **2017**, *45*, 222–238. [CrossRef]
42. Arpan, L.; Rissetto, R.; Yan, Z.; Roetzel, A.; Azar, E.; Jazizadeh, F.; Morandi, F.; Zhu, Y.; Heydarian, A.; Bourikas, L.; et al. The hopeful expect to be comfortable: Exploring emotion and personal norms related to sustainable buildings in the United States. *Energy Res. Soc. Sci.* **2022**, *93*, 102846. [CrossRef]
43. Banham, R. *The Architecture of the Well-Tempered Environment*; The Architecture Press: London, UK, 1969.
44. Hameed, I.; Khan, K. An extension of the goal-framing theory to predict consumer's sustainable behavior for home appliances. *Energy Effic.* **2020**, *13*, 1441–1455. [CrossRef]
45. Gerhardsson, K.M.; Laike, T.; Johansson, M. Residents' lamp purchasing behaviour, indoor lighting characteristics and choices in Swedish homes. *Indoor Built Environ.* **2019**, *28*, 964–983. [CrossRef]
46. Wall, R.; Crosbie, T. Potential for reducing electricity demand for lighting in households: An exploratory socio-technical study. *Energy Policy* **2009**, *37*, 1021–1031. [CrossRef]
47. Bamberg, S.; Möser, G. Twenty years after Hines, Hungerford, and Tomera: A new meta-analysis of psycho-social determinants of pro-environmental behaviour. *J. Environ. Psychol.* **2007**, *27*, 14–25. [CrossRef]
48. Lindenberg, S.; Steg, L. Normative, Gain and Hedonic Goal Frames Guiding Environmental Behavior. *J. Soc. Issues* **2007**, *63*, 117–137. [CrossRef]
49. Schweiker, M.; Brasche, S.; Hawighorst, M.; Bischof, W.; Wagner, A. Presenting LOBSTER, an innovative climate chamber, and the analysis of the effect of a ceiling fan on the thermal sensation and performance under summer conditions in an office-like setting. In Proceedings of the 8th Windsor Conference: Counting the Cost of Comfort in a Changing World, Windsor, UK, 10–13 April 2014.
50. *ISO 7730:2005*; Ergonomics of the Thermal Environment—Analytical Determination and Interpretation of Thermal Comfort Using Calculation of the PMV and PPD Indices and Local Thermal Comfort Criteria. ISO: Geneva, Switzerland, 2005.
51. LimeSurvey Project Team/Carsten Schmitz. LimeSurvey: An Open Source Survey Tool. 2012. Available online: http://www.limesurvey.org (accessed on 14 January 2024).
52. Rissetto, R.; Schweiker, M.; Wagner, A. Personalized ceiling fans: Effects of air motion, air direction and personal control on thermal comfort. *Energy Build.* **2021**, *235*, 110721. [CrossRef]
53. Schakib-Ekbatan, K.; Wagner, A.; Lussac, C. Occupant satisfaction as an indicator for the socio-cultural dimension of sustainable office buildings—Development of an overall building index. In Proceedings of the Adapting to Change: New Thinking on Comfort Cumberland Lodge, Windsor, UK, 9–11 April 2010.
54. *ASHRAE Standard 55*; Thermal Environmental Conditions for Human Occupancy. American Society of Heating, Refrigerating, and Air-Conditioning Engineers: Atlanta, GA, USA, 2017.
55. Carbonare, N.; Pflug, T.; Wagner, A. Clustering the occupant behavior in residential buildings: A method comparison. *Bauphysik* **2018**, *40*, 427–433. [CrossRef]
56. Raschka, S. *Python Machine Learning: Unlock Deeper Insights into Machine Learning with This Vital Guide to Cutting-Edge Predictive Analytics*; Community Experience Distilled, Packt Publishing: Birmingham, UK, 2015.
57. Vapnik, V. *The Nature of Statistical Learning Theory*; Springer: New York, NY, USA, 1995.
58. Zhai, Y.; Zhang, Y.; Zhang, H.; Pasut, W.; Arens, E.; Meng, Q. Human comfort and perceived air quality in warm and humid environments with ceiling fans. *Build. Environ.* **2015**, *90*, 178–185. [CrossRef]
59. Huang, L.; Ouyang, Q.; Zhu, Y.; Jiang, L. A study about the demand for air movement in warm environment. *Build. Environ.* **2013**, *61*, 27–33. [CrossRef]
60. R Core Team. *R: A Language and Environment for Statistical Computing*; R Core Team: Vienna, Austria, 2022.
61. Faul, F.; Erdfelder, E.; Buchner, A.; Lang, A.G. Statistical power analyses using G*Power 3.1: Tests for correlation and regression analyses. *Behav. Res. Methods* **2009**, *41*, 1149–1160. [CrossRef] [PubMed]
62. WHO. World Health Organization, 2020. Available online: https://www.who.int/ (accessed on 14 January 2024).
63. Schweiker, M.; Fuchs, X.; Becker, S.; Shukuya, M.; Dovjak, M.; Hawighorst, M.; Kolarik, J. Challenging the assumptions for thermal sensation scales. *Build. Res. Inf.* **2017**, *45*, 572–589. [CrossRef]
64. Christensen, R.H.B. Cumulative Link Models for Ordinal Regression with the R Package Ordinal; 2015 . Available online: https://cran.r-project.org/ (accessed on 14 January 2024).
65. Hayes, A.F.; Montoya, A.K. A Tutorial on Testing, Visualizing, and Probing an Interaction Involving a Multicategorical Variable in Linear Regression Analysis. *Commun. Methods Meas.* **2017**, *11*, 1–30. [CrossRef]
66. McIntyre, D.A. Chapter 13 Design Requirements for a Comfortable Environment. In *Bioengineering, Thermal Physiology and Comfort*; Studies in Environmental Science; Elsevier: Amsterdam, The Netherlands, 1981; Volume 10, pp. 195–220. [CrossRef]
67. Chappells, H.; Shove, E. Debating the future of comfort: Environmental sustainability, energy consumption and the indoor environment. *Build. Res. Inf.* **2005**, *33*, 32–40. [CrossRef]
68. Anderson, R.E. Consumer Dissatisfaction_ The Effect of Disconfirmed Expectancy on Perceived Product Performance—3149407. *J. Mark. Res.* **1973**, *10*, 38–44. [CrossRef]

69. Rupp, R.F.; Kim, J.; de Dear, R.; Ghisi, E. Associations of occupant demographics, thermal history and obesity variables with their thermal comfort in air-conditioned and mixed-mode ventilation office buildings. *Build. Environ.* **2018**, *135*, 1–9. [CrossRef]
70. Ahima, R.S.; Lazar, M.A. Physiology. The health risk of obesity–better metrics imperative. *Science* **2013**, *341*, 856–858. [CrossRef] [PubMed]

buildings

MDPI

Article

Skin Heat Transfer and Thermal Sensation Coupling Model under Steady Stimulation

Yijia Zhou, Hang Yu *, Maohui Luo and Xiang Zhou

School of Mechanical Engineering, Tongji University, Shanghai 201804, China; 2130278@tongji.edu.cn (Y.Z.); luomaohui@tongji.edu.cn (M.L.); zhouxiang@tongji.edu.cn (X.Z.)
* Correspondence: yuhang@tongji.edu.cn

Abstract: Thermal sensation prediction models can help to evaluate complex thermal environments and guide the environment conditioning strategy. However, most existing models are established basing on the thermal status of the entire human body or local body parts, failing to reflect thermal sensation generating mechanism or micro-scale (centimeter-scale) thermal sensation. This study put forward a new thermal sensation predicting approach by coupling the skin heat transfer and the thermoreceptor impulse signals. The micro-scale thermal sensitivity data under steady stimuli were applied to bridging the objective heat transfer model and the subjective sensation model. The model contains a one-dimensional skin heat transfer equation and three sensation-generating equations: the thermoreceptor impulse equation, the psychosensory intensity equation, and the thermal sensation equation. The dimension of the skin heat transfer equation was determined through a skin temperature diffusion experiment, and the coefficients of the static/dynamic impulse in the thermoreceptor impulse equation and the thermal sensation equation were obtained through polynomial fitting using thermal sensitivity data. The validated mean absolute percentage error (MAPE) was 0.08 and 0.1 under cooling and heating stimuli, respectively. This new model can predict thermal sensation on the centimeter scale and be applied under different boundary conditions. In the future, the new model can be further developed by testing dynamic stimuli and other boundary conditions so that it can be applied to more complex thermal exposures.

Keywords: thermal sensitivity; thermal sensation; skin heat transfer; thermoreceptor

Citation: Zhou, Y.; Yu, H.; Luo, M.; Zhou, X. Skin Heat Transfer and Thermal Sensation Coupling Model under Steady Stimulation. *Buildings* **2024**, *14*, 547. https://doi.org/10.3390/buildings14020547

Academic Editor: Constantinos A. Balaras

Received: 15 January 2024
Revised: 9 February 2024
Accepted: 14 February 2024
Published: 19 February 2024

1. Introduction

Nowadays, people tend to have high demands on thermally comfortable environments. To meet this requirement, researchers have constructed thermal sensation models to predict and evaluate an occupant's thermal comfort status, so that the optimal thermal environment combinations could be determined. In the classic prediction method, thermal sensation or thermal comfort is generally estimated based on the thermal status of the human body or local body parts. Fanger [1] firstly introduced six parameters, namely, air temperature, relative humidity, air velocity, mean radiant temperature, clothing insulation, and metabolic rate, to paraphrase the heat balance equation, that is the classic PMV (predicted mean vote) model. However, the PMV model and its derivatives are not accurate when applied in spatially ununiform and temporally unsteady thermal conditions due to complex physiological and psychological parameters. For the prediction for dynamic thermal sensation and comfort, based on the multi-node thermoregulation model by Stolwijk and Hardy [2], Fiala [3] incorporated the controlled passive system and controlling active system model of the human body to construct the comfort prediction model. Modified models, with a 65-node thermoregulation model and a radiation model by Tanabe [4], and the model constructed by UCB [5] made further improvements. However, these models were constructed mainly by exploring the potential influencing factors of thermal sensation (skin temperature, clothing insulation, etc.) and obtaining thermal sensation directly through simple regression. They failed to reflect the mechanism of how thermal sensation is generated. Thus, there can be great discrepancies between their predictive results and actual thermal sensation, especially under complex thermal exposures [6–8]. Additionally, these models were constructed on a whole-body or local-body-part scale, while the model at micro-scale, i.e., the centimeter scale, is lacking. The micro-scale thermal sensation data can be beneficial for the development of personal comfort system (PCS) devices which can provide cooling or heating directly to the small spots on the skin, and for guiding the controlling temperature of PCS devices that still remains unanswered [9,10].

When the skin is exposed to external thermal stimuli, the temperature signal transfers from the skin surface to a thermoreceptor beneath the surface, and the thermoreceptor releases an impulse through nerve fibers to the central nervous system, thus generating thermal sensation [11]. This is the mechanism of the generation of thermal sensation, which can be paraphrased as the process of 'external thermal stimuli–temperature change at thermoreceptor–thermoreceptor impulse–thermal sensation'. Based on this process, thermal sensation can be determined by thermoreceptor temperature and impulse, so the form of the model constructed under this mechanism will not change under different conditions. To date, few models have been constructed based on the full thermal sensation generation process, while there are some research studies on how humans respond to external stimuli based on the partial process. For the process 'external thermal stimuli–temperature change at thermoreceptor', researchers have constructed skin heat transfer models. Pennes [12] firstly proposed the BHTE (bioheat transfer equation) to calculate the temperature distribution within the skin layers. considering the skin as an integral consisting of three parts: cells, blood vessels (vascular region), and the extravascular region. To simplify the blood diffusion in BHTE, Vafai [13] overlooked the boundary between the vascular and extravascular region, and proposed the skin structure as a two-part porous medium structure. Based on the porous medium model, Khaled and Vafai [14] investigated the transport application in porous tissue, and Xuan and Roetzel [15] explored the heat transfer between the blood and peripheral tissue using a two-energy equation. Both the BHTE and porous medium model consider the skin as the mono-layer structure, while the real skin is composed of different layers with different thermophysical properties. To simulate heat transfer within the skin precisely, Mahjoob and Vafai [16,17] firstly explored the analytical solution for the tissue and blood temperature profiles and heat transfer correlations to develop the BHTE in a dual-layer skin structure and a multi-layer skin structure. Researchers further explored the multi-layer model in transient heat transfer [18], and extended the model to a two-dimensional [19] or three-dimensional model [20].

For the process 'temperature change at thermoreceptor–thermoreceptor impulse–thermal sensation', Hensel [21] discovered that a thermoreceptor converts the temperature signal into an impulse signal. In a steady-state condition, the thermoreceptor only releases a static impulse, while in a transient condition, the thermoreceptor releases a static and a dynamic impulse at the same time. The magnitude of the static impulse depends on the temperature at the thermoreceptor, and that of the dynamic impulse depends on the change in thermoreceptor temperature. Based on the impulse property of the thermoreceptor, Ring and de Dear [22] discovered that the integral of a cold or warm receptor impulse frequency was proportional to PSI (psychosensory intensity), and constructed the thermoreceptor impulse model for the prediction of PSI. Researchers further validated the model under sinusoidal stimuli [23], and compared it with other prediction models [24].

To quantify the skin temperature and thermal sensation at micro-scale, thermal sensitivity, defined as the thermal sensation change caused by a certain thermal stimulus [25], has been used to describe the perception of small regional parts to surrounding thermal environments. Donaldson [26] figured the distribution of cold and heat spots on the back of the left and right hand. Luo [27] mapped the high-density thermal sensitivity of the human body, and explored the variance among different body parts and gender. Our previous studies [28,29] also obtained the thermal sensitivity data of the hand, arm, and face, and discovered the effect of the initial thermal states, stimulus intensity, and stimulating time on its distribution.

This study further developed Hensel's 'temperature change at thermoreceptor–thermoreceptor impulse–thermal sensation' model by coupling skin heat transfer and the thermoreceptor impulse model. The micro-scale thermal sensitivity data obtained from our previous studies [28,29] were applied to bridging the objective heat transfer model and the subjective sensation model. By doing this, the new model can reflect the full process of the mechanism of how thermal sensation was generated, and predict micro-scale (centimeter-scale) thermal sensation.

The framework of the new model and its comparison with classic thermal sensation prediction models are shown in Figure 1. This paper will firstly construct the model and obtain the unknown coefficients in the model, and then it will discuss its accuracy, potential application, and limitations.

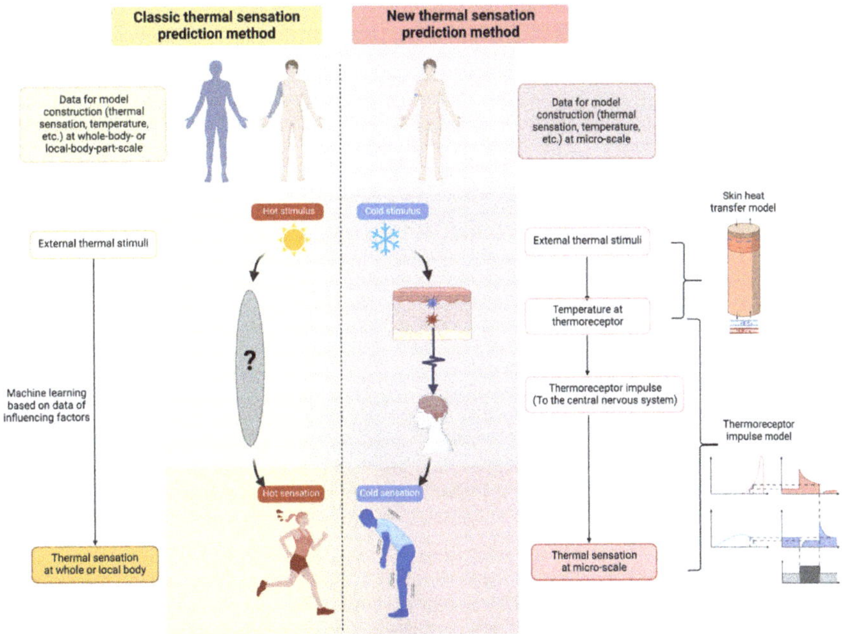

Figure 1. The classic thermal sensation prediction model (**left**) and the new construction of this study (**right**) (created with BioRender.com with permission, accessed on 15 February 2024). In the figure of 'Thermoreceptor impulse model', the left part represents static impulse with temperature, the right part represents dynamic impulse with time, and the lowest part represents ambient temperature with time. The red shadow represents impulse at warm receptor, and the blue one represents impulse at cold receptor. In the lowest figure, the darker color represents higher ambient temperature.

2. Methods

The flowchart of the construction of the model is shown in Figure 2. The coupling model includes a skin heat transfer model and a thermoreceptor impulse model. The skin heat transfer model is used to calculate the temperature at the thermoreceptor, and the thermoreceptor impulse model is used to calculate thermal sensation based on thermoreceptor temperature. The dataset for the construction of the model came from the thermal sensation and thermal sensitivity data obtained in the experiments of our previous studies [28,29].

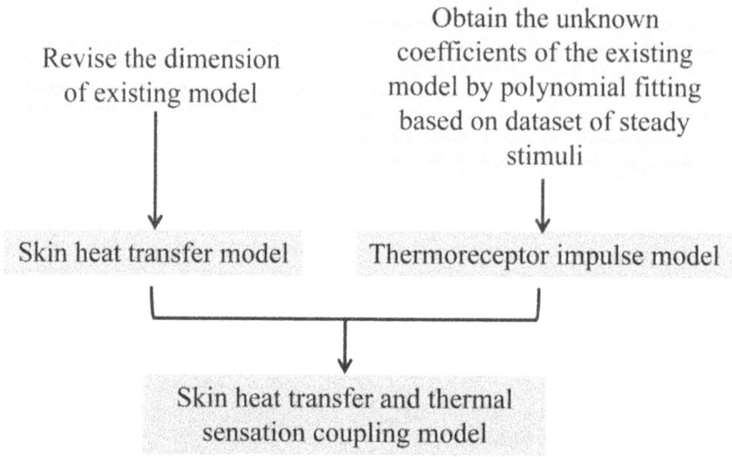

Figure 2. The flowchart of the construction of the model.

2.1. Construction of Skin Heat Transfer Model

Based on the heat balance within the skin, Pennes suggested the BHTE to reflect the heat transfer in each layer of the skin. Since the thermoreceptor is far away from the blood vessel, the equation could overlook the heat from blood flow and metabolism [30]. Thus, the original skin heat transfer equation is described as Equation (1).

$$\lambda(\frac{\partial^2 T_{sk}}{\partial x^2} + \frac{\partial^2 T_{sk}}{\partial y^2} + \frac{\partial^2 T_{sk}}{\partial z^2}) = c\rho(\frac{\partial T_{sk}}{\partial t}) \tag{1}$$

where λ is the thermal conductivity (W/m/K). c is the heat capacity (J/kg/K). ρ is density (g/cm^3). t is time (s). T_{sk} is the skin temperature at a certain layer (K). x, y, and z represent the length in a certain direction in three dimensions (m). The original equation considers the heat transfer in three dimensions, while under the short-period stimulation of 10 s [28], the heat transfer in some dimensions may not be so evident. To verify the dimension of the skin heat transfer equation, this paper carried out the skin temperature diffusion experiment to explore the extent of heat transfer after the stimulation in three dimensions.

The apparatus and test spots of the experiment are shown in Figure 3. A shows the stimulation equipment (accuracy of ±0.1 °C, temperature adjustment between 0 °C and 50 °C, NTE-2A, Physitemp Instruments Inc., Hawthorne, NJ, USA), with its temperature controller (left) and stimulating probe (right). The area of the probe was 1.54 cm^2 with a diameter of 1.4 cm. B shows the infrared camera (accuracy of ±3 °C, temperature measurement between 0 °C and 400 °C, FLIR ONE Pro, Teledyne FLIR, Wilsonville, OR, USA). C shows the distribution of test spots. The test spots were the dorsum of the hand, the forearm, the upper arm, and the left side of the face. They were the spots with high sensitivity in our previous study [28], and as the center of three body parts, respectively, they could represent the thermal response of each body part. During the experiment, after the subject adapted to the experimental condition, the researcher firstly set the probe temperature to the reference temperature of 32 °C, which could cause local sensation close to 0 [28], conducted neutral stimulation on one test spot for 1 min, and then photographed the temperature distribution around one test spot with the infrared camera. Secondly, the researcher set the probe temperature to −5 °C from the reference temperature to conduct cooling stimulation for 1 min, and then photographed the temperature distribution at the same spot. After the cooling stimulation of all test spots, the researcher conducted heating stimulation following the same process. The infrared photos before and after the stimulation and the skin temperature data were processed to obtain the diffusivity in three dimensions.

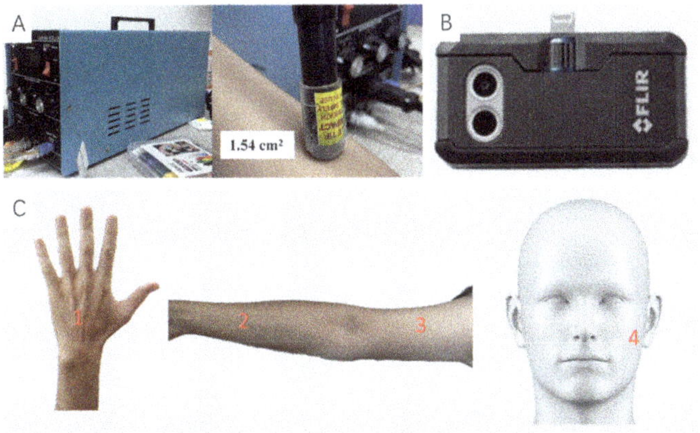

Figure 3. (**A**) Stimulation equipment. (**B**) Infrared camera. (**C**) Test spots. The '1-4' represents the number of each test spot.

The results are shown in Figures 4 and 5. The infrared photos in Figure 4 show the temperature distribution before and after the cooling or heating stimulation of four test spots. It can be concluded that after stimulation, the temperature at the stimulated area would change evidently, while the stimulation hardly diffused to the surrounding area. After cooling stimulation, the skin area with a temperature change would diffuse with an increase in diameter of 0.05 cm, that is, with a diffusivity of 1.039, and after heating stimulation, the area would diffuse with an increase in diameter of 0.04 cm, that is, with a diffusivity of 1.026. Compared with the transverse temperature diffusion on the skin surface, the longitudinal diffusion along with the skin depth could reach the core of the skin, that is, at least 0.54 cm from the skin surface, as shown in Figure 5. Since the transverse diffusivity was only 10% of the longitudinal one, the skin heat transfer model could be considered as a one-dimensional model in the longitudinal direction.

Based on the experimental conclusion, the skin heat transfer equation of this study could be described as Equation (2).

$$\lambda\left(\frac{\partial^2 T_{sk}(x,t)}{\partial x^2}\right) = c\rho\left(\frac{\partial T_{sk}(x,t)}{\partial t}\right) \tag{2}$$

where x represents the depth from the skin surface (m).

To ensure the boundary condition of the skin heat transfer equation, the skin structure model was constructed as shown in Figure 6. The idealized skin structure consists of three parts: epidermis, dermis, and hypodermis [31]. The thermophysical properties of each part are shown in Table 1 [32,33]. Therefore, the depth from the skin surface to the skin core is 5.4 mm. The cold thermoreceptors are mainly distributed at a depth of 0.2 mm and the warm thermoreceptors are at a depth of 0.5 mm [21].

Table 1. Thermophysical properties of epidermis, dermis, and hypodermis.

Skin Layer	Thickness (mm)	Heat Capacity (J/kg/K)	Thermal Conductivity (W/m/K)	Density (g/cm^3)
Epidermis	0.15	3590	0.24	1.2
Dermis	1.5	3300	0.45	1.2
Hypodermis	3.75	2675	0.19	1

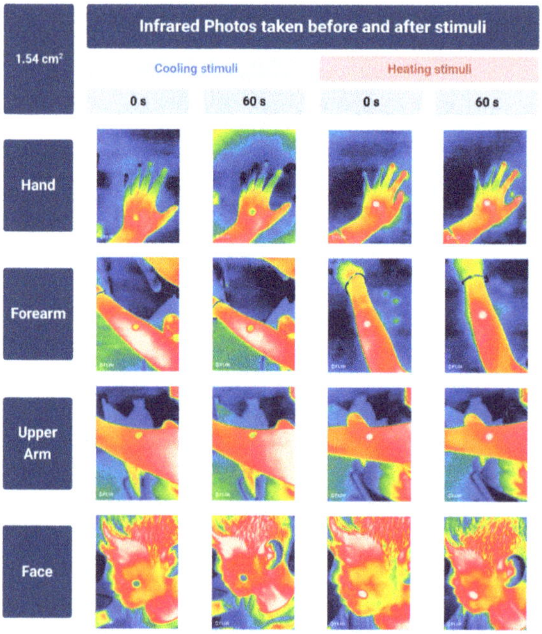

Figure 4. The infrared photos of four test spots before and after the cooling and heating stimuli (created with BioRender.com with permission).

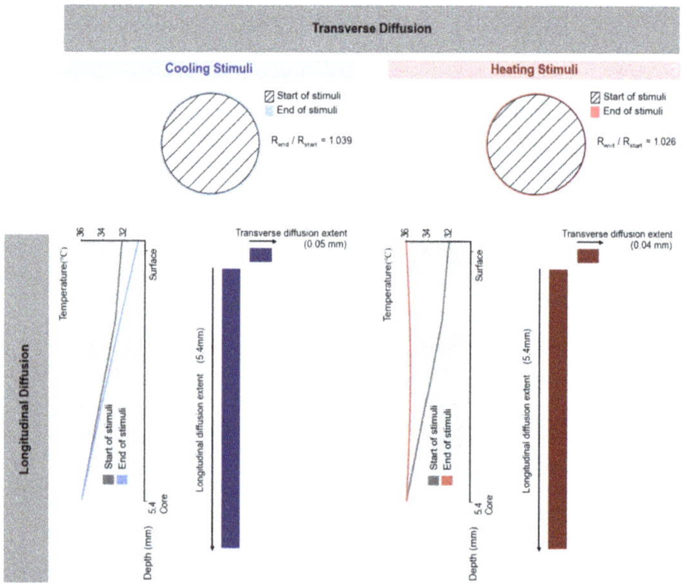

Figure 5. The temperature diffusion before and after the cooling and heating stimuli in the transverse (skin surface) and longitudinal (along skin depth) dimension. In the transverse part, the circle with a slanted shadow represents the skin area with a temperature change at the beginning of stimuli, the one with a blue shadow represents the area after cooling stimuli, and the one with a red shadow represents the area after heating stimuli. In the longitudinal part, the gray line is the temperature distribution before stimuli, and the blue or red line is the distribution after the stimuli.

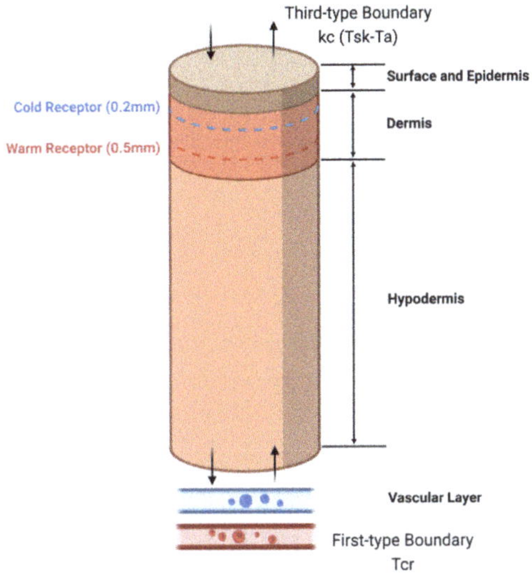

Figure 6. The skin structure model (created with BioRender.com with permission).

This model uses the third-type boundary condition at the skin surface. The boundary condition is described as Equation (3).

$$-\lambda(\frac{\partial T_{sk}}{\partial x}) = k_c(T_{sk} - T_a) \quad (3)$$

where Ta is the temperature of stimulation (K). k_c is the heat transfer coefficient between the stimulation probe and the skin surface (W/m^2/K), and it can be calculated using Equation (4) [33].

$$k_c = 1.25\lambda_c \frac{\Delta a_c}{\varepsilon_c} (\frac{P_c}{H_c})^{0.95} \quad (4)$$

where λ_c is the harmonic mean thermal conductivity (W/m/K). Δa_c is the average surface asperity slope of the contact interface (rad). ε_c is the average surface roughness of the contact interface (µm). The three parameters were calculated using Equations (5)–(7). The subscript 1 and 2 represent the parameters of the stimulation probe and skin surface, respectively. P_c is the contact pressure (kPa). H_c is the microhardness of the contact material (MPa). The data of the parameters of the stimulation probe and skin surface are listed in Table 2 [33].

$$\Lambda_c = \frac{2\lambda_1\lambda_2}{\lambda_1 + \lambda_2} \quad (5)$$

$$\Delta a_c = \sqrt{\Delta a_1{}^2 + \Delta a_2{}^2} \quad (6)$$

$$\varepsilon_c = \sqrt{\varepsilon_1{}^2 + \varepsilon_2{}^2} \quad (7)$$

Table 2. Thermophysical properties of the skin surface and stimulation probe.

Material	Thermal Conductivity (W/m/K)	Surface Roughness (µm)	Surface Asperity Slope (rad)	Contact Pressure (kPa)	Microhardness (Mpa)
Skin surface	0.24	21.69	0.3	0.8	0.1225
Stimulation probe (stainless steel 304)	14.986	0.008	0.009	--	--

At the skin core, the model uses the first-type boundary condition. The core temperatures of the hand, arm, and face are shown in Table 3 [34].

Table 3. Core temperatures of the hand, arm, and face.

Body Part	Core Temperature (°C)
Hand	35.05
Arm	35.87
Face	37.46

2.2. Construction of Thermoreceptor Impulse Model

Based on the thermoreceptor impulse theory by Hensel [21], Ring and de Dear [22] explored the mathematical relationship between the thermoreceptor impulse and thermal sensation. Under the external thermal stimulation, the thermoreceptor releases both a static and a dynamic impulse. Thus, the total impulse is described as Equation (8).

$$R(t) = K_s(T_{rec}(t) - T_{rec}(0)) + K_d\frac{\partial T_{rec}(t)}{\partial t} \tag{8}$$

where R is the total thermoreceptor impulse (HZ). T_{rec} is the temperature at the thermoreceptor (K). K_s is the static coefficient, and K_d is the dynamic coefficient. According to existing research [23], K_s is related to the initial thermoreceptor temperature before stimulation, and K_d is related to stimulus intensity and the change rate of stimulus intensity. Since the dataset for the steady model was obtained in the experiment with only one type of stimulus intensity (−5 °C under cooling stimulation and 5 °C under heating stimulation) and only one type of initial stimulation temperature (32 °C), K_s and K_d of the steady model were both constants. To reflect the difference of thermal sensory intensity at different body parts, the equation should multiply the thermal sensitivity as the areal summation factor [35]. The micro-scale thermal sensitivity in the dataset can also reflect the thermal response of micro-scale test spots. Thus, the total impulse equation of the steady model can be described as Equation (9).

$$R(t) = TS(C_1(T_{rec}(t) - T_{rec}(0)) + C_2\frac{\partial T_{rec}(t)}{\partial t}) \tag{9}$$

where TS is the thermal sensitivity of each test spot (1/K). C_1 and C_2 are constants representing K_s and K_d, respectively.

The PSI is calculated by integrating the total impulse for the first 20 s of stimulation, and thermal sensation is proportional to PSI [36]. The correlation can be described as Equations (10) and (11), where TSV is the thermal sensation vote, and C_3 is the constant coefficient. Thus, the thermoreceptor impulse model is the combination of Equations (9)–(11). To obtain the coefficients C_1, C_2, and C_3, the training set of the thermal sensation and thermal sensitivity data will be used to train the model.

$$PSI = \int_0^{20} R(t)dt \tag{10}$$

$$TSV = C_3 PSI \tag{11}$$

2.3. Dataset for the Model Construction

The data of thermal sensation and thermal sensitivity for the construction of the model under steady stimulation were obtained from the experiment on the hand, arm, and face [29]. A total of 50 subjects (25 male, 25 female) were invited to participate in the experiments under thermally neutral conditions. They were healthy college students, about 20 years old, living in Shanghai for more than 3 months before the experiment. To ensure that the body surface area of the male and female subjects could match to avoid its influence on the thermal sensation votes [37], the weights and heights of the subjects were filtered using Equation (12), where A is the body surface area (m^2), H is the height (m), and m is the weight (kg).

$$A = 0.202m^{0.425}H^{0.725} \tag{12}$$

During the experiment, 57 test spots of the hand, 25 test spots of the arm, and 25 test spots of the face were stimulated, firstly under a reference temperature of 32 °C, that is, the temperature producing a thermal sensation vote of 0, and then under a cooling stimulation of 27 °C (−5 °C from the reference) or a heating stimulation of 37 °C (+5 °C from the reference). In the meantime, the subjects recorded thermal sensation before and after the stimulation at the test spot, and the researcher recorded the change in skin temperature of the test spot. Thermal sensitivity was calculated as the change in thermal sensation vote divided by the change in skin temperature. The bar charts of the thermal sensation and thermal sensitivity of the hand, arm, and face are shown in Figure 7. A total of 80% of the data were used for the training set, and 20% were used for the test set.

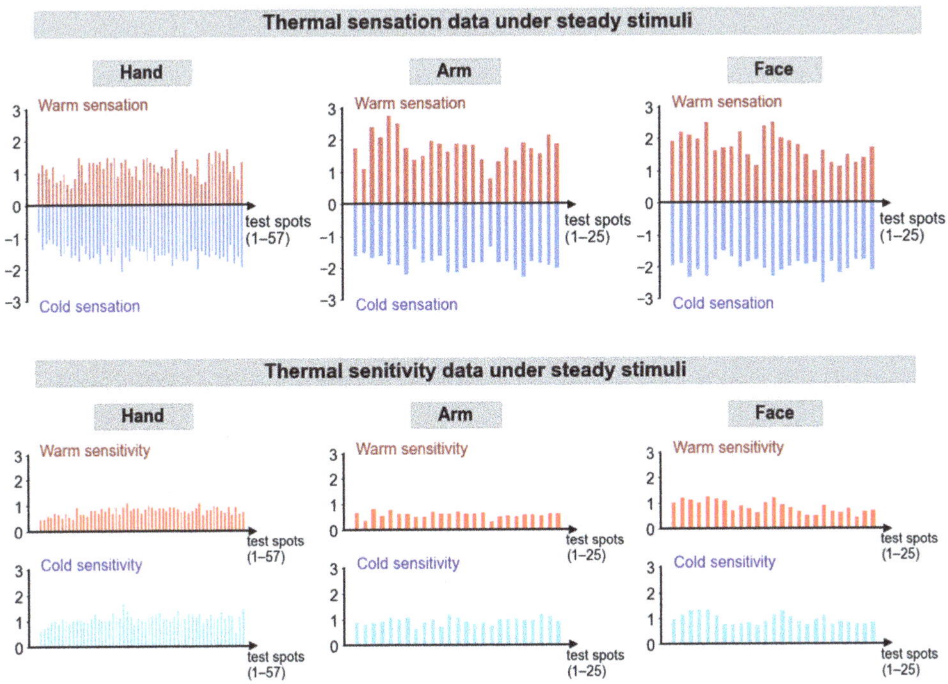

Figure 7. The data of the thermal sensation and thermal sensitivity of the hand, arm, and face. In the upper figures, the red represents warm sensation, and the blue represents cold sensation. In the lower figures, the red represents warm sensitivity, and the blue represents cold sensitivity.

2.4. Data Processing

The thermal sensation predicting process of this model is shown in the flowchart in Figure 8. This paper analyzed the skin temperature distribution under stimulation through the calculation of the skin heat transfer model, obtained the unknown coefficients of the thermoreceptor impulse equations to construct the complete skin heat transfer coupling model based on the training set, and validated the model based on the test set. MATLAB (R2020b, The MathWorks, Inc., Natick, MA, USA) and Python were used for the construction and validation of the model. In our programming, the initial skin temperature of each test spot was input to simulate the temperature distribution within the skin, and then the stimulating temperature was input to simulate how temperature changes in 20 s at each depth of the skin using the finite difference method.

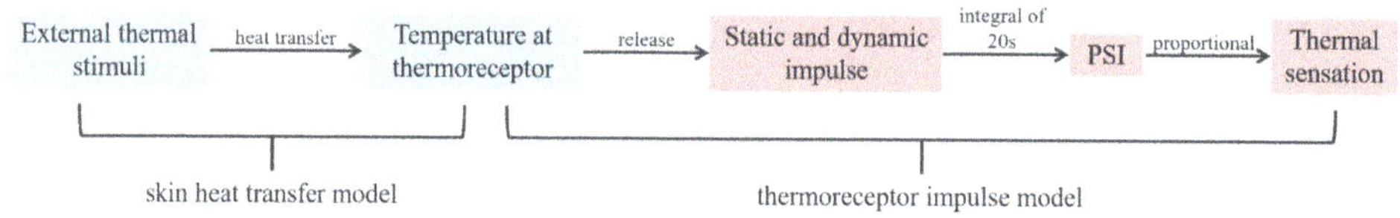

Figure 8. The predicting process of the model.

3. Results

3.1. Skin Temperature Distribution after Stimulation

The average skin temperature distribution after the steady cooling and heating stimulation is shown in Figure 9. The depth of 0 mm is the skin surface, and the depth of 5.4 mm is the skin core. The core temperature was the average of the hand, arm, and face at 36.1 °C. Since the skin heat transfer model was a one-dimensional model, the temperature remained the same at the same skin layer, and changed only with the depth. After cooling stimulation, the average skin surface temperature was 31.5 °C, and increased with the depth almost linearly to the core temperature. After heating stimulation, the average skin temperature at the surface was 35 °C, and firstly decreased to the lowest point at the depth of 1.9 mm, then increased to the core temperature. This may be because in a stimulation lasting only 20 s, the temperature within the skin was not sufficiently heated to the stable state as the linear distribution, and the heating rate of the skin layers near the heat source (stimulating probe at the skin surface and blood flow at the skin core) was higher than the ones far from the heat source.

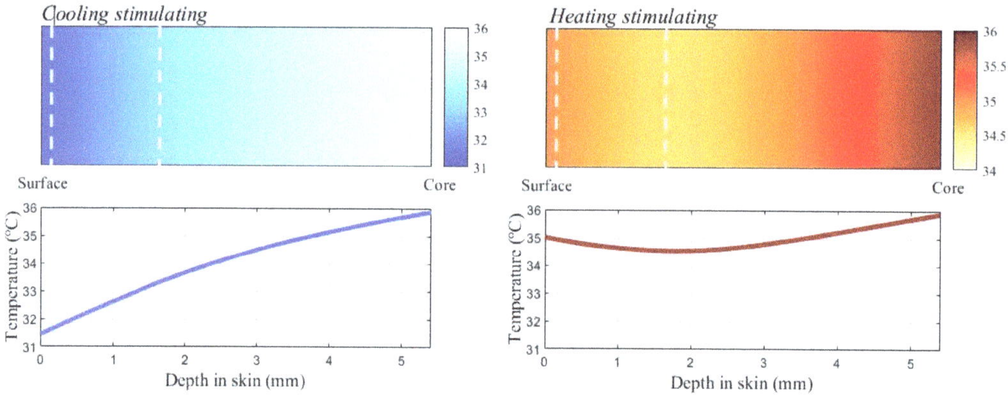

Figure 9. The skin temperature distribution with the depth of the skin.

3.2. Thermoreceptor Temperature Change with Time

The change in thermoreceptor temperature with time under cooling and heating stimulation is shown in Figure 10. The left part of the figure shows the temperature of the cold receptor, and the right part shows that of the warmth receptor. Before the stimulation, since the cold receptor was located at a shallower depth than the warmth receptor, the initial cold receptor temperature was 32 °C, slightly lower than the warmth receptor which was 32.2 °C. During stimulation, the cold receptor temperature decreased gradually, then remained stable after 15 s from the beginning of stimulation, while the warmth receptor temperature continued to increase throughout the stimulation. This may also be because of the insufficient heating during the short-time stimulation, lasting only 20 s.

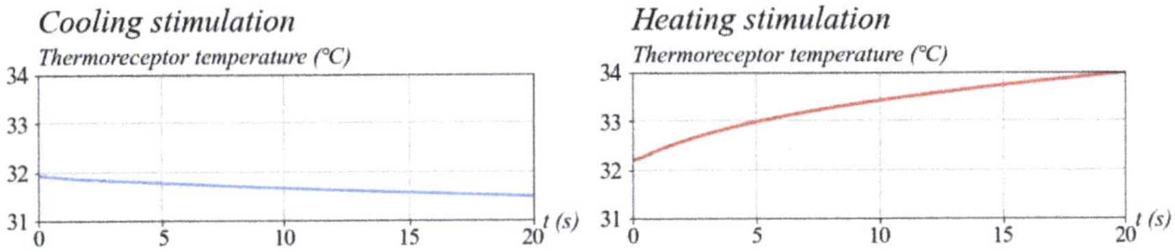

Figure 10. Temperature change with time of the cold and warmth receptor.

3.3. Fitting of the Model

To obtain the unknown coefficients C_1, C_2, and C_3 to complete the steady thermoreceptor impulse model, the training set of thermal sensation and thermal sensitivity data and the simulated thermoreceptor temperature data from the skin heat transfer model were used for nonlinear polynomial fitting. The fitting results are shown in Figure 11. The fitting curves were represented by the cold and warm sensation calculated by the fitted skin heat transfer coupling model based on the training set. For cooling stimulation, the coefficients C_1, C_2, and C_3 were −3, −12.6, and 0.0499, respectively, and for heating stimulation, they were 2.4, 6.7, and 0.0466, respectively. The determination coefficient (R_2) of the model under cooling stimulation was 0.85, and under heating stimulation it was 0.88, indicating a good fitting result.

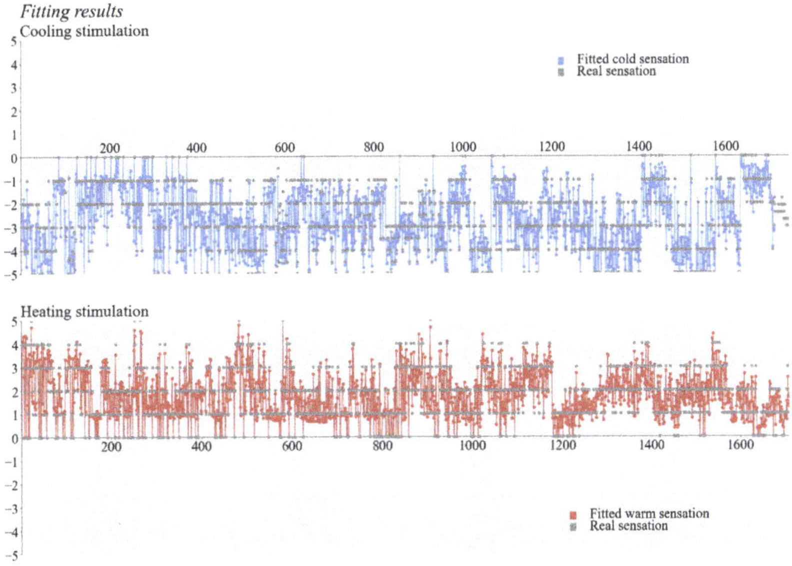

Figure 11. Model fitting results.

3.4. Validation of the Model

To explore the accuracy of the skin heat transfer model, the test set of the data was used to validate the steady and dynamic model. The validation results are shown in Figure 12. Each type of stimulation took 20% of the experimental data as the test set. The test thermal sensitivity data were applied to calculate the predictive thermal sensation, and the test thermal sensation data were applied for the comparison between the predictive and real thermal sensation. As shown in the figure, the MAPE (mean absolute percentage error) was 0.03 under cooling stimulation, and 0.05 under heating stimulation. Thus, the model could be considered to be accurate in predicting thermal sensation under steady and dynamic stimulation.

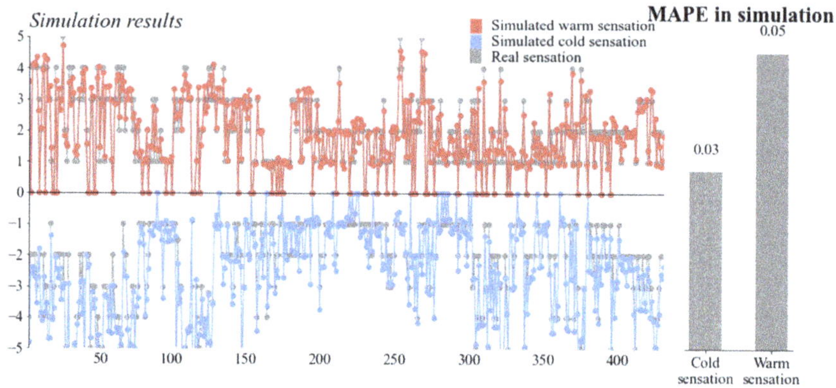

Figure 12. Model validation results.

4. Discussion

4.1. Potential Application

The skin heat transfer coupling model constructed in this study can be applied to the micro-scale thermal sensation prediction under different complex exposures. The existing thermal sensation prediction models are mostly constructed using regression under only one type of exposure, e.g., a room with air conditioning. Therefore, when there are various exposures, e.g., an air-conditioned room with a fan, the influencing factors for thermal sensation change and the regression functions are unavailable. However, the model in this study can be applied in different thermal conditions, since its input only contains initial skin temperature and external stimulating temperature. Additionally, it can be applied to the development of small-scale personal comfort system (PCS) devices. Such devices can be worn by the user to regulate skin temperature in the small areas where they are worn, like wrist-bands and neck-worn fans for smaller scales [38,39], to improve thermal comfort. By using this model, the user can better regulate the stimulating temperature of the device to enhance micro-scale satisfaction.

4.2. Limitations

This paper constructed a skin heat transfer model and a thermoreceptor impulse model, and validated their performance in thermal sensation prediction under steady stimuli. However, whether the skin temperature simulated by the skin heat transfer model and the impulse simulated by the thermoreceptor impulse model were accurate is unknown. To validate either of the two models, the data of the real temperature within the skin and the real impulse released by the thermoreceptor were required, but the current apparatus for these measurements was unavailable. Additionally, this study only considered steady and contact stimuli without considering dynamic changing-temperature stimuli or non-contact stimuli such as the radiative and convective stimuli. Since the model was constructed using the third-type boundary condition at the skin surface, it has the potential to be developed in a transient and non-contact environment, such as draught, radiation, or showering comfort studies.

5. Conclusions

This study constructed the skin heat transfer and thermal sensation coupling model under steady stimulation by reconstructing the skin heat transfer model and the thermoreceptor impulse model, and validated the model using the test set of experimental data. The model used the fixed temperature boundary condition at the skin surface, and thus the input for the model only requires external stimulation. This study obtained conclusions as follows.

First, in the stimulation time of 1 min, skin temperature diffused with an increase in diameter of 0.05 cm under cooling stimulation and 0.04 under heating stimulation at the skin surface, far less than the diffusing depth of 0.54 cm in the longitudinal direction.

Second, the temperature distribution of the skin was stable at the end of stimulation under cooling stimulation, while it still changed under heating stimulation. The temperature of the cold receptor remained stable after 15 s from the beginning of stimulation, while the warmth receptor temperature continued to increase.

Third, the R^2 of the model indicates a good fitting result, and the MAPE of the model indicates high accuracy in the prediction.

Author Contributions: Methodology, Y.Z. and M.L.; Software, Y.Z.; Resources, Y.Z. and M.L.; Data curation, Y.Z.; Writing—original draft, Y.Z.; Writing—review and editing, M.L., H.Y. and X.Z.; Funding acquisition, M.L. and H.Y. All authors have read and agreed to the published version of the manuscript.

Funding: The research was supported by the National Natural Science Foundation of China (No. 52078355, No. 52108086, No. 51578386), the China National Key R&D Program during the 14th Five-year Plan Period (No. 2022YFC3801500), the National Engineering Research Center of New Energy Vehicles and Power Systems, and the Shanghai Key Lab of Vehicle Aerodynamics and Vehicle Thermal Management Systems.

Data Availability Statement: The data presented in this study are available upon request from the corresponding author. The original data are not publicly available.

Conflicts of Interest: The authors declare no conflicts of interest.

References

1. Fanger, P.O. *Thermal Comfort, Analysis and Applications in Environmental Engineering*; Danish Technical Press: Copenhagen, Denmark, 1970.
2. Stolwijk, J.A.J.; Hardy, J.D. Temperature regulation in man—A theoretical study. *Pflüger's Arch. Gesamte Physiol. Menschen Tiere* **1966**, *291*, 129–162. [CrossRef]
3. Fiala, D. *Dynamic Simulation of Human Heat Transfer and Thermal Comfort*; De Montfort University: Leicester, UK, 1998.
4. Tanabe, S.-i.; Kobayashi, K.; Nakano, J.; Ozeki, Y.; Konishi, M. Evaluation of thermal comfort using combined multi-node thermoregulation (65MN) and radiation models and computational fluid dynamics (CFD). *Energy Build.* **2002**, *34*, 637–646. [CrossRef]
5. Huizenga, C.; Hui, Z.; Arens, E. A model of human physiology and comfort for assessing complex thermal environments. *Build. Environ.* **2001**, *36*, 691–699. [CrossRef]
6. Du, H.; Lian, Z.; Lai, D.; Duanmu, L.; Zhai, Y.; Cao, B.; Zhang, Y.; Zhou, X.; Wang, Z.; Zhang, X.; et al. Evaluation of the accuracy of PMV and its several revised models using the Chinese thermal comfort Database. *Energy Build.* **2022**, *271*, 112334. [CrossRef]
7. Xie, Y.; Huang, T.; Li, J.; Liu, J.; Niu, J.; Mak, C.M.; Lin, Z. Evaluation of a multi-nodal thermal regulation model for assessment of outdoor thermal comfort: Sensitivity to wind speed and solar radiation. *Build. Environ.* **2018**, *132*, 45–56. [CrossRef]
8. Zhou, X.; Liu, S.; Liu, X.; Lin, X.; Qing, K.; Zhang, W.; Li, J.; Dong, J.; Lai, D.; Chen, Q. Evaluation of Four Models for Predicting Thermal Sensation in Chinese Residential Kitchen. In *E3S Web Conference*; EDP Sciences: Les Ulis, France, 2019; Volume 111.
9. Luo, M.; Arens, E.; Zhang, H.; Ghahramani, A.; Wang, Z. Thermal comfort evaluated for combinations of energy-efficient personal heating and cooling devices. *Build Environ.* **2018**, *143*, 206–216. [CrossRef]
10. Wang, Z.; Warren, K.; Luo, M.; He, X.; Zhang, H.; Arens, E.; Chen, W.; He, Y.; Hu, Y.; Jin, L.; et al. Evaluating the comfort of thermally dynamic wearable devices. *Build Environ.* **2020**, *167*, 106443. [CrossRef]
11. Djongyang, N.; Tchinda, R.; Njomo, D. Thermal comfort: A review paper. *Renew. Sustain. Energy Rev.* **2010**, *14*, 2626–2640. [CrossRef]
12. Pennes, H.H. Analysis of Tissue and Arterial Blood Temperatures in the Resting Human Forearm. *J. Appl. Physiol.* **1948**, *1*, 93–122. [CrossRef] [PubMed]
13. Vafai, K. *Porous Media: Applications in Biological Systems and Biotechnology*; Taylor & Francis Group: Milton, UK, 2010.
14. Khaled, A.R.A.; Vafai, K. The role of porous media in modeling flow and heat transfer in biological tissues. *Int. J. Heat Mass Transf.* **2003**, *46*, 4989–5003. [CrossRef]
15. Xuan, Y.; Roetzel, W. Bioheat equation of the human thermal system. *Chem. Eng. Technol.* **1997**, *20*, 268–276. [CrossRef]
16. Mahjoob, S.; Vafai, K. Analytical characterization of heat transport through biological media incorporating hyperthermia treatment. *Int. J. Heat Mass Transf.* **2009**, *52*, 1608–1618. [CrossRef]
17. Mahjoob, S.; Vafai, K. Analysis of Bioheat Transport Through a Dual Layer Biological Media. *J. Heat Transf.* **2009**, *132*, 031101. [CrossRef]

18. Becker, S.M. One-Dimensional Transient Heat Conduction in Composite Living Perfuse Tissue. *J. Heat Transf.* **2013**, *135*, 071002. [CrossRef]
19. Sarkar, D.; Haji-Sheikh, A.; Jain, A. Temperature distribution in multi-layer skin tissue in presence of a tumor. *Int. J. Heat Mass Transf.* **2015**, *91*, 602–610. [CrossRef]
20. Ma, J.; Yang, X.; Sun, Y.; Yang, J. Thermal damage in three-dimensional vivo bio-tissues induced by moving heat sources in laser therapy. *Sci. Rep.* **2019**, *9*, 10987. [CrossRef]
21. Hensel, H. Thermoreception and temperature regulation. *Monogr. Physiol. Soc.* **1981**, *38*, 1–321.
22. Ring, J.W.; de Dear, R. Temperature Transients: A Model for Heat Diffusion through the Skin, Thermoreceptor Response and Thermal Sensation. *Indoor Air* **1991**, *1*, 448–456. [CrossRef]
23. Ring, J.W.; de Dear, R.; Melikov, A. Human thermal sensation: Frequency response to sinusoidal stimuli at the surface of the skin. *Energy Build.* **1993**, *20*, 159–165. [CrossRef]
24. Parkinson, T.; Zhang, H.; Arens, E.; He, Y.; de Dear, R.; Elson, J.; Parkinson, A.; Maranville, C.; Wang, A. Predicting thermal pleasure experienced in dynamic environments from simulated cutaneous thermoreceptor activity. *Indoor Air* **2021**, *31*, 2266–2280. [CrossRef]
25. McIntyre, D.A. *Indoor Climate*; Applied Science Publishers Ltd.: London, UK, 1980; Volume 443.
26. Donaldson, H.H. On the temperature-sense. *Mind* **1885**, *10*, 398–416. [CrossRef]
27. Luo, M.; Wang, Z.; Zhang, H.; Arens, E.; Filingeri, D.; Jin, L.; Ghahramani, A.; Chen, W.; He, Y.; Si, B. High-density thermal sensitivity maps of the human body. *Build Environ.* **2020**, *167*, 106435. [CrossRef]
28. Zhou, Y.; Yu, H.; Xu, S.; Luo, M.; Zhou, X. High-density thermal sensitivity of the hand under different thermal states and stimulus intensities. *Indoor Air* **2022**, *32*, e13089. [CrossRef]
29. Zhou, Y.; Luo, M.; Yu, H.; Zhou, X. Evaluating the dynamic thermal sensitivities of human body with high-dense contacting thermal stimuli. *Indoor Built Environ.* **2023**, 1420326X231201847. [CrossRef]
30. Dufour, A.; Després, O.; Pebayle, T.; Lithfous, S. Thermal sensitivity in humans at the depth of thermal receptor endings beneath the skin: Validation of a heat transfer model of the skin using high-temporal resolution stimuli. *Eur. J. Appl. Physiol.* **2020**, *120*, 1509–1518. [CrossRef] [PubMed]
31. Wang, S.; Yu, R.-X.; Fan, W.; Li, C.-X.; Fei, W.-M.; Li, S.; Zhou, J.; Hu, R.; Liu, M.; Xu, F.; et al. Detection of skin thickness and density in healthy Chinese people by using high-frequency ultrasound. *Ski. Res. Technol.* **2023**, *29*, e13219. [CrossRef] [PubMed]
32. Gowrishankar, T.R.; Stewart, D.A.; Martin, G.T.; Weaver, J.C. Transport lattice models of heat transport in skin with spatially heterogeneous, temperature-dependent perfusion. *BioMedical Eng. OnLine* **2004**, 3, 42. [CrossRef] [PubMed]
33. Zhang, A.; Pang, D.; Wang, B.; Wang, J. Dynamic responses of wearable thermoelectric generators used for skin waste heat harvesting. *Energy* **2023**, *262*, 125621. [CrossRef]
34. Takahashi, Y.; Nomoto, A.; Yoda, S.; Hisayama, R.; Ogata, M.; Ozeki, Y.; Tanabe, S.-i. Thermoregulation model JOS-3 with new open source code. *Energy Build.* **2021**, *231*, 110575. [CrossRef]
35. de Dear, R.J.; Ring, J.W.; Fanger, P.O. Thermal Sensations Resulting From Sudden Ambient Temperature Changes. *Indoor Air* **1993**, *3*, 181–192. [CrossRef]
36. Molinari, H.H.; Greenspan, J.D.; Krenshalo, D.R. The effects of rate of temperature change and adapting temperature on thermal sensitivity. *Sens. Process.* **1977**, *1*, 354–362.
37. *ASHRAE 55:2020*; Thermal Environmental Conditions for Human Occupancy. American Society of Heating Refrigerating and Air-Conditioning Engineers (ASHRAE): Atlanta, GA, USA, 2020.
38. Reon Pocket. Available online: https://popbee.com/tag/reon-pocket/ (accessed on 10 January 2024).
39. Embr Wave. Available online: https://embrlabs.com/embr-wave/ (accessed on 10 January 2024).

buildings

MDPI

Article

A Field Investigation to Quantify the Correlation between Local and Overall Thermal Comfort in Cool Environments

Xiaohong Liang [1], Yingdong He [2,*], Nianping Li [2,*], Yicheng Yin [1] and Jinhua Hu [3]

1 School of Architecture, Hunan University, Changsha 410082, China; liangxiaohong@hnu.edu.cn (X.L.), yinyicheng@hnu.edu.cn (Y.Y.)
2 College of Civil Engineering, Hunan University, Changsha 410082, China
3 College of Civil Engineering, Hunan University of Science and Technology, Xiangtan 411201, China; hujinhua@hnust.edu.cn
* Correspondence: heyingdong2022@hnu.edu.cn (Y.H.); linianping@hnu.edu.cn (N.L.)

Abstract: The thermal comfort of local body parts is the essential factor that affects people's health and comfort as well as a buildings' energy. This study aims to (1) investigate the characteristics of the local thermal comfort of different body parts of occupants in real buildings in winter, (2) quantify the correlation between the amount of local body parts with coolness or discomfort and the overall subjective thermal responses, and (3) validate an easy-to-use local–overall thermal comfort model. A field investigation in the office and study rooms of a university was conducted in winter. The results indicate that the top five percentages of local coolness appeared in the feet (41.02%), the hands (26.58%), the calves (25.18%), the thighs (13.99%), and the head (9.72%) and that the top five percentages of local discomfort appeared in the feet (44.99%), the palms (28.2%), the calves (24.74%), the head (19.66%), and the thighs (16.35%). Moreover, when the whole body felt cool, at least four local body parts had cool sensations; when the whole body felt thermally uncomfortable, at least three local body parts had cool sensations; and when the whole body felt that the ambient environment was thermally unacceptable, at least seven local body parts had cool sensations. Meanwhile, the correlation between local discomfort and whole-body responses was different: when the whole body felt thermal uncomfortable, at least three local body parts had discomfort; and when the whole body felt that the ambient environment was thermally unacceptable, at least four local body parts had discomfort. Further, the local–overall thermal comfort model proposed by the authors exerted high accuracy in predicting overall thermal comfort.

Keywords: local thermal sensation; local thermal comfort; thermal environment

Citation: Liang, X.; He, Y.; Li, N.; Yin, Y.; Hu, J. A Field Investigation to Quantify the Correlation between Local and Overall Thermal Comfort in Cool Environments. *Buildings* **2024**, *14*, 1171. https://doi.org/10.3390/buildings14041171

Academic Editors: Kian Jon Chua and Vincenzo Costanzo

Received: 25 February 2024
Revised: 4 April 2024
Accepted: 19 April 2024
Published: 21 April 2024

1. Introduction

As defined in the ASHRAE Standard 55 [1], thermal comfort is a subjective state of awareness in which the human body evaluates the ambient thermal environment; it is affected by the heat transfer of convection, radiation, and evaporation between the human body and the ambient environment, which are mainly decided by indoor air temperature, radiant temperature, air humidity, air velocity, clothing insulation, and human metabolic rate. The detailed heat transfer between the human body and the ambient environment can be calculated by the predicted mean vote (PMV) model, which is also incorporated in the ASHRAE Standard 55 [1]. Thermal comfort is the ultimate criterion for evaluating the quality of buildings' thermal environments since it significantly affects the health and work performance of the occupants of buildings [2]. To improve the thermal comfort demands of occupants and work performance in buildings, heating, ventilation, and air conditioning (HVAC) systems have been widely adopted in modern buildings and, at the same time, have resulted in massive energy consumption.

In China, the total energy consumption for heating, ventilation, and air conditioning systems accounts for up to half of the total building energy consumption [3]. One of the primary reasons for this is that a large amount of energy is consumed to create uniform thermal environments, which, however, are unable to accommodate people's comfort demands, as the local thermal states of body parts can result in individual differences [4–6]. For example, females usually feel cool discomfort in their hands and feet in the winter even when the indoor temperature is moderate and comfortable for men [7]. Numerous studies have found that local thermal comfort has a significant impact on the overall thermal comfort [8–10]. Local heating that warms one or several body parts allows a person to maintain comfort when the ambient environment is cool or cold. For example, in cold environments, warming the lower body parts (feet and legs) can make occupants feel that the environment is thermally acceptable, even if it has been lowered to 14 °C indoors [11], which helps to reduce the energy consumption needed for heating building spaces to a relatively high temperature like 20 °C. To energy efficiently improve local thermal comfort, personal comfort systems (PCSs) have been widely investigated, including seats [12,13], foot warmers [14], wrist warmers [15], radiant panels [16], garments [17], local ventilation devices [18], etc.

In recent years, to understand well how local body parts affect overall thermal responses, many researchers have conducted human experiments in climate chambers. In 2003, Zhang et al. [19] found that the local thermal sensations of different body parts had various influences on the overall thermal sensations. They proposed the UCB (University of California, Berkeley) model for predicting overall thermal sensations and further improved the model in 2009 [8]. Fang et al. [20] proposed that the local thermal sensations of body parts are sparsely distributed differently. Khiavi et al. [21] combined local thermal responses with the UCB model to form a new integrated assessment method to predict overall thermal sensations. Jian et al. [22] set up three experimental scenarios to explore how the coldest local thermal sensation affects the overall thermal sensation, and they found that the feet had the coldest sensations, with the most obvious skin temperature drops. Liu et al. [23] heated 15 body parts and found that heating the waist was better than heating the feet for overall thermal comfort. Sun et al. [24] studied wrist skin temperature during sleeping through several human experiments and found that wrist skin temperature was sufficient for predicting human thermal status. Li et al. [25] proposed that the head, the chest, the back, and the hands were easy to be warm, which led to a warm overall thermal sensation, and the arms, the hands, the legs, and the feet were easily cool, which led to a cool overall feeling during cold exposure. Du et al. [26] suggested that the occupants in their study had a preference for heating their feet and lower-body areas, with a significant increase in the thermal sensations in the feet and the whole body when local heating is provided. Cheng et al. [27] established non-contact measurements of skin temperatures and deep learning to reflect the human thermal comfort state, which can be used to control HVAC systems. Jia et al. [28] claimed that overall thermal sensation is significantly correlated with facial skin temperature.

In addition to laboratory experiments, some scholars have also explored the correlation between local and overall thermal comfort in real environments. Shahzad et al. [29] found that people's local thermal comfort in residential buildings may be improved by heated chairs, which, in turn, enhance people's overall thermal comfort. Kim et al. [30] suggested that 99% of occupants' thermal satisfaction can be obtained with PCSs that locally cool or warm human bodies. Chen et al. [31], after a questionnaire survey conducted in six areas of Harbin (with a cold climate), pointed out that the actual overall thermal sensation is decided by the thermal status of the coldest body part.

The authors have also carried out several field studies on local and overall thermal comfort in the past few years: He et al. [32], based on a field study in an office in Zhuhai, suggested that the head, the hands, the legs, and the feet affected the overall thermal sensation more than other body segments. He et al. [33] also conducted an on-site investigation in air-conditioned university dormitories. They found that, in the summer, the overall thermal comfort was determined by the two most extreme body parts, which were usually the head and the chest. Hu et al. [7] carried out a winter field survey at a university to analyze the gender differences in thermal comfort, work performance, and sick building syndromes in real-world environments, and they found that the fingers easily had very low skin temperatures in the winter. He et al. [34] proposed a new method based on infrared thermography to predict the thermal state of an occupant based on the local skin temperatures of the hands, the cheeks, and the nose.

Although the above-stated literature review indicates the importance of local thermal comfort, some critical scientific gaps remain unfilled. First, most of the above-mentioned studies are based on laboratory experiments, and the obtained results may deviate from those of realistic environments; the existing studies seldom analyze the percentages of local discomfort of different body parts in actual buildings. Second, the existing studies seldom indicate how the amount of uncomfortable local body parts affects the overall thermal comfort. For instance, when a person feels discomfort, how many local body parts are experiencing discomfort? Third, most of the existing local–overall thermal comfort models have not been validated in real buildings. An easy-to-use model with full validation in real buildings can help further the investigation of local thermal comfort in the future.

The primary purposes of this study are the following: (1) explore the local thermal comfort of occupants through an investigation in real buildings in winter, as well as determine the percentages of different local parts with cool sensations and discomfort under different indoor temperatures; (2) quantify the effects of the amount of local cool sensations and discomfort on one's overall feelings, including overall thermal sensation, comfort, and acceptability; and (3) validate an easy-to-use local–overall comfort model proposed by the authors using data collected in real office buildings.

2. Materials and Methods

2.1. Geography and Climate

The investigation was carried out between November 2019 and January 2020 (the winter period in the Hunan province includes November, December, January, and February), and all the surveys were conducted on weekdays from 8:00 a.m. to 6:00 p.m. when there were many occupants indoors. Each participant filled out the questionnaire only once, which could usually be completed within ten minutes.

The field investigation was performed at Hunan University in Changsha, which is located in central China (latitude 27°51′–28°40′ N and longitude 11°53′–114°15′ E), as shown in Figure 1, with abundant precipitation. Changsha has a warm climate in the summer (the average temperature in July is about 29 °C) and a cold climate in the winter (the average temperature in January is about 5 °C) [35]. Meanwhile, the outdoor relative humidity is often higher than 70% throughout the year.

Figure 1. The location of Changsha [36].

2.2. Investigated Buildings

The investigation was performed in the office and study rooms of three buildings of Hunan University (as shown in Figure 2), including the main hall of the College of Civil Engineering (Building A), the Research Center for Green Building Energy Efficiency and Green Buildings (Building B), and the Library of Hunan University (Building C). All three buildings had reinforced concrete structures, with a total of six floors in Building A, three floors in Building B, and nine floors in Building C. Each room had several operable single-glazed windows and split-type air conditioners, and each air conditioner had a plug-in controller panel for the occupants to freely change the indoor set-point temperature. Each window was closed, and each air conditioner was running during the investigation. In general, the air conditioning was turned on at around 8 a.m. on weekdays, when the earliest occupants entered the office, and it was turned off when the last occupant left the office, depending on when the occupants finished their work.

(a) (b) (c)

Figure 2. The investigated buildings and rooms: (**a**) Building A, (**b**) Building B, and (**c**) Building C.

2.3. Measurement

This study measured the same environmental and physiological parameters with the same types of instruments as those in the previous work of the researchers [6]. The main topic, scenarios, locations, and findings of this study are different from those of our previous work, which focused on the gender differences in thermal comfort, work performance, and sick building syndrome in naturally ventilated and air-conditioned classrooms [7].

The indoor temperature and humidity recorder were placed 0.6 m above the floor. An oximeter was used to measure oxygen saturation (SpO_2) and heart rate and was attached to a participant's left middle finger for each measurement. The instruments recorded the environmental parameters after having been placed in a room for 30 min and displayed a stable value. Portable thermocouples were adopted to record the skin temperatures of the participants' foreheads, cheeks, necks, backs of the hands, and fingertips. Since the participants generally moved and used their right hand frequently, the measurement was conducted on their left body side to reduce disturbance on the participants. Detailed information about the instruments used in this study is presented in Table 1. All the instruments were calibrated by the manufacturer before the investigation, and the measurement was conducted when the questionnaire was distributed to the participants.

Table 1. Experimental instruments used in this study.

Type	Parameter	Instrument	Models	Range	Accuracy
Environmental	Air temperature	Temp. and humidity recorder	TR-72Ui	−10~60°C	±0.3 °C
	Relative humidity	Temp. and humidity recorder	TR-72Ui	10–90%	±5%
Physiological	Skin temperature	Thermocouple	905 12	−50~350 °C	0.1 °C
	Heart rate/SpO_2	Oximeter	YX303	70–100%	±2%

2.4. Survey

Before conducting the survey, each student who participated gave their oral consent. Second, the questionnaire was administered to subjects who had been seated in the office for more than 30 min. Third, the questionnaire was explained orally to each participant by the researchers. Figure 3 shows the main questions of the used questionnaire. The participants selected the most proper thermal sensation and comfort levels from an integer seven-point scale (from "cold" to "hot" and from "very uncomfortable" to "very comfortable", with the corresponding values ranging from −3 to 3). In addition, the participants were asked to choose whether the indoor thermal environment was "acceptable" or "unacceptable" (the corresponding values were 1 and 0, respectively). The specific scales are shown in Table 2.

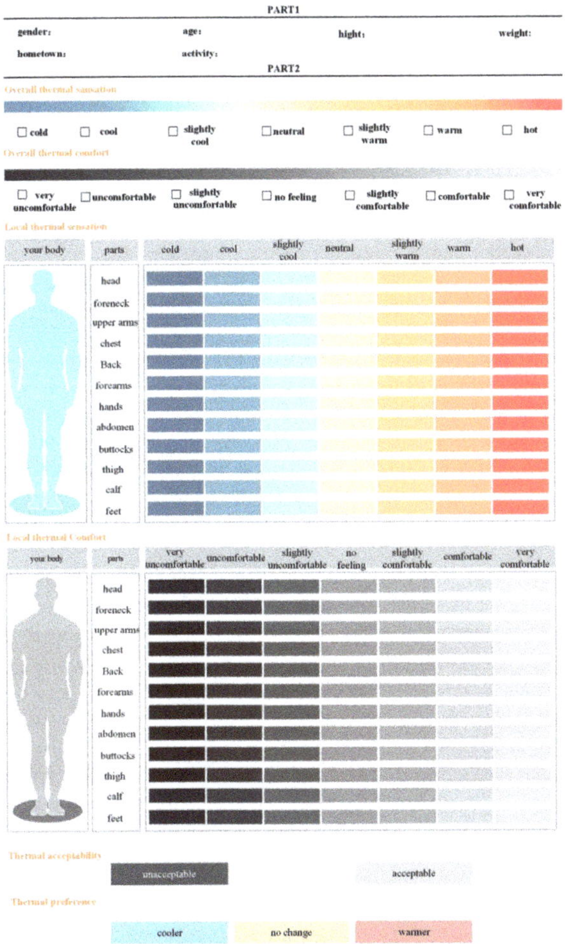

Figure 3. The main questions of the questionnaire used in this study.

Table 2. Scales of subjective responses.

Voting Scale	Thermal Sensation	Thermal Comfort	Thermal Acceptability
3	Hot	Very comfortable	
2	Warm	Comfortable	
1	Slightly warm	Slightly comfortable	Acceptable
0	Neutral	Neutral	Unacceptable
−1	Slightly cool	Slightly uncomfortable	
−2	Cool	Uncomfortable	
−3	Cold	Very uncomfortable	

2.5. Participants

All the participants were graduate or undergraduate students at Hunan University, and they were working or studying in the investigated rooms from 8:00 a.m. to 8.00 p.m. during the investigation. All the students were in good health, with no disabilities or illnesses, and they did not have alcohol or smoke before or during the survey. In addition, there were no restrictions on students' clothes in the university, and the students wore thick clothes in different rooms due to the cold weather during the investigation. To be specific, each of the students mainly wore a thick jacket, a sweater, and long trousers.

2.6. Data Filtration and Analysis

In total, 1479 questionnaires were assigned, and 1356 valid questionnaires were obtained (656 men (48.4%) and 700 women (51.6%)) after eliminating the questionnaires with incomplete key data.

First, the overall percentages of cool sensations (including "slightly cool", "cool", "cold", and "very cold"), discomfort (including "slightly uncomfortable", "uncomfortable", and "very uncomfortable"), and thermal unacceptability were calculated. Then, the bin method was adopted for data analysis. Specifically, the data under different air temperatures were assigned to different integer temperature bins with an interval of 1 °C, and the integer temperature value was used as the standard temperature for each bin, based on which the key data for our analysis such as the thermal discomfort percentages were calculated at different integer temperature values. For example, the thermal comfort values under indoor air temperatures of 18.7 °C, 18.9 °C, and 19.1 °C were within the "18.5–19.5 °C" range and then were assigned to the temperature bin of 19.0 °C, which was taken as the standard temperature for the above-mentioned thermal comfort values. Afterward, the relationship between the amount of cool and uncomfortable local body parts and the overall subjective responses was analyzed using the above-mentioned percentage data.

Further, with the survey data, an easy-to-use local–overall thermal comfort model proposed in the previous experimental work of the authors [37] was validated. To be specific, the local cool sensation and discomfort data under different indoor temperatures in this study were substituted into the model to obtain the predicted overall cool sensation and overall discomfort data. The differences between the actual and predicted overall cool sensations and discomfort under different indoor temperatures were calculated as the absolute deviation (the absolute value of the difference between the actual and predicted values) to determine whether the model was applicable in real buildings. The model was expressed as follows:

(1) For thermal sensations, the association between the overall thermal sensation (*OTS*) and the local thermal sensation (*LTS*) could be represented as follows:

$$OTS = 0.5 \times (LTS_{1st,\ warm} + LTS_{2nd,\ warm} + LTS_{1st,\ cool} + LTS_{2nd,\ cool}) \quad (1)$$

where $LTS_{1st,\,warm}$ is the local thermal sensation of the warmest body segment on the warm side; $LTS_{2nd,\,warm}$ is the local thermal sensation of the second-warmest body segment on the warm side; $LTS_{1st,\,cool}$ is the local thermal sensation of the coolest body segment on the cool side; and $LTS_{2nd,\ cool}$ is the local thermal sensation of the second-coolest body segment on the cool side.

When the amount of body parts on the warm or cool side was less than two, the calculation would be based on the actual number of body parts on the warm or cool side.

(2) For thermal comfort, the association between the overall thermal comfort (*OTC*) and the local thermal comfort (*LTC*) was determined through two rules below:

(1) Rule 1: The amount of local body parts on the comfortable side (n+) was less than that on the uncomfortable side (n−). In this case, the relationship between the overall thermal comfort (*OTC*) and the local thermal comfort (*LTC*) was modeled as follows:

$$OTC = 0.5 \times \left(LTC_{1st,\ \ discomf} + LTC_{2nd,\ \ discomf} + LTC_{1st,\ \ comf}\right) \tag{2}$$

where $LTC_{1st,\ discomfort}$ is the local thermal comfort of the most uncomfortable body segment on the uncomfortable side; $LTC_{2nd,\ discomf}$ is the local thermal comfort of the second most uncomfortable body segment on the uncomfortable side; and $LTC_{1st,\ comf}$ is the local thermal comfort of the most comfortable body segment on the comfortable side.

(2) Rule 2: The amount of local body parts on the comfortable side (n+) was greater than that of the body parts on the discomfort side (n−). In this case, the relationship between the overall thermal comfort (*OTC*) and the local thermal comfort (*LTC*) was modeled as follows:

$$OTC = 0.5 \times \left(LTC_{1st,\ \ comf} + LTC_{2nd,\ \ comf} + LTC_{1st,\ \ discomf}\right) \tag{3}$$

where $LTS_{2nd,\ comf}$ is the local thermal comfort of the second most comfortable body segment on the comfortable side.

If no body segment had discomfort, Model (3) could be written as follows:

$$OTC = 0.5 \times \left(LTC_{1st,\ \ comf} + LTC_{2nd,\ \ comf}\right) \tag{4}$$

3. Results

3.1. Overview of the Subjects

Table 3 shows the information of the participants in this study. It can be found that 92.7% of the participants were 16–24 years old; 30.0% of the female participants were in the height range of 160–170 cm, while 29.6% of the male participants were in the height range of 170–180 cm; 27.8% of the female participants weighed 50–60 kg, while 20.5% of the male participants weighed 60–70 kg. In addition, 70.1% of the participants had BMI values of 18.5–23.9 kg/m^2.

Table 3. The distributions of participants' parameters.

Information	Range	Percentage (Overall %)	Percentage (Male %)	Percentage (Female %)
Age	16–18	42.4	23.1	19.4
	19–21	41.3	19.9	21.4
	22–24	9.0	4.3	4.7
	>24	7.3	2.8	4.5
Hight	<160	16.4	0.1	16.4
	160–170	40.8	10.8	30.0
	170–180	33.1	29.6	3.5
	180–190	9.3	9.2	0.1
	>190	0.3	0.3	0.0

Table 3. *Cont.*

Information	Range	Percentage (Overall %)	Percentage (Male %)	Percentage (Female %)
Weight	<50	16.0	1.0	15.0
	50–60	43.0	15.2	27.8
	60–70	26.5	20.5	6.1
	70–80	11.1	10.3	0.8
	>80	1.5	1.3	0.2
BMI	<18.4	19.4	9.0	10.4
	18.5–23.9	70.1	32.9	37.2
	24.0–27.9	9.7	7.7	2.1
	>28	0.8	0.6	0.2

3.2. Environmental Parameters

During the survey period, the main outdoor temperatures in Changsha were 5.8–14.5 °C, and the outdoor humidity was mainly 60–100%. Figure 4 illustrates the indoor temperatures and indoor humidity during the survey period. The average indoor temperature and humidity were 20.4 °C and 39.7%, respectively, and most of the rooms had temperatures higher than 18 °C.

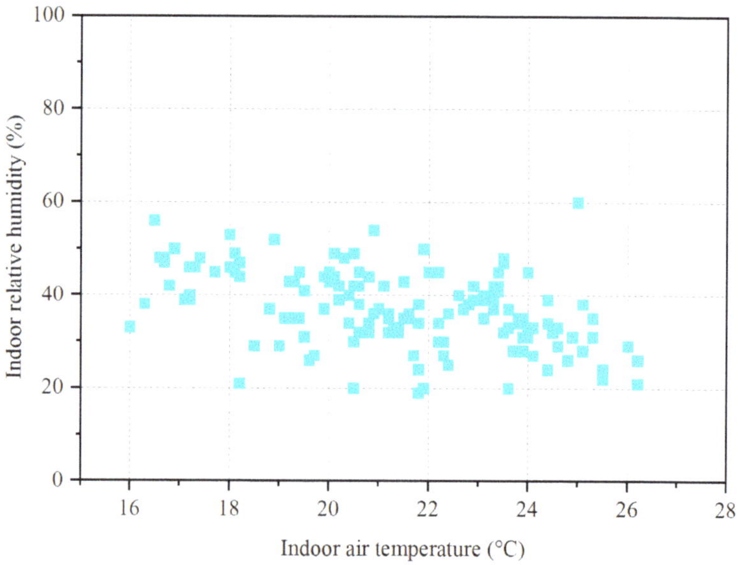

Figure 4. Environmental parameters: indoor temperature and relative humidity.

3.3. Physiological Parameters

Figure 5 illustrates the local skin temperatures under different indoor temperatures. Noticeably, local skin temperature gradually increased as the room temperature rose. The fingertip temperatures were the lowest under different indoor temperatures, and the second-lowest one was the back of the hand.

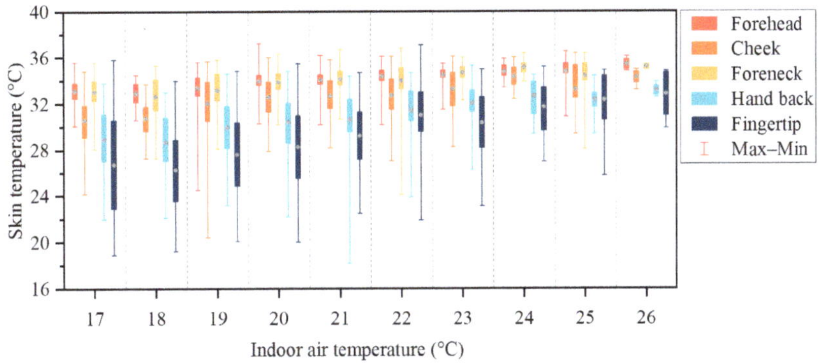

Figure 5. Distributions of participants' local skin temperatures.

Figure 6 shows the regression results of local skin temperatures and indoor temperatures. Local body parts had larger skin temperature differences as the indoor environment became cool. For example, when it was 17 °C indoors, the difference between the forehead and the fingertips was 6.5 °C, and, when it was 26 °C indoors, the skin temperature difference was merely 2.6 °C.

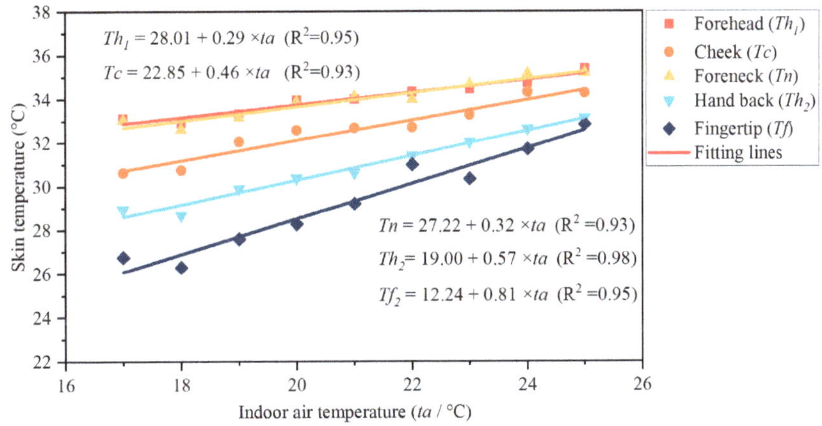

Figure 6. Average skin temperatures of different body parts.

In addition, at different temperatures, the SpO_2 values of the participants mainly varied between 97% and 98%, and the heart rate mainly varied between 60 and 80 bpm, which had no significant correlation with the indoor temperature.

3.4. Ranking of Local Cool Sensations

Figure 7 represents the percentages of different local body parts with cool sensations. It shows that body extremities are prone to cool sensations. Specifically, the top five percentages of local cool sensations appeared in the feet (41.02%), the hands (26.58%), the calves (25.18%), the thighs (13.99%), and the head (9.72%). It should also be noted that the head could also have noticeable cool sensations in the winter despite the indoor temperatures being usually higher than 18 °C (see Figure 4), which could be attributed to the fact that no participants wore hats or had their heads covered when staying indoors.

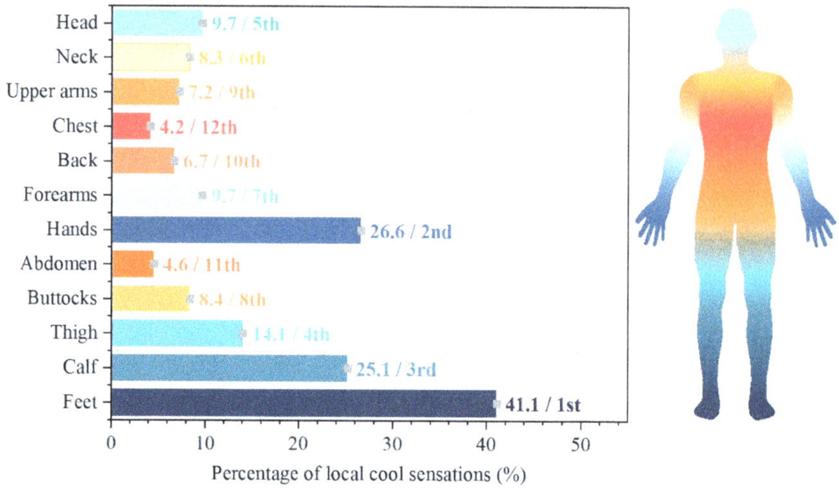

Figure 7. Percentages of the top five discomfort body parts under different indoor temperatures.

Figure 8 shows the change in the proportions of local coolness in the top-five-ranked body parts under different indoor temperatures. It was found that, as the room temperature decreased, the percentages of cool sensations increased rapidly for body extremities (feet, hands, calves). To be specific, when the temperature reached 19 °C, nearly half of the participants had cold sensations in their feet, and, when the temperature reached 17 °C, close to 50% of the participants had cool hands, and 40% of the participants had cool calves.

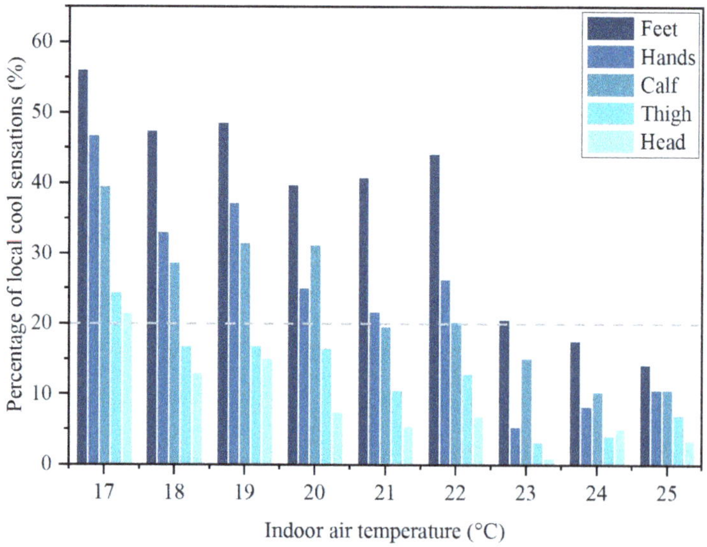

Figure 8. Percentages of the top five cool body parts under different indoor temperatures.

Furthermore, it was found that the top three body parts with the largest percentages of cool sensations remained unchanged under different indoor temperatures, with small differences in the order of the local body parts. Specifically, under different indoor temperatures, the top three body parts with the largest percentages of cool sensations were the feet, the hands, and the calves, and the cool sensation of the feet was much higher than that of the other two parts. In addition, it should be noted that, even when it was

20 °C or higher indoors, some body parts (feet, calves, and hands) still had noticeable cool sensations (higher than 20%), which indicates that using conventional air conditioning systems is insufficient to effectively reduce local thermal discomfort but may cause high energy costs.

3.5. *Ranking of Local Discomfort*

Figure 9 shows the percentages of different local body parts that had discomfort. The body extremities were prone to experience discomfort. Specifically, the higher percentages of local discomfort appeared in the feet (44.99%), the palms (28.2%), the calves (24.74%), the head (19.66%), and the thighs (16.35%). It also should be noted that the head could also have notable discomfort in the winter despite the indoor temperatures being usually higher than 18 °C (see Figure 4).

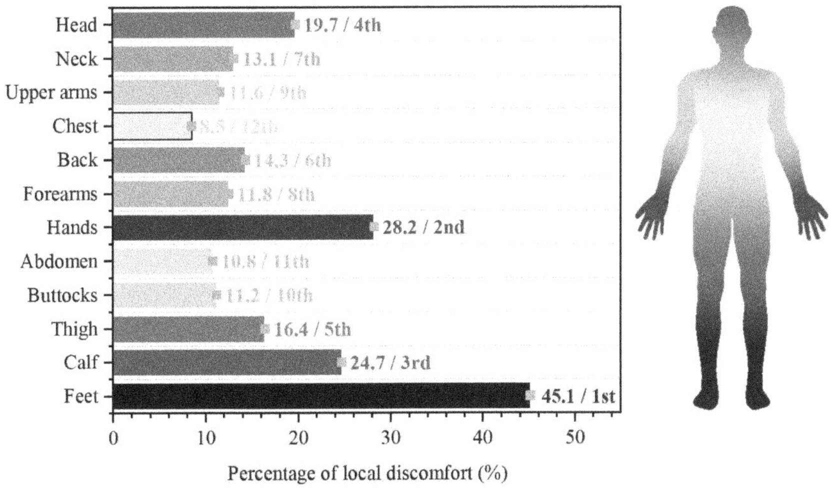

Figure 9. Percentages of local discomfort of different body parts.

Figure 10 illustrates the percentages of local discomfort of the top-five-ranked body parts under different indoor temperatures. It was found that, as the room temperature decreased, the percentages of local discomfort increased rapidly for body extremities (feet, hands, calves). To be specific, when the temperature reached 22 °C, more than 50% of the participants had discomfort in their feet, and, when the temperature reached 17 °C, nearly 40% of the participants had discomfort in their hands, and more than 30% of the participants had discomfort in their calves.

In addition, it could be seen that the top three local uncomfortable body parts remained unchanged under different indoor temperatures, with small differences in the order of the local parts. Specifically, under different indoor temperatures, the top three local discomforts were also felt in the feet, hands, and calves, the discomfort of the feet was much higher than the others, and the highest percentage of local discomfort was always in the feet regardless of the temperature. Moreover, it should be noted that, even when it was 22 °C or higher indoors, some body parts (feet, calves, and hands) still felt noticeable discomfort (higher than 20%), which indicates that using conventional air conditioning systems is insufficient to reduce local discomfort but may cause high energy costs.

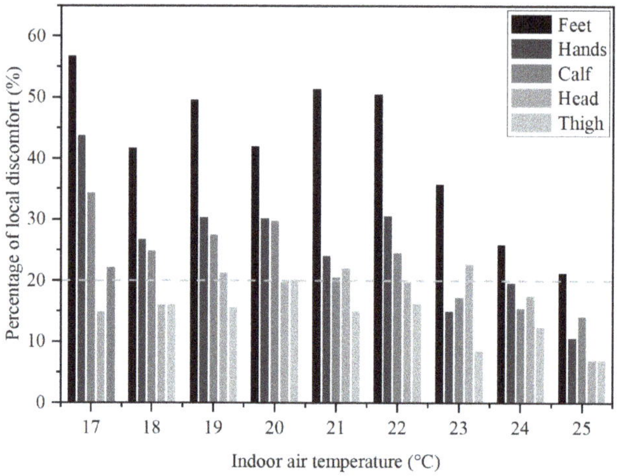

Figure 10. Percentages of the top five uncomfortable body parts under different indoor temperatures.

3.6. Correlation between the Percentages of Local and Overall Thermal Status

Figure 11 represents the percentages of participants having overall cool sensations (voting "slightly cool", "cool", or "cold"), uncomfortable feelings (voting "slightly uncomfortable", "uncomfortable", or "very uncomfortable"), and thermal unacceptability (voting "unacceptable") under the different indoor temperatures in this study. It is obvious that, under the same indoor temperature, the percentage of overall discomfort was higher than that of the overall cool sensation, and the percentage of unacceptability was usually less than 10%, regardless of what the indoor temperature was. Moreover, less than 30% of the participants was cool and uncomfortable when it was 17 °C or higher indoors. When it was 19 °C or higher indoors, less than 20% of the participants felt cool sensations, and, when it was 22.3 °C or higher indoors, less than 20% of the participants felt discomfort. The above-stated results also indicate that considering only thermal acceptability may lead to a large overestimation of occupant thermal comfort in real buildings.

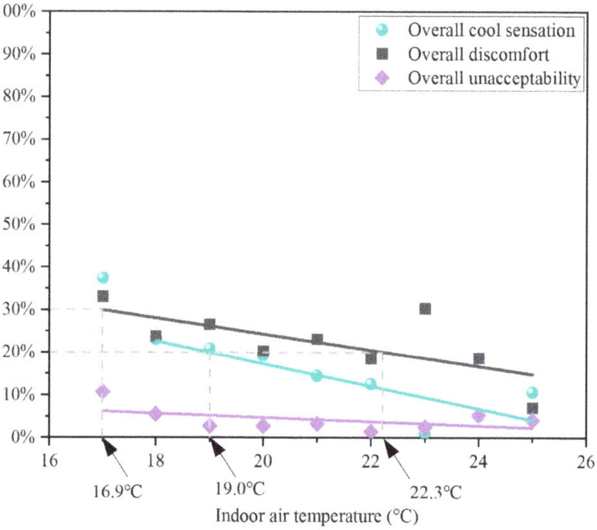

Figure 11. Percentages of overall cool sensation, discomfort, and unacceptability under different indoor temperatures.

3.6.1. Amount of Cool Body Parts and Overall Subjective Responses

Figure 12 shows the percentages of overall subjective responses and different amounts of cool body parts (local body parts had cool sensations) under different indoor temperatures. The results indicate the following: when the whole body has a cool sensation, at least four local body parts have cool sensations; when the whole body has discomfort, at least three local body parts have cool sensations; and when the whole body feels that the ambient environment is thermally unacceptable, at least seven local body parts have cool sensations.

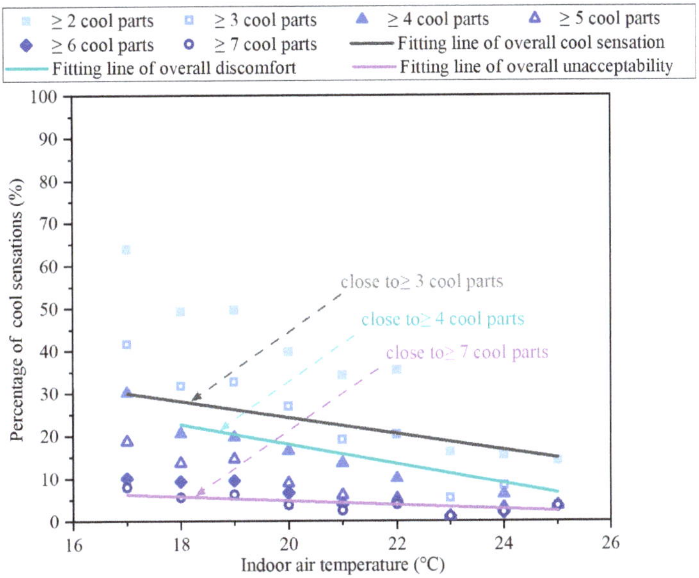

Figure 12. Percentages of overall subjective responses and different amounts of cool body parts.

Specifically, the percentage of at least four cool body parts was similar to that of the overall cool sensation, with an average absolute difference of 1.1%. The percentage of at least three cool body parts was similar to that of the overall discomfort, with an average absolute difference of 2.3%. At the same time, the percentage of at least seven cool body parts was similar to that of the overall unacceptability, with an average absolute difference of 0.2%.

The above-stated results also indicate that, if overall subjective responses are obtained, they can be used to estimate the amount of cool body parts.

3.6.2. Number of Uncomfortable Body Parts and Overall Subjective Responses

Figure 13 shows the percentages of the overall subjective responses and the different amounts of uncomfortable body parts (local body parts which had discomfort) under different indoor temperatures. The results indicate that, when the whole body feels discomfort, at least three local body parts have discomfort and that, when the whole body feels that the ambient environment is thermally unacceptable, at least four local body parts have discomfort.

Specifically, the percentage of at least three uncomfortable body parts was similar to that of the overall discomfort, with an average absolute difference of 2.9%, and the percentage of at least four uncomfortable body parts was similar to that of the overall unacceptability, with an average absolute difference of 0.09%. The above-stated results indicate that, if only overall subjective responses are obtained, they can be used to estimate the amounts of uncomfortable body parts.

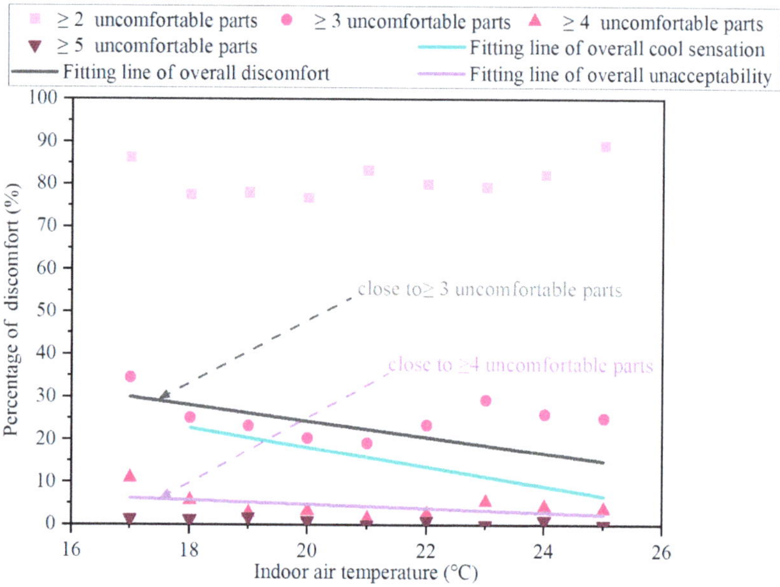

Figure 13. Percentages of overall subjective responses and different amounts of uncomfortable body parts.

3.7. Validation of Local–Overall Thermal Comfort Model

3.7.1. Local and Overall Thermal Sensations

Figure 14 shows the actual and predicted overall thermal sensation values under different indoor temperatures, which were obtained using the bin method, as described in Section 2.6, while the detailed results are represented in Table 4. It can be seen that the average deviation of the model is 0.13, and most of the deviations are lower than 0.2, which demonstrates that the local–overall thermal sensation model proposed by the authors (model (1)) has a high accuracy level.

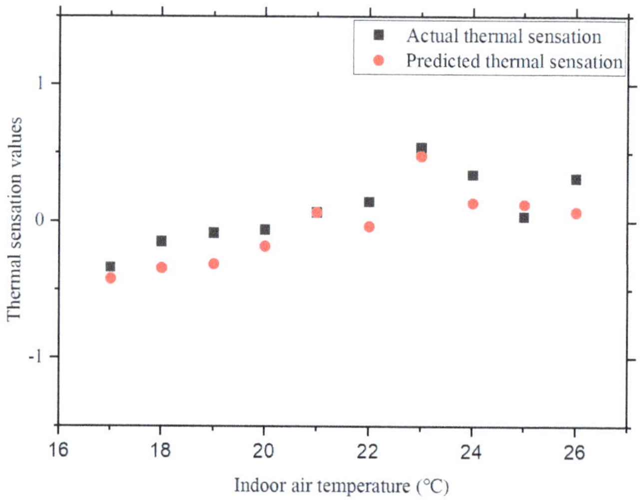

Figure 14. Comparison between the actual and predicted overall thermal sensations.

Table 4. The deviation between the actual and predicted overall thermal sensation values.

Air Temperature (°C)	Actual Value	Predicted Value
17	−0.34	−0.42
18	−0.15	−0.35
19	−0.09	−0.32
20	−0.06	−0.18
21	0.07	0.07
22	0.14	−0.04
23	0.54	0.48
24	0.34	0.14
25	0.04	0.13
26	0.32	0.07

3.7.2. Local and Overall Thermal Comfort

Figure 15 represents the actual and predicted overall thermal comfort values under different indoor temperatures, and the detailed results are represented in Table 5. It can be seen that the average absolute deviation of the model is 0.07, and most of the deviations are lower than 0.3, which demonstrates that the local–overall thermal comfort model proposed by the authors (models (2), (3), and (4)) has a high accuracy level.

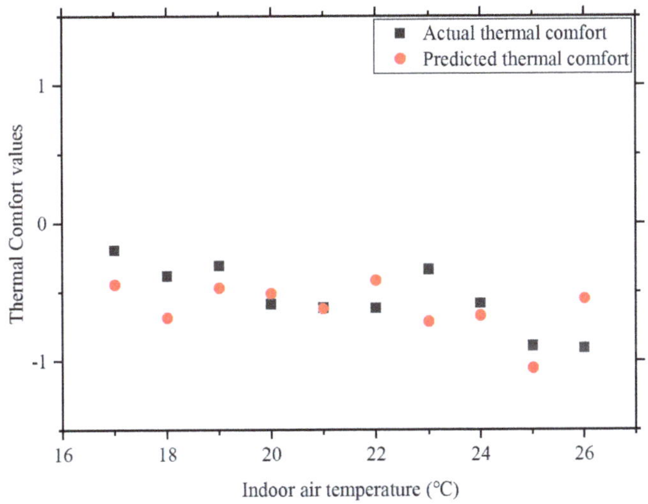

Figure 15. Comparison between the actual and predicted overall thermal comfort.

Table 5. The deviation between the actual and predicted overall thermal comfort values.

Air Temperature (°C)	Actual Value	Predicted Value
17	−0.19	−0.45
18	−0.38	−0.69
19	−0.31	−0.47
20	−0.59	−0.51
21	−0.62	−0.62
22	−0.62	−0.42
23	−0.34	−0.72
24	−0.58	−0.68
25	−0.89	−1.05
26	−0.91	−0.55

4. Discussion

Applications and Limitations of This Study

This study may contribute to current academia as follows:

(1) This study gave detailed information on the local thermal sensations and comfort of different body parts in the winter, especially under different indoor temperatures, which provided guiding suggestions for the future design and deployment of PCSs. For example, this study found that, in addition to the well-known high rates of coldness and discomfort in extremities like the hands and the feet, the head is also susceptible to coldness and discomfort. Thus, special attention should be paid to designing wearable head-heating systems like neck-warmer devices [38] and heated helmets [39]. Using the percentages of local cool sensations and discomfort rather than the average values shed some light on the future work on local–overall thermal comfort and reflected the real comfort needs of indoor occupants.

(2) This study established a quantitative correlation between the amount of local cool or uncomfortable body parts and the overall subjective thermal responses, which provided a novel method to quickly estimate how many local body parts had cool sensations or discomfort according to the overall subjective responses. Such a method can help simplify the questionnaire in the future work by reducing the metrics for evaluating the local thermal comfort of different body parts. For instance, in the existing studies, when investigating local body status, subjects are usually required to answer questions for each body part repeatedly, and then researchers can analyze the amount of cool or warm body parts [32], which is time-consuming.

(3) This study demonstrated a previous local–overall thermal comfort model proposed by the authors [37]. This model simplifies the calculations of overall thermal comfort since it only needs the most extreme local status as the input. With the understanding of local thermal comfort under different indoor temperatures, this model is convenient to use in real environments.

This study also has some limitations:

(1) In this study, it was found that the indoor temperatures were within or close to the comfortable range, and, thus, the gender difference was not as obvious as that found in our previous study [6]. However, if an indoor environment was out of the comfort range, the findings may differ significantly due to gender differences. In the future, the local thermal comfort of different groups (male and female participants) will be explored in environments deviating from the comfortable range.

(2) The participants of this study were healthy young students, who may have been more resilient to the indoor environments in our study than other groups like patients, children, and elders. The key body parts that should be heated as a priority may be different for other groups, a matter to which the development of PCSs should pay careful attention to. For example, for a patient, body parts with an injury may be sensitive to cool or warm stimuli and could easily have cool or warm sensations.

(3) The investigated buildings in this study were all air-conditioned. The applicability of the results of this study in natural-ventilated environments remains to be investigated. In naturally ventilated environments, which are usually much colder than air-conditioned environments in the winter, people can wear more clothes to cover their bodies, such as hats and gloves, which may reduce the cool sensations of the head and hands.

5. Conclusions

A field investigation was conducted to study the correlation between local and overall thermal comfort in the winter, including the measurements of indoor environmental parameters (e.g., air temperature and relative humidity) and participants' physiological parameters (e.g., skin temperature), in addition to 1356 questionnaires from the participants (656 men (48.4%) and 700 women (51.6%)). The main conclusions are listed below.

(1) In terms of local cool sensations, body extremities were the main areas prone to have cool sensations, and the top-five-ranked local cool sensation percentages were those of

the feet (41.02%), the hands (26.58%), the calves (25.18%), the thighs (13.99%), and the head (9.72%). The highest percentage of local cool sensation was always in the feet, regardless of the indoor temperature. Moreover, the percentage of local coolness in the feet, hands, and lower legs increased rapidly as the room temperature decreased. When the indoor temperature decreased by 1 °C, the percentages of local cool sensations of the feet, hands, and lower legs increased by 3–14%. Moreover, there was a significant cool sensation in the head when the temperature was higher than 18 °C indoors.

(2) In terms of local discomfort, body extremities were also more likely to feel uncomfortable, and the top-five-ranked local discomfort percentages laid in the feet (44.99%), the palms (28.2%), the calves (24.74%), the head (19.66%), and the thighs (16.35%), and the highest percentage of local discomfort was always in the feet, regardless of the indoor temperature. At the same time, with the decrease in the indoor temperature, the percentages of local discomfort of the feet, hands, and calves increased rapidly. It is worth noting that the head also had notable discomfort when the temperature was higher than 18 °C indoors.

(3) Moreover, it was found that the percentage of overall thermal discomfort was higher than the percentage of overall cool sensation at the same indoor temperature, while the percentage of thermal unacceptability was usually less than 10%, regardless of the indoor temperature. And, when the whole body felt cool, at least four local body parts had local cool sensations; when the whole body felt uncomfortable, at least three local body parts had local cool sensations; and when the whole body felt that the ambient environment was thermally unacceptable, at least seven local body parts had local cool sensations. In addition, at least three local body parts felt uncomfortable when the whole body felt uncomfortable, and at least four local body parts felt uncomfortable when the whole body felt that the ambient environment was thermally unacceptable.

(4) Further, the mean absolute difference between the actual and predicted overall thermal sensation values was 0.13, indicating that the local–overall thermal sensation model (Equation (1)) proposed by the authors has a high accuracy level. The mean absolute difference between the actual and predicted overall comfort values was 0.07, indicating that the local–overall thermal comfort model (Equations (2)–(4)) proposed by the authors also has a high accuracy level.

This study serves as a reference for prioritizing which local body parts should be warmed in actual cool environments, which may help develop energy-efficient personal comfort systems. For example, Figure 8 shows the percentages of different body parts experiencing cool sensations. The results show that body extremities are prone to coolness and that the head also has noticeable cool sensations in the winter even when the indoor temperatures is above 18 °C. When designing PCSs, the results of this paper can help determine which local part needs to be prioritized for heating, while the consumed energy can be lowered by reducing the heating energy spent on the body parts which are resilient to coolness when the indoor temperature is low. Also, this study contributes to quickly predicting the amount of local body parts with cool discomfort by merely investigating the overall subjective responses, which helps with the simplification of thermal comfort questionnaires for future studies.

Author Contributions: Conceptualization, Y.H. and N.L.; Methodology, X.L., Y.H., Y.Y. and J.H.; Formal analysis, X.L.; Investigation, X.L., Y.H. and J.H.; Resources, N.L.; Writing—original draft, X.L.; Writing—review & editing, Y.H., N.L. and Y.Y.; Funding acquisition, Y.H. All authors have read and agreed to the published version of the manuscript.

Funding: This research is financially supported by the National Natural Science Foundation of China (Number: 52308092) and the Fundamental Research Funds for the Central Universities of China (Number: 531118010826).

Data Availability Statement: All data generated or analyzed during this study are included in this article.

Acknowledgments: The authors would like to thank all the participants of this study.

Conflicts of Interest: The authors declare no conflicts of interest.

References

1. *ASHRAE Standard 55*; Thermal Environment Conditions for Human Occupancy. American Society of Heating, Refriger ation and Air-conditioning Engineers: Atlanta, GA, USA, 2017.
2. Yang, H.; Deng, Y.; Cao, B.; Zhu, Y. Study on the local and overall thermal perceptions under nonuniform thermal exposure using a cooling chair. *Build. Environ.* **2020**, *176*, 106864. [CrossRef]
3. Xiong, W.; Wang, Y.; Mathiesen, B.V.; Lund, H.; Zhang, X. Heat roadmap China: New heat strategy to reduce energy consumption towards 2030. *Energy* **2015**, *81*, 274–285. [CrossRef]
4. Nilsson, H.O. Thermal comfort evaluation with virtual manikin methods. *Build. Environ.* **2007**, *42*, 4000–4005. [CrossRef]
5. Park, S.; Hellwig, R.T.; Grün, G.; Holm, A. Local and overall thermal comfort in an aircraft cabin and their interrelations. *Build. Environ.* **2011**, *46*, 1056–1064. [CrossRef]
6. Hu, J.; He, Y.; Wang, Q.; Wang, B.; Hao, X.; Li, N.; Yin, W.; Liu, L. Resilient or resistant to non-neutral environments? A comparative study on occupant thermal needs in buildings under natural ventilation, fee-free heating, and fee-charged heating modes. *J. Build. Eng.* **2023**, *72*, 106651. [CrossRef]
7. Hu, J.; He, Y.; Hao, X.; Li, N.; Su, Y.; Qu, H. Optimal temperature ranges considering gender differences in thermal comfort, work performance, and sick building syndrome: A winter field study in university classrooms. *Energy Build.* **2022**, *254*, 111554. [CrossRef]
8. Zhang, H.; Arens, E.; Huizenga, C.; Han, T. Thermal sensation and comfort models for non-uniform and transient environments: Part I: Local sensation of individual body parts. *Build. Environ.* **2010**, *45*, 380–388. [CrossRef]
9. Arens, E.; Zhang, H.; Huizenga, C. Partial- and whole-body thermal sensation and comfort—Part II: Non-uniform environmental conditions. *J. Therm. Biol.* **2006**, *31*, 60–66. [CrossRef]
10. Jin, Q.; Li, X.; Duanmu, L.; Shu, H.; Sun, Y.; Ding, Q. Predictive model of local and overall thermal sensations for non-uniform environments. *Build. Environ.* **2012**, *51*, 330–344. [CrossRef]
11. He, Y.; Wang, X.; Li, N.; He, M.; He, D. Heating chair assisted by leg-warmer: A potential way to achieve better thermal comfort and greater energy conservation in winter. *Energy Build.* **2018**, *158*, 1106–1116. [CrossRef]
12. Yang, H.; Cao, B.; Zhu, Y. Study on the effects of chair heating in cold indoor environments from the perspective of local thermal sensation. *Energy Build.* **2018**, *180*, 16–28. [CrossRef]
13. He, Y.; Wang, X.; Li, N.; He, M.; He, D.; Wang, K. Cooling ceiling assisted by desk fans for comfort in hot-humid environment. *Build. Environ.* **2017**, *122*, 23–34. [CrossRef]
14. Ren, Z.; Gao, X.; Xiao, Y.; Liu, Y. Thermal comfort and energy conservation of a four-sided enclosed local heating device in a cold environment. *Build. Environ.* **2023**, *228*, 109837. [CrossRef]
15. Aryal, A.; Becerik-Gerber, B. A comparative study of predicting individual thermal sensation and satisfaction using wrist-worn temperature sensor, thermal camera and ambient temperature sensor. *Build. Environ.* **2019**, *160*, 106223. [CrossRef]
16. Echarri-Iribarren, V.; Wong, N.H.; Sánchez-Ostiz, A. Radiant Floors versus Radiant Walls Using Ceramic Thermal Panels in Mediterranean Dwellings: Annual Energy Demand and Cost-Effective Analysis. *Sustainability* **2021**, *13*, 588. [CrossRef]
17. Tang, J.; Liu, Y.; Du, H.; Lan, L.; Sun, Y.; Wu, J. The effects of portable cooling systems on thermal comfort and work performance in a hot environment. *Build. Simul.* **2021**, *14*, 1667–1683. [CrossRef]
18. Zhang, J.; Zhou, X.; Lei, S.; Luo, M. Energy and comfort performance of occupant-centric air conditioning strategy in office buildings with personal comfort devices. *Build. Simul.* **2021**, *15*, 899–911. [CrossRef]
19. Zhang, H.; Huizenga, C.; Arens, E.; Wang, D. Thermal sensation and comfort in transient non-uniform thermal environments. *Eur. J. Appl. Physiol.* **2004**, *92*, 728–733. [CrossRef] [PubMed]
20. Fang, Z.; Liu, H.; Li, B.; Tan, M.; Olaide, O.M. Experimental investigation on thermal comfort model between local thermal sensation and overall thermal sensation. *Energy Build.* **2018**, *158*, 1286–1295. [CrossRef]
21. Khiavi, N.M.; Maerefat, M.; Zolfaghari, S.A. Assessment of overall body thermal sensation based on the thermal response of local cutaneous thermoreceptors. *J. Therm. Biol.* **2019**, *83*, 187–194. [CrossRef]
22. Jian, Y.; Hou, Y.; Liu, W.; Chang, X. How the coldest local thermal sensation affects overall thermal sensation after turning on the air conditioning—Evidence from chamber experiments. *Build. Environ.* **2022**, *191*, 107589. [CrossRef]
23. Liu, F.; Huang, Y.; Zhang, L.; Li, G. Marine environmental pollution, aquatic products trade and marine fishery Economy—An empirical analysis based on simultaneous equation model. *Ocean Coast. Manag.* **2022**, *222*, 106096. [CrossRef]
24. Sun, Y.; Zhang, H.; Yan, Y.; Lan, L.; Cao, T.; Lian, Z.; Fan, X.; Wargocki, P.; Zhu, J.; Xu, X. Comparison of wrist skin temperature with mean skin temperature calculated with Hardy and Dubois's seven-point method while sleeping. *Energy Build.* **2022**, *259*, 111894. [CrossRef]
25. Li, W.; Chen, J.; Lan, F.; Xie, H. Human thermal sensation and its algorithmic modelization under dynamic environmental thermal characteristics of vehicle cabin. *Indoor Air* **2022**, *32*, e13168. [CrossRef] [PubMed]
26. Du, C.; Liu, H.; Li, C.; Xiong, J.; Li, B.; Li, G.; Xi, Z. Demand and efficiency evaluations of local convective heating to human feet and low body parts in cold environments. *Build. Environ.* **2020**, *171*, 106662. [CrossRef]
27. Cheng, X.; Yang, B.; Tan, K.; Isaksson, E.; Li, L.; Hedman, A.; Olofsson, T.; Li, H. A Contactless Measuring Method of Skin Temperature based on the Skin Sensitivity Index and Deep Learning. *Appl. Sci.* **2019**, *9*, 1375. [CrossRef]

28. Jia, M.; Choi, J.-H.; Liu, H.; Susman, G. Development of facial-skin temperature driven thermal comfort and sensation modeling for a futuristic application. *Build. Environ.* **2022**, *207*, 108479. [CrossRef]
29. Shahzad, S.; Calautit, J.K.; Aquino, A.I.; Nasir, D.S.; Hughes, B.R. A user-controlled thermal chair for an open plan workplace: CFD and field studies of thermal comfort performance. *Appl. Energy* **2017**, *207*, 283–293. [CrossRef]
30. Kim, J.; Bauman, F.; Raftery, P.; Arens, E.; Zhang, H.; Fierro, G.; Andersen, M.; Culler, D. Occupant comfort and behavior: High-resolution data from a 6-month field study of personal comfort systems with 37 real office workers. *Build. Environ.* **2019**, *148*, 348–360. [CrossRef]
31. Chen, X.; Xue, P.; Gao, L.; Du, J.; Liu, J. Physiological and thermal response to real-life transient conditions during winter in severe cold area. *Build. Environ.* **2019**, *157*, 284–296. [CrossRef]
32. He, Y.; Li, N.; Huang, Q. A field study on thermal environment and occupant local thermal sensation in offices with cooling ceiling in Zhuhai, China. *Energy Build.* **2015**, *102*, 277–283. [CrossRef]
33. He, Y.; Li, N.; Zhang, W.; Peng, J. Overall and local thermal sensation & comfort in air-conditioned dormitory with hot-humid climate. *Build. Environ.* **2016**, *101*, 102–109. [CrossRef]
34. He, Y.; Zhang, H.; Arens, E.; Merritt, A.; Huizenga, C.; Levinson, R.; Wang, A.; Ghahramani, A.; Alvarez-Suarez, A. Smart detection of indoor occupant thermal state via infrared thermography, computer vision, and machine learning. *Build. Environ.* **2023**, *228*, 109811. [CrossRef]
35. Zhang, Q.; Yang, H. *Typical Meteorological Database Handbook for Buildings*; China Building Industry Press: Beijing, China, 2012.
36. 0OpenStreetMap (CCBY-SA2.0), Licences: Commons Attribution ShareAlike 2.0. Available online: https://www.openstreetmap.org (accessed on 20 March 2024).
37. He, Y. Research on Characteristics of Thermal Comfort and Energy-Use Behaviors of Occupants with Personal Comfort Systems. Ph.D. Thesis, Hunan University, Changsha, Hunan, April 2019.
38. Cao, Y.; Lei, T.-H.; Wang, F.; Yang, B.; Mündel, T. Head, Face and Neck Cooling as Per-cooling (Cooling During Exercise) Modalities to Improve Exercise Performance in the Heat: A Narrative Review and Practical Applications. *Sports Med.-Open* **2022**, *8*, 16. [CrossRef]
39. Lv, S.; He, W.; Wang, L.; Li, G.; Ji, J.; Chen, H.; Zhang, G. Design, fabrication and feasibility analysis of a thermo-electric wearable helmet. *Appl. Therm. Eng.* **2016**, *109*, 138–146. [CrossRef]

MDPI
St. Alban-Anlage 66
4052 Basel
Switzerland
www.mdpi.com

Buildings Editorial Office
E-mail: buildings@mdpi.com
www.mdpi.com/journal/buildings

www.ingramcontent.com/pod-product-compliance
Lightning Source LLC
Chambersburg PA
CBHW061834210926
51850CB00006B/887
9783725812257